An Overview of Biometrics

" The New Age of Personal Identification "

Edited by Paul F. Kisak

Contents

1	**Biometrics**	**1**
1.1	Biometric functionality	1
1.2	Multimodal biometric system	2
1.3	Performance	3
1.4	History of biometrics	3
1.5	Adaptive biometric systems	4
1.6	India's national ID program	4
1.7	Recent advances in emerging biometrics	4
	1.7.1 Operator signatures	4
	1.7.2 Proposed requirement for certain public networks	5
1.8	Issues and concerns	5
	1.8.1 Human dignity	5
	1.8.2 Privacy and discrimination	5
	1.8.3 Danger to owners of secured items	6
	1.8.4 Cancelable biometrics	6
	1.8.5 Soft biometrics	6
	1.8.6 International sharing of biometric data	7
	1.8.7 Likelihood of full governmental disclosure	7
1.9	Countries applying biometrics	7
1.10	See also	7
1.11	Notes	8
1.12	References	8
1.13	Further reading	10
1.14	External links	11
2	**Access control**	**12**
2.1	Physical security	12
	2.1.1 Access control system operation	13
	2.1.2 Credential	14
	2.1.3 Access control system components	14

		2.1.4	Access control topology	14
		2.1.5	Types of readers	15
		2.1.6	Access control system topologies	15
		2.1.7	Security risks	19
	2.2	Computer security		20
	2.3	Access control models		21
	2.4	Telecommunication		21
	2.5	In object-oriented programming		21
		2.5.1	Comparison of use of access modifier keywords in different OOP languages	22
		2.5.2	Attribute accessors	22
	2.6	Public policy		22
	2.7	See also		22
	2.8	References		22
	2.9	External links		23
3	**Fingerprint**			**24**
	3.1	Biology		24
	3.2	Types		24
	3.3	Dactyloscopy		25
	3.4	Types		26
		3.4.1	Exemplar	26
		3.4.2	Latent	26
		3.4.3	Patent	26
		3.4.4	Plastic	27
		3.4.5	Electronic recording	27
		3.4.6	Footprints	27
	3.5	Capture and detection		27
		3.5.1	Live scan devices	27
		3.5.2	Scanning dead or unconscious people	28
		3.5.3	Latent detection	28
		3.5.4	Laboratory techniques	29
	3.6	Research		29
	3.7	Disappearance of children's latent prints		30
	3.8	Detection of drug use		31
	3.9	United States databases and compression		31
	3.10	Validity		31
		3.10.1	Criticism	31
		3.10.2	Track record	32
	3.11	Professional certification		32

CONTENTS

- 3.12 Errors 33
  - 3.12.1 Brandon Mayfield and the Madrid bombing 33
  - 3.12.2 René Ramón Sánchez 33
  - 3.12.3 Shirley McKie 33
  - 3.12.4 Stephan Cowans 33
  - 3.12.5 Craig D. Harvey 33
- 3.13 History 34
  - 3.13.1 Antiquity and the medieval period 34
  - 3.13.2 Europe in the 17th and 18th centuries 34
  - 3.13.3 Modern era 34
- 3.14 Privacy 36
  - 3.14.1 Fingerprinting of children 36
- 3.15 Other uses 37
  - 3.15.1 Welfare claimants 37
  - 3.15.2 Log-in authentication and other locks 37
  - 3.15.3 Electronic registration and library access 37
- 3.16 Absence or mutilation of fingerprints 37
- 3.17 Fingerprint recognition 37
  - 3.17.1 Patterns 38
  - 3.17.2 Fingerprint processing 38
  - 3.17.3 Minutiae features 38
  - 3.17.4 Defeats 38
- 3.18 Fingerprint recognition in electronic devices 38
- 3.19 Fingerprint sensors 38
  - 3.19.1 Optical 39
  - 3.19.2 Ultrasonic 39
  - 3.19.3 Capacitance 39
- 3.20 Algorithms 39
  - 3.20.1 Pre-processing 39
  - 3.20.2 Pattern-based (or image-based) algorithms 40
- 3.21 In other species 40
- 3.22 In fiction 40
  - 3.22.1 Mark Twain 40
  - 3.22.2 Crime fiction 40
  - 3.22.3 Film and television 40
- 3.23 Other reliable identifiers 40
- 3.24 See also 41
- 3.25 References 41

		3.26 Further reading	45
		3.27 External links	46

4 Facial recognition system — 47

- 4.1 Techniques for face acquisition . . . 47
  - 4.1.1 Traditional . . . 47
  - 4.1.2 3-dimensional recognition . . . 48
  - 4.1.3 Skin texture analysis . . . 48
  - 4.1.4 Thermal cameras . . . 48
- 4.2 Notable users and deployments . . . 48
  - 4.2.1 Additional uses . . . 49
- 4.3 Advantages and disadvantages . . . 49
  - 4.3.1 Compared to other technologies . . . 49
  - 4.3.2 Weaknesses . . . 50
  - 4.3.3 Effectiveness . . . 50
  - 4.3.4 Privacy issues . . . 50
- 4.4 History . . . 51
- 4.5 Emotion detection . . . 52
- 4.6 Anti facial recognition systems . . . 52
- 4.7 See also . . . 52
- 4.8 References . . . 53
- 4.9 Further reading . . . 55
- 4.10 External links . . . 55

5 DNA — 56

- 5.1 Properties . . . 57
  - 5.1.1 Nucleobase classification . . . 58
  - 5.1.2 Non canonical bases . . . 58
  - 5.1.3 Listing of non canonical bases found in DNA . . . 59
  - 5.1.4 Grooves . . . 60
  - 5.1.5 Base pairing . . . 60
  - 5.1.6 Sense and antisense . . . 60
  - 5.1.7 Supercoiling . . . 61
  - 5.1.8 Alternative DNA structures . . . 61
  - 5.1.9 Alternative DNA chemistry . . . 61
  - 5.1.10 Quadruplex structures . . . 62
  - 5.1.11 Branched DNA . . . 62
- 5.2 Chemical modifications and altered DNA packaging . . . 62
  - 5.2.1 Base modifications and DNA packaging . . . 62

	5.2.2 Damage	63
5.3	Biological functions	64
	5.3.1 Genes and genomes	64
	5.3.2 Transcription and translation	64
	5.3.3 Replication	65
	5.3.4 Extracellular nucleic acids	65
5.4	Interactions with proteins	65
	5.4.1 DNA-binding proteins	65
	5.4.2 DNA-modifying enzymes	66
5.5	Genetic recombination	67
5.6	Evolution	68
5.7	Uses in technology	68
	5.7.1 Genetic engineering	68
	5.7.2 DNA profiling	69
	5.7.3 DNA enzymes or catalytic DNA	69
	5.7.4 Bioinformatics	69
	5.7.5 DNA nanotechnology	70
	5.7.6 History and anthropology	70
	5.7.7 Information storage	70
5.8	History of DNA research	70
5.9	See also	72
5.10	References	72
5.11	Further reading	81
5.12	External links	82

6 Palm print — 83
6.1 References — 83

7 Hand geometry — 84
7.1 See also — 84
7.2 References — 84

8 Iris recognition — 86
8.1 History — 86
8.2 Visible vs near infrared imaging — 87
8.3 Operating principle — 87
8.4 Advantages — 88
8.5 Shortcomings — 88
8.6 Security considerations — 89

8.7	Deployed applications	89
8.8	Iris recognition in television and movies	91
8.9	See also	91
8.10	References	92
8.11	Further reading	93
8.12	External links	93

9 Retinal scan — 94

9.1	Introduction	94
9.2	History	94
9.3	In popular culture	94
9.4	Uses	95
9.5	Pros and cons	95
9.6	See also	95
9.7	References	95

10 Keystroke dynamics — 96

10.1	Science of Keystroke Dynamics	96
10.2	Origin of Keystroke Dynamics	96
10.3	Use as Biometric Data	96
10.4	Authentication versus identification	97
10.5	Temporal variation	97
10.6	Legal and regulatory issues	98
10.7	Other uses	98
10.8	See also	98
10.9	References	98
	10.9.1 Other references	98

11 Gait analysis — 100

11.1	History	100
11.2	Process and equipment	100
11.3	Factors and parameters	101
11.4	Techniques	101
	11.4.1 Temporal / spatial	101
	11.4.2 Kinematics	102
	11.4.3 Markerless Gait Capture	102
	11.4.4 Pressure Measurement	102
	11.4.5 Kinetics	102
	11.4.6 Dynamic electromyography	102

- 11.5 Applications 103
  - 11.5.1 Medical diagnostics 103
  - 11.5.2 Chiropractic and Osteopathic Utilizations 103
  - 11.5.3 Biometric identification and forensics 103
  - 11.5.4 Comparative biomechanics 103
- 11.6 Popular media 103
- 11.7 Textbooks 103
- 11.8 See also 104
- 11.9 References 104
- 11.10 External links 105

12 Speaker recognition — 106
- 12.1 Verification versus identification 106
- 12.2 Variants of speaker recognition 106
- 12.3 Technology 107
- 12.4 Applications 107
- 12.5 See also 108
- 12.6 Notes 108
- 12.7 References 108
- 12.8 Bibliography 108
- 12.9 External links 108
  - 12.9.1 Software 109

13 Electroencephalography — 110
- 13.1 History 110
- 13.2 Medical use 111
- 13.3 Research use 112
  - 13.3.1 Advantages 112
  - 13.3.2 Disadvantages 113
  - 13.3.3 With other neuroimaging techniques 113
- 13.4 Mechanisms 114
- 13.5 Method 114
  - 13.5.1 Limitations 116
  - 13.5.2 EEG vs fMRI, fNIRS and PET 116
  - 13.5.3 EEG vs MEG 116
- 13.6 Normal activity 117
  - 13.6.1 Wave patterns 118
- 13.7 Artifacts 119
  - 13.7.1 Biological artifacts 119

13.7.2 Environmental artifacts 120

13.7.3 Artifact correction 120

13.8 Abnormal activity 121

13.8.1 Remote communication 121

13.9 Economics 121

13.10 Future research 122

13.11 See also 122

13.12 References 123

13.13 Further reading 127

13.14 External links 127

14 Electrocardiography 128

14.1 History 128

14.2 Medical uses 129

14.3 Electrocardiographs 130

14.4 Electrodes and leads 131

14.4.1 Limb leads 132

14.4.2 Augmented limb leads 132

14.4.3 Precordial leads 132

14.4.4 Specialized leads 133

14.4.5 Lead locations on an ECG report 133

14.4.6 Contiguity of leads 133

14.5 Electrophysiology 133

14.6 Interpretation 133

14.6.1 Theory 134

14.6.2 Electrocardiogram grid 134

14.6.3 Rate and rhythm 135

14.6.4 Axis 135

14.6.5 Amplitudes and intervals 136

14.6.6 Ischemia and infarction 136

14.6.7 Artifacts 136

14.7 Diagnosis 136

14.8 See also 137

14.9 Notes 137

14.10 References 138

14.11 External links 139

15 BioAPI 140

15.1 Origins 140

15.2 What and why?	140
15.3 The basic architecture	141
15.4 Procurement issues	141
15.5 Distributed systems	141
15.6 References	141

16 Biometric passport — 142

16.1 Availability	142
16.2 Data protection	142
16.3 Inspection process	143
16.4 Attacks	143
16.5 Opposition	144
16.6 Countries using biometric passports	145
16.6.1 European Union	145
16.6.2 Albania	146
16.6.3 Algeria	146
16.6.4 Argentina	147
16.6.5 Armenia	147
16.6.6 Australia	147
16.6.7 Azerbaijan	147
16.6.8 Bangladesh	147
16.6.9 Bosnia and Herzegovina	147
16.6.10 Botswana	147
16.6.11 Brazil	147
16.6.12 Brunei	147
16.6.13 Cambodia	148
16.6.14 Canada	148
16.6.15 Cape Verde	148
16.6.16 Chile	148
16.6.17 China	148
16.6.18 Colombia	148
16.6.19 Dominican Republic	148
16.6.20 Egypt	148
16.6.21 Gabon	148
16.6.22 Ghana	149
16.6.23 Hong Kong	149
16.6.24 Iceland	149
16.6.25 India	149
16.6.26 Indonesia	149

- 16.6.27 Iran 149
- 16.6.28 Iraq 149
- 16.6.29 Ireland 150
- 16.6.30 Israel 150
- 16.6.31 Japan 150
- 16.6.32 Kazakhstan 150
- 16.6.33 Kosovo 150
- 16.6.34 Kuwait 150
- 16.6.35 Laos 150
- 16.6.36 Lebanon 150
- 16.6.37 Lesotho 151
- 16.6.38 Macau 151
- 16.6.39 Macedonia 151
- 16.6.40 Madagascar 151
- 16.6.41 Malaysia 151
- 16.6.42 Maldives 151
- 16.6.43 Sovereign Military Order of Malta 151
- 16.6.44 Moldova 151
- 16.6.45 Montenegro 151
- 16.6.46 Mongolia 152
- 16.6.47 Mauritania 152
- 16.6.48 Morocco 152
- 16.6.49 Mozambique 152
- 16.6.50 New Zealand 152
- 16.6.51 Nigeria 152
- 16.6.52 Norway 152
- 16.6.53 Pakistan 153
- 16.6.54 Panama 153
- 16.6.55 Peru 153
- 16.6.56 Philippines 153
- 16.6.57 Qatar 153
- 16.6.58 Russia 154
- 16.6.59 Saudi Arabia 154
- 16.6.60 Serbia 154
- 16.6.61 Singapore 154
- 16.6.62 Somalia 154
- 16.6.63 South Korea 154
- 16.6.64 South Sudan 154

CONTENTS   xi

        16.6.65 Slovakia 154

        16.6.66 Sri Lanka 154

        16.6.67 Sudan 155

        16.6.68 Switzerland 155

        16.6.69 Taiwan 155

        16.6.70 Tajikistan 155

        16.6.71 Thailand 155

        16.6.72 Togo 155

        16.6.73 Tunisia 155

        16.6.74 Turkey 155

        16.6.75 Turkmenistan 156

        16.6.76 Ukraine 156

        16.6.77 United Arab Emirates 156

        16.6.78 United States 156

        16.6.79 Uruguay 156

        16.6.80 Uzbekistan 156

        16.6.81 Venezuela 156

        16.6.82 Zimbabwe 157

  16.7 See also 157

  16.8 References 157

  16.9 External links 161

17 Biometric voter registration   **162**

  17.1 Countries with biometric voter registration 162

  17.2 Advocacy and criticism 162

  17.3 See also 163

  17.4 References 163

18 Biometrics in schools   **166**

  18.1 Types of Biometrics Used in Schools 166

  18.2 United Kingdom 166

  18.3 Belgium 167

  18.4 Early applications 167

  18.5 Applications 167

  18.6 Ages 168

  18.7 Current usage 168

  18.8 Security concerns 168

  18.9 Advantages 168

  18.10 See also 168

18.11 References	168
18.12 External links	169
18.12.1 Legislation	169
18.12.2 Non statutory advice	170

19 BioSlimDisk — 171

- 19.1 Signature . . . 171
- 19.2 References . . . 171
- 19.3 External links . . . 171

20 Handwritten biometric recognition — 172

- 20.1 Static and Dynamic recognition . . . 172
- 20.2 Difference from OCR . . . 172
- 20.3 References . . . 172

21 Private biometrics — 174

- 21.1 Comparison with handling computer passwords . . . 174
- 21.2 References . . . 174

22 Signature recognition — 176

- 22.1 Related techniques . . . 176
- 22.2 Databases . . . 176
- 22.3 References . . . 177

23 Vein matching — 178

- 23.1 History of Vein Matching . . . 178
- 23.2 Commercial applications . . . 179
- 23.3 Forensic identification . . . 179
- 23.4 Other applications . . . 179
- 23.5 See also . . . 179
- 23.6 References . . . 179
- 23.7 Further reading . . . 180
- 23.8 External links . . . 180

24 Voice analysis — 181

- 24.1 Typical voice problems . . . 181
- 24.2 Analysis methods . . . 181
- 24.3 See also . . . 181
- 24.4 External links . . . 182

25 Identity Cards Act 2006 — 183

- 25.1 Development 183
  - 25.1.1 Reasons for introduction 183
  - 25.1.2 Legislative progress 184
  - 25.1.3 Timescale and implementation progress 185
  - 25.1.4 Pilot schemes and partial rollouts 186
  - 25.1.5 2010 general election 186
  - 25.1.6 Ending of the scheme 187
- 25.2 Historical and international comparisons 187
  - 25.2.1 ID cards during the World Wars 187
  - 25.2.2 International comparisons 188
- 25.3 System 188
  - 25.3.1 Legal requirements 188
  - 25.3.2 National Identity Register 188
  - 25.3.3 Identity Registration Number 189
  - 25.3.4 Types of cards 189
- 25.4 Use as travel document 189
- 25.5 Reaction 190
  - 25.5.1 Public reaction 190
  - 25.5.2 Terrorism and crime 190
- 25.6 Objections to the scheme 190
  - 25.6.1 Costs 190
  - 25.6.2 Effectiveness 191
  - 25.6.3 Ethnic minorities 191
  - 25.6.4 Concerns raised by the Information Commissioner 191
  - 25.6.5 Human rights 191
  - 25.6.6 Feature creep 192
  - 25.6.7 Database extent and access 192
  - 25.6.8 Vulnerable individuals 192
  - 25.6.9 Identity theft 192
  - 25.6.10 Technology 192
- 25.7 Opposition campaigns 192
  - 25.7.1 Scotland 193
  - 25.7.2 Northern Ireland 193
- 25.8 See also 193
- 25.9 Notes 193
- 25.10 References 193
- 25.11 External links 197
  - 25.11.1 News stories 197

  25.11.2 Guides 197
  25.11.3 Opposition groups 198

26 International Identity Federation   199
 26.1 Sources 199
 26.2 References 199
 26.3 Text and image sources, contributors, and licenses 200
  26.3.1 Text 200
  26.3.2 Images 208
  26.3.3 Content license 215

Chapter 1

Biometrics

For the academic journal of statistics in biology, see Biometrics (journal). For the application of statistics to topics in biology, see Biostatistics. For the fraudulently marketed "thigh contour treatment", see Peter Foster.

At Walt Disney World in Lake Buena Vista, Florida, biometric measurements are taken from the fingers of guests to ensure that a ticket is used by the same person from day to day

Biometrics refers to metrics related to human characteristics. Biometrics authentication (or realistic authentication)[*][note 1] is used in computer science as a form of identification and access control.[*][1] It is also used to identify individuals in groups that are under surveillance.

Biometric identifiers are then distinctive, measurable characteristics used to label and describe individuals.[*][2] Biometric identifiers are often categorized as physiological versus behavioral characteristics.[*][3] Physiological characteristics are related to the shape of the body. Examples include, but are not limited to fingerprint, palm veins, face recognition, DNA, palm print, hand geometry, iris recognition, retina and odour/scent. Behavioral characteristics are related to the pattern of behavior of a person, including but not limited to typing rhythm, gait,[*][4] and voice.[*][5][*][note 2] Some researchers have coined the term behaviometrics to describe the latter class of biometrics.[*][6]

More traditional means of access control include token-based identification systems, such as a driver's license or passport, and knowledge-based identification systems, such as a password or personal identification number.[*][2] Since biometric identifiers are unique to individuals, they are more reliable in verifying identity than token and knowledge-based methods; however, the collection of biometric identifiers raises privacy concerns about the ultimate use of this information.[*][2][*][7]

1.1 Biometric functionality

Many different aspects of human physiology, chemistry or behavior can be used for biometric authentication. The selection of a particular biometric for use in a specific application involves a weighting of several factors. Jain *et al.* (1999)[*][8] identified seven such factors to be used when assessing the suitability of any trait for use in biometric authentication.

- Universality means that every person using a system should possess the trait.

- Uniqueness means the trait should be sufficiently different for individuals in the relevant population such that they can be distinguished from one another.

- Permanence relates to the manner in which a trait varies over time. More specifically, a trait with 'good' permanence will be reasonably invariant over time with respect to the specific matching algorithm.

- Measurability (collectability) relates to the ease of acquisition or measurement of the trait. In addition, acquired data should be in a form that permits subsequent processing and extraction of the relevant feature sets.

- Performance relates to the accuracy, speed, and robustness of technology used (see performance section for more details).

- Acceptability relates to how well individuals in the relevant population accept the technology such that they are willing to have their biometric trait captured and assessed.
- Circumvention relates to the ease with which a trait might be imitated using an artifact or substitute.

Proper biometric use is very application dependent. Certain biometrics will be better than others based on the required levels of convenience and security.*[9] No single biometric will meet all the requirements of every possible application.*[8]

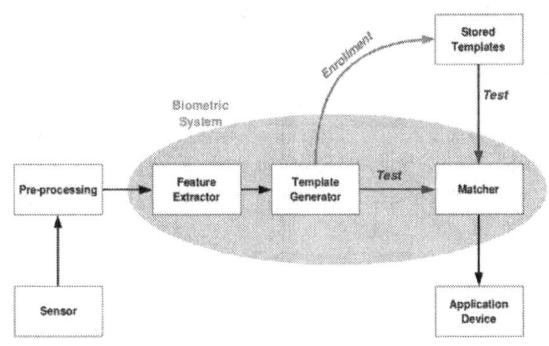

The block diagram illustrates the two basic modes of a biometric system.*[3] First, in verification (or authentication) mode the system performs a one-to-one comparison of a captured biometric with a specific template stored in a biometric database in order to verify the individual is the person they claim to be. Three steps are involved in the verification of a person.*[10] In the first step, reference models for all the users are generated and stored in the model database. In the second step, some samples are matched with reference models to generate the genuine and impostor scores and calculate the threshold. Third step is the testing step. This process may use a smart card, username or ID number (e.g. PIN) to indicate which template should be used for comparison.*[note 3] 'Positive recognition' is a common use of the verification mode, "where the aim is to prevent multiple people from using the same identity".*[3]

Second, in identification mode the system performs a one-to-many comparison against a biometric database in an attempt to establish the identity of an unknown individual. The system will succeed in identifying the individual if the comparison of the biometric sample to a template in the database falls within a previously set threshold. Identification mode can be used either for 'positive recognition' (so that the user does not have to provide any information about the template to be used) or for 'negative recognition' of the person "where the system establishes whether the person is who she (implicitly or explicitly) denies to be".*[3] The latter function can only be achieved through biometrics since other methods of personal recognition such as passwords, PINs or keys are ineffective.

The first time an individual uses a biometric system is called *enrollment*. During the enrollment, biometric information from an individual is captured and stored. In subsequent uses, biometric information is detected and compared with the information stored at the time of enrollment. Note that it is crucial that storage and retrieval of such systems themselves be secure if the biometric system is to be robust. The first block (sensor) is the interface between the real world and the system; it has to acquire all the necessary data. Most of the times it is an image acquisition system, but it can change according to the characteristics desired. The second block performs all the necessary pre-processing: it has to remove artifacts from the sensor, to enhance the input (e.g. removing background noise), to use some kind of normalization, etc. In the third block necessary features are extracted. This step is an important step as the correct features need to be extracted in the optimal way. A vector of numbers or an image with particular properties is used to create a *template*. A template is a synthesis of the relevant characteristics extracted from the source. Elements of the biometric measurement that are not used in the comparison algorithm are discarded in the template to reduce the filesize and to protect the identity of the enrollee.

During the enrollment phase, the template is simply stored somewhere (on a card or within a database or both). During the matching phase, the obtained template is passed to a matcher that compares it with other existing templates, estimating the distance between them using any algorithm (e.g. Hamming distance). The matching program will analyze the template with the input. This will then be output for any specified use or purpose (e.g. entrance in a restricted area). Selection of biometrics in any practical application depending upon the characteristic measurements and user requirements.*[10] In selecting a particular biometric, factors to consider include, performance, social acceptability, ease of circumvention and/or spoofing, robustness, population coverage, size of equipment needed and identity theft deterrence. Selection of a biometric based on user requirements considers sensor and device availability, computational time and reliability, cost, sensor size and power consumption.

1.2 Multimodal biometric system

Multimodal biometric systems use multiple sensors or biometrics to overcome the limitations of unimodal biometric systems.*[11] For instance iris recognition systems can be compromised by aging irises*[12] and finger scanning systems by worn-out or cut fingerprints. While unimodal biometric systems are limited by the integrity of their identifier, it is unlikely that several unimodal systems will suffer from identical limitations. Multimodal biometric systems can obtain sets of information from the same marker (i.e., multiple images of an iris, or scans of the same finger) or information from different biometrics (requiring fingerprint scans and, using voice recognition, a spoken passcode).[*][13][*][14]

Multimodal biometric systems can fuse these unimodal systems sequentially, simultaneously, a combination thereof, or in series, which refer to sequential, parallel, hierarchical and serial integration modes, respectively. Fusion of the biometrics information can occur at different stages of a recognition system. In case of feature level fusion, the data itself or the features extracted from multiple biometrics are fused. Matching-score level fusion consolidates the scores generated by multiple classifiers pertaining to different modalities. Finally, in case of decision level fusion the final results of multiple classifiers are combined via techniques such as majority voting. Feature level fusion is believed to be more effective than the other levels of fusion because the feature set contains richer information about the input biometric data than the matching score or the output decision of a classifier. Therefore, fusion at the feature level is expected to provide better recognition results.[*][11]

Spoof attacks consist in submitting fake biometric traits to biometric systems, and are a major threat that can curtail their security. Multi-modal biometric systems are commonly believed to be intrinsically more robust to spoof attacks, but recent studies[*][15] have shown that they can be evaded by spoofing even a single biometric trait.

1.3 Performance

The following are used as performance metrics for biometric systems:[*][16]

- **False match rate** (FMR, also called FAR = False Accept Rate): the probability that the system incorrectly matches the input pattern to a non-matching template in the database. It measures the percent of invalid inputs that are incorrectly accepted. In case of similarity scale, if the person is an imposter in reality, but the matching score is higher than the threshold, then he is treated as genuine. This increases the FMR, which thus also depends upon the threshold value.[*][10]

- **False non-match rate** (FNMR, also called FRR = False Reject Rate): the probability that the system fails to detect a match between the input pattern and a matching template in the database. It measures the percent of valid inputs that are incorrectly rejected.

- **Receiver operating characteristic** or relative operating characteristic (ROC): The ROC plot is a visual characterization of the trade-off between the FMR and the FNMR. In general, the matching algorithm performs a decision based on a threshold that determines how close to a template the input needs to be for it to be considered a match. If the threshold is reduced, there will be fewer false non-matches but more false accepts. Conversely, a higher threshold will reduce the FMR but increase the FNMR. A common variation is the *Detection error trade-off (DET)*, which is obtained using normal deviation scales on both axes. This more linear graph illuminates the differences for higher performances (rarer errors).

- **Equal error rate** or crossover error rate (EER or CER): the rate at which both acceptance and rejection errors are equal. The value of the EER can be easily obtained from the ROC curve. The EER is a quick way to compare the accuracy of devices with different ROC curves. In general, the device with the lowest EER is the most accurate.

- **Failure to enroll rate** (FTE or FER): the rate at which attempts to create a template from an input is unsuccessful. This is most commonly caused by low quality inputs.

- **Failure to capture rate** (FTC): Within automatic systems, the probability that the system fails to detect a biometric input when presented correctly.

- **Template capacity**: the maximum number of sets of data that can be stored in the system.

1.4 History of biometrics

An early cataloging of fingerprints dates back to 1891 when Juan Vucetich started a collection of fingerprints of criminals in Argentina. The History of Fingerprints. Josh Ellenbogen and Nitzan Lebovic argued that Biometrics is originated in the identificatory systems of criminal activity developed by Alphonse Bertillon (1853–1914) and developed by Francis Galton's theory of fingerprints and physiognomy.[*][17] According to Lebovic, Galton's work "led to the application of mathematical models to fingerprints, phrenology, and facial characteristics", as part of "absolute identification" and "a key to both inclusion and exclusion" of populations.[*][18] Accordingly, "the biometric system is the absolute political weapon of our era" and a form of "soft control".[*][19] The theoretician David Lyon showed that during the past two decades biometric systems have penetrated the civilian market, and blurred the lines between governmental forms of control and private corporate control.[20] Kelly A. Gates identified 9/11 as the turning point for the cultural language of our present: "in the language of cultural studies, the aftermath of 9/11 was a moment of articulation, where objects or events that have no necessary connection come together and a new discourse formation is established: automated facial recognition as a homeland security technology." Kelly A. Gates, *Our Biometric Future: Facial Recognition Technology and the Culture of Surveillance* (New York, 2011), p. 100.

1.5 Adaptive biometric systems

Adaptive biometric Systems aim to auto-update the templates or model to the intra-class variation of the operational data.[21] The two-fold advantages of these systems are solving the problem of limited training data and tracking the temporal variations of the input data through adaptation. Recently, adaptive biometrics have received a significant attention from the research community. This research direction is expected to gain momentum because of their key promulgated advantages. First, with an adaptive biometric system, one no longer needs to collect a large number of biometric samples during the enrollment process. Second, it is no longer necessary to re-enroll or retrain the system from scratch in order to cope with the changing environment. This convenience can significantly reduce the cost of maintaining a biometric system. Despite these advantages, there are several open issues involved with these systems. For mis-classification error (false acceptance) by the biometric system, cause adaptation using impostor sample. However, continuous research efforts are directed to resolve the open issues associated to the field of adaptive biometrics. More information about adaptive biometric systems can be found in the critical review by Rattani *et al*.

1.6 India's national ID program

India's national ID program called Aadhaar is the largest biometric database in the world. It is a biometrics-based digital identity assigned for a person's lifetime, verifiable online instantly in the public domain, at any time, from anywhere, in a paperless way. It is designed to enable government agencies to deliver a retail public service, securely based on biometric data (fingerprint, iris scan and face photo), along with demographic data (name, age, gender, address, parent/spouse name, mobile phone number) of a person. The data is transmitted in encrypted form over the internet for authentication, aiming to free it from the limitations of physical presence of a person at a given place.

About 550 million residents have been enrolled and assigned 480 million Aadhaar national identification numbers as of 7 November 2013.[22] It aims to cover the entire population of 1.2 billion in a few years.[23]

1.7 Recent advances in emerging biometrics

In recent times, biometrics based on brain (electroencephalogram) and heart (electrocardiogram) signals have emerged.[24][25] The research group at University of Kent led by Ramaswamy Palaniappan has shown that people have certain distinct brain and heart patterns that are specific for each individual. The advantage of such 'futuristic' technology is that it is more fraud resistant compared to conventional biometrics like fingerprints. However, such technology is generally more cumbersome and still has issues such as lower accuracy and poor reproducibility over time. This new generation of biometrical systems is called biometrics of intent and it aims to scan *intent*. The technology will analyze physiological features such as eye movement, body temperature, breathing etc. and predict dangerous behaviour or hostile intent before it materializes into action.

On the portability side of biometric products, more and more vendors are embracing significantly-miniaturized Biometric Authentication Systems (BAS) thereby driving elaborate cost savings especially for large scale deployments. This is also complimented by Kenneth Okereafor's innovative integrated liveness detection technique.,[26] which is a novel design of a framework that performs biometric liveness detection using / multi-modal random trait technique; and was first presented at the 2017 UKSim-AMSS 19th International Conference on Modelling & Simulation [27] and published by the IEEE International Journal of Simulation, Systems, Science and Technology. The innovative framework tackles biometric spoofing from a randomized trait analysis approach that guarantees more secure authentication.

1.7.1 Operator signatures

An operator signature is a biometric mode where the manner in which a person using a device or complex system is recorded as a verification template.[28] One potential use for this type of biometric signature is to distinguish among remote users of telerobotic surgery systems that utilize public networks for communication.[28]

1.7.2 Proposed requirement for certain public networks

John Michael (Mike) McConnell, a former vice admiral in the United States Navy, a former Director of U.S. National Intelligence, and Senior Vice President of Booz Allen Hamilton promoted the development of a future capability to require biometric authentication to access certain public networks in his keynote speech*[29] at the 2009 Biometric Consortium Conference.

A basic premise in the above proposal is that the person that has uniquely authenticated themselves using biometrics with the computer is in fact also the agent performing potentially malicious actions from that computer. However, if control of the computer has been subverted, for example in which the computer is part of a botnet controlled by a hacker, then knowledge of the identity of the user at the terminal does not materially improve network security or aid law enforcement activities.*[30]

Recently, another approach to biometric security was developed, this method scans the entire body of prospects to guarantee a better identification of this prospect. This method is not globally accepted because it is very complex and prospects are concerned about their privacy.

1.8 Issues and concerns

1.8.1 Human dignity

Biometrics have been considered also instrumental to the development of state authority*[31] (to put it in Foucauldian terms, of discipline and biopower*[32]). By turning the human subject into a collection of biometric parameters, biometrics would dehumanize the person,*[33] infringe bodily integrity, and, ultimately, offend human dignity.*[34]

In a well-known case,*[35] Italian philosopher Giorgio Agamben refused to enter the United States in protest at the United States Visitor and Immigrant Status Indicator (US-VISIT) program's requirement for visitors to be fingerprinted and photographed. Agamben argued that gathering of biometric data is a form of bio-political tattooing, akin to the tattooing of Jews during the Holocaust. According to Agamben, biometrics turn the human persona into a bare body. Agamben refers to the two words used by Ancient Greeks for indicating "life", *zoe*, which is the life common to animals and humans, just life; and *bios*, which is life in the human context, with meanings and purposes. Agamben envisages the reduction to bare bodies for the whole humanity.*[36] For him, a new bio-political relationship between citizens and the state is turning citizens into pure biological life (*zoe*) depriving them from their humanity (*bios*); and biometrics would herald this new world.

In Dark Matters: On the Surveillance of Blackness, surveillance scholar Simone Browne formulates a similar critique as Agamben, citing a recent study*[37] relating to biometrics R&D that found that the gender classification system being researched "is inclined to classify Africans as males and Mongoloids as females." *[37] Consequently, Browne argues that the conception of an objective biometric technology is difficult if such systems are subjectively designed, and are vulnerable to cause errors as described in the study above. The stark expansion of biometric technologies in both the public and private sector magnifies this concern. The increasing commodification of biometrics by the private sector adds to this danger of loss of human value. Indeed, corporations value the biometric characteristics more than the individuals value them.*[38] Browne goes on to suggest that modern society should incorporate a "biometric consciousness" that "entails informed public debate around these technologies and their application, and accountability by the state and the private sector, where the ownership of and access to one's own body data and other intellectual property that is generated from one's body data must be understood as a right." *[39]

Other scholars*[40] have emphasized, however, that the globalized world is confronted with a huge mass of people with weak or absent civil identities. Most developing countries have weak and unreliable documents and the poorer people in these countries do not have even those unreliable documents.*[41] Without certified personal identities, there is no certainty of right, no civil liberty.*[42] One can claim her rights, including the right to refuse to be identified, only if she is an identifiable subject, if she has a public identity. In such a sense, biometrics could play a pivotal role in supporting and promoting respect for human dignity and fundamental rights.*[43]

The biometrics of intent poses further risks. In his paper in Harvard International Review, Prof Nayef Al-Rodhan cautions about the high risks of miscalculations, wrongful accusations and infringements of civil liberties. Critics in the US have also signalled a conflict with the 4th Amendment.

1.8.2 Privacy and discrimination

It is possible that data obtained during biometric enrollment may be used in ways for which the enrolled individual has not consented. For example, most biometric features could disclose physiological and/or pathological medical conditions (e.g., some fingerprint patterns are related to chromosomal diseases, iris patterns could reveal genetic sex, hand vein patterns could reveal vascular diseases, most behavioral biometrics could reveal neurological diseases, etc.).*[44] Moreover, second generation biometrics, notably behavioral and electro-physiologic biometrics (e.g., based on electrocardiography, electroencephalography, electromyography), could be also used for emotion detection.*[45]

There are three categories of privacy concerns:*[46]

1. Unintended functional scope: The authentication goes further than authentication, such as finding a tumor.

2. Unintended application scope: The authentication process correctly identifies the subject when the subject did not wish to be identified.

3. Covert identification: The subject is identified without seeking identification or authentication, i.e. a subject's face is identified in a crowd.

1.8.3 Danger to owners of secured items

When thieves cannot get access to secure properties, there is a chance that the thieves will stalk and assault the property owner to gain access. If the item is secured with a biometric device, the damage to the owner could be irreversible, and potentially cost more than the secured property. For example, in 2005, Malaysian car thieves cut off the finger of a Mercedes-Benz S-Class owner when attempting to steal the car.*[47]

1.8.4 Cancelable biometrics

One advantage of passwords over biometrics is that they can be re-issued. If a token or a password is lost or stolen, it can be cancelled and replaced by a newer version. This is not naturally available in biometrics. If someone's face is compromised from a database, they cannot cancel or reissue it. If the electronic biometric identifier is stolen, it is nearly impossible to change a biometric feature. This renders the person's biometric feature questionable for future use in authentication, such as the case with the hacking of security-clearance-related background information from the Office of Personnel Management (OPM) in the United States.*[48]

Cancelable biometrics is a way in which to incorporate protection and the replacement features into biometrics to create a more secure system. It was first proposed by Ratha et al.*[49]

"Cancelable biometrics refers to the intentional and systematically repeatable distortion of biometric features in order to protect sensitive user-specific data. If a cancelable feature is compromised, the distortion characteristics are changed, and the same biometrics is mapped to a new template, which is used subsequently. Cancelable biometrics is one of the major categories for biometric template protection purpose besides biometric cryptosystem." *[50] In biometric cryptosystem, "the error-correcting coding techniques are employed to handle intraclass variations." *[51] This ensures a high level of security but has limitations such as specific input format of only small intraclass variations.

Several methods for generating new exclusive biometrics have been proposed. The first fingerprint-based cancelable biometric system was designed and developed by Tulyakov et al.*[52] Essentially, cancelable biometrics perform a distortion of the biometric image or features before matching. The variability in the distortion parameters provides the cancelable nature of the scheme. Some of the proposed techniques operate using their own recognition engines, such as Teoh et al.*[53] and Savvides et al.,*[54] whereas other methods, such as Dabbah et al.,*[55] take the advantage of the advancement of the well-established biometric research for their recognition front-end to conduct recognition. Although this increases the restrictions on the protection system, it makes the cancellable templates more accessible for available biometric technologies

1.8.5 Soft biometrics

Soft biometrics traits are physical, behavioral or adhered human characteristics that have been derived from the way human beings normally distinguish their peers (e.g. height, gender, hair color). They are used to complement the identity information provided by the primary biometric identifiers . Although soft biometric characteristics lack the distinctiveness and permanence to recognize an individual uniquely and reliably, and can be easily faked, they provide some evidence about the users identity that could be beneficial. In other words, despite the fact they are unable to individualize a subject, they are effective in distinguishing between people. Combinations of personal attributes like gender, race, eye color, height and other visible identification marks can be used to improve the performance of traditional biometric systems.*[56] Most soft biometrics can be easily collected and are actually collected during enrollment.Two main ethical issues are raised by soft biometrics.*[57] First, some of soft biometric traits are strongly cultural based; e.g., skin colors for determining ethnicity risk to support racist approaches, biometric sex recognition at the best recognizes gender from tertiary sexual characters, being unable to determine genetic and chromosomal sexes; soft biometrics for aging recognition are often deeply influenced by ageist stereotypes, etc. Second, soft biometrics have strong potential for categorizing and profiling people, so risking of supporting processes of stigmatization and exclusion.*[58]

1.8.6 International sharing of biometric data

Many countries, including the United States, are planning to share biometric data with other nations.

In testimony before the US House Appropriations Committee, Subcommittee on Homeland Security on "biometric identification" in 2009, Kathleen Kraninger and Robert A Mocny*[59] commented on international cooperation and collaboration with respect to biometric data, as follows:

According to an article written in 2009 by S. Magnuson in the National Defense Magazine entitled "Defense Department Under Pressure to Share Biometric Data" the United States has bilateral agreements with other nations aimed at sharing biometric data.*[60] To quote that article:

1.8.7 Likelihood of full governmental disclosure

Certain members of the civilian community are worried about how biometric data is used but full disclosure may not be forthcoming. In particular, the Unclassified Report of the United States' Defense Science Board Task Force on Defense Biometrics states that it is wise to protect, and sometimes even to disguise, the true and total extent of national capabilities in areas related directly to the conduct of security-related activities.*[61] This also potentially applies to Biometrics. It goes on to say that this is a classic feature of intelligence and military operations. In short, the goal is to preserve the security of 'sources and methods'.

1.9 Countries applying biometrics

Main article: Countries applying biometrics

Countries using biometrics include Australia, Brazil, Canada, Cyprus, Greece, China, Gambia, Germany, India, Iraq, Israel, Italy, Malaysia, Netherlands, New Zealand, Nigeria, Norway, Pakistan, South Africa, Saudi Arabia, Tanzania,*[62] Ukraine, United Arab Emirates, United Kingdom, United States and Venezuela.

Among low to middle income countries, roughly 1.2 billion people have already received identification through a biometric identification program.*[63]

There are also numerous countries applying biometrics for voter registration and similar electoral purposes. According to the International IDEA's ICTs in Elections Database,*[64] some of the countries using (2017) Biometric Voter Registration (BVR) are Armenia, Angola, Bangladesh, Bhutan, Bolivia, Brazil, Burkina Faso, Cambodia, Cameroon, Chad, Colombia, Comoros, Congo (Democratic Republic of), Costa Rica, Ivory Coast, Dominican Republic, Fiji, Gambia, Ghana, Guatemala, India, Iraq, Kenya, Lesotho, Liberia, Malawi, Mali, Mauritania, Mexico, Morocco, Mozambique, Namibia, Nepal, Nicaragua, Nigeria, Panama, Peru, The Philippines, Senegal, Sierra Leone, Solomon Islands, Somaliland, Swaziland, Tanzania, Uganda, Uruguay, Venezuela, Yemen, Zambia, and Zimbabwe.*[65]*[66]*[67]

1.10 See also

- Aadhaar
- Access control
- AFIS
- AssureSign
- BioAPI
- Biometric passport
- Biometric voter registration
- Biometrics in schools
- BioSlimDisk
- Facial recognition system
- Fingerprint recognition
- Fuzzy extractor
- Gait analysis
- Government database
- Hand geometry
- Handwritten biometric recognition
- Identity Cards Act 2006
- International Identity Federation
- Iris recognition
- Keystroke dynamics
- Private biometrics
- Retinal scan
- Signature recognition
- Smart city

- Speaker recognition
- Surveillance
- Vein matching
- Voice analysis

1.11 Notes

[1] As Jain and Ross (2008, footnote 4 on page 1) point out, "the term *biometric authentication* is perhaps more appropriate than *biometrics* since the latter has been historically used in the field of statistics to refer to the analysis of biological (particularly medical) data [36]" (wikilink added to original quote).

[2] Strictly speaking, *voice* is also a physiological trait because every person has a different vocal tract, but voice recognition is classed as behavioural as it is affected by a person's mood. Biometric voice recognition is separate and distinct from speech recognition with the latter being concerned with accurate understanding of speech content rather than identification or recognition of the person speaking.

[3] Systems can be designed to use a template stored on media like an e-Passport or smart card, rather than a remote database.

1.12 References

[1] "Biometrics: Overview". Biometrics.cse.msu.edu. 6 September 2007. Archived from the original on 7 January 2012. Retrieved 2012-06-10.

[2] Jain, A.; Hong, L. and Pankanti, S. (2000). "Biometric Identification". *Communications of the ACM*, 43(2), p. 91–98. DOI 10.1145/328236.328110

[3] Jain, Anil K.; Ross, Arun (2008). "Introduction to Biometrics". In Jain, AK; Flynn; Ross, A. *Handbook of Biometrics*. Springer. pp. 1–22. ISBN 978-0-387-71040-2.

[4] Damaševičius, R.; Maskeliūnas, R.; Venčkauskas, A.; Woźniak, M. Smartphone User Identity Verification Using Gait Characteristics, Symmetry 2016, 8, 100.

[5] Sahidullah, Md (2015). "Enhancement of Speaker Recognition Performance Using Block Level, Relative and Temporal Information of Subband Energies". PhD Thesis (Indian Institute of Technology Kharagpur).

[6] "Biometrics for Secure Authentication" (PDF). Archived from the original (PDF) on 25 March 2012. Retrieved 29 July 2012.

[7] Weaver, A. C. (2006). "Biometric Authentication". *Computer*, 39 (2), p. 96–97. DOI 10.1109/MC.2006.47

[8] Jain, A. K.; Bolle, R.; Pankanti, S., eds. (1999). *Biometrics: Personal Identification in Networked Society*. Kluwer Academic Publications. ISBN 978-0-7923-8345-1.

[9] Bleicher, Paul (2005). "Biometrics comes of age: despite accuracy and security concerns, biometrics are gaining in popularity". *Applied Clinical Trials*.

[10] Sahoo, SoyujKumar; Mahadeva Prasanna, SR (1 January 2012). Mahadeva Prasanna, SR, Choubisa, Tarun. "Multimodal Biometric Person Authentication : A Review". *IETE Technical Review*. **29** (1): 54. doi:10.4103/0256-4602.93139. Archived from the original on 17 September 2013. Retrieved 23 February 2012. Missing |last2= in Authors list (help)

[11] M. Haghighat, M. Abdel-Mottaleb, & W. Alhalabi (2016). Discriminant Correlation Analysis: Real-Time Feature Level Fusion for Multimodal Biometric Recognition. IEEE Transactions on Information Forensics and Security, 11(9), 1984–1996.

[12] "Questions Raised About Iris Recognition Systems". *Science Daily*. 12 July 2012.

[13] Saylor, Michael (2012). *The Mobile Wave: How Mobile Intelligence Will Change Everything*. Perseus Books/Vanguard Press. p. 99.

[14] Bill Flook (3 October 2013). "This is the 'biometric war' Michael Saylor was talking about". *Washington Business Journal*.

[15] Zahid Akhtar, "Security of Multimodal Biometric Systems against Spoof Attacks", Department of Electrical and Electronic Engineering, University of Cagliari, Cagliari, Italy, 6 March 2012.

[16] "Characteristics of Biometric Systems". Cernet. Archived from the original on 17 October 2008.

[17] Josh Ellenbogen, *Reasoned and Unreasoned Images: The Photography of Bertillon, Galton, and Marey* (University Park, PA, 2012)

[18] Nitzan Lebovic, "Biometrics or the Power of the Radical Center", in *Critical Inquiry* 41:4 (Summer, 2015), 841–868.

[19] Nitzan Lebovic, "Biometrics or the Power of the Radical Center", in *Critical Inquiry* 41:4 (Summer, 2015), p. 853.

[20] David Lyon, *Surveillance Society: Monitoring Everyday Life* (Philadelphia, 2001).

[21] A. Rattani, "Adaptive Biometric System based on Template Update Procedures", PhD thesis, University of Cagliari, Italy, 2010

[22] "Aadhaar scheme does not violate fundamental rights, says UIDAI". Zee News. October 22, 2013.

1.12. REFERENCES

[23] "Building a Biometric National ID: Lessons for Developing Countries from India's Universal ID Program", Alan Gelb and Julia Clark, The Center for Global Development, October 2012, http://www.cgdev.org/doc/full_text/GelbClarkUID/1426583.html

[24] [R. Palaniappan, "Electroencephalogram signals from imagined activities: A novel biometric identifier for a small population", published in E. Corchado *et al.* (eds): *Intelligent Data Engineering and Automated Learning – IDEAL 2006*, Lecture Notes in Computer Science, vol. 4224, pp. 604–611, Springer-Verlag, Berlin Heidelberg, 2006. DOI: 10.1007/11875581_73]

[25] R. Palaniappan, and S. M. Krishnan, "Identifying individuals using ECG signals", *Proceedings of International Conference on Signal Processing and Communications*, Bangalore, India, pp. 569–572, 11–14 December 2004. DOI: 10.1109/SPCOM.2004.1458524]

[26] [5] K. U. Okereafor, C. Onime and O. E. Osuagwu, "Multibiometric Liveness Detection - A New Perspective," West African Journal of Industrial and Academic Research, vol. 16, no. 1, pp. 26 - 37, 2016 (https://www.ajol.info/index.php/wajiar/article/view/145878)

[27] K. U. Okereafor, C. Onime and O. E. Osuagwu, "Enhancing Biometric Liveness Detection Using Trait Randomization Technique," 2017 UKSim-AMSS 19th International Conference on Modelling & Simulation, University of Cambridge, Conference Proceedings, pp. 28 – 33, 2017 (http://uksim.info/uksim2017/CD/data/2735a028.pdf)

[28] Langston, Jennifer (8 May 2015). "Researchers hack Teleoperated Surgical Robot to Reveal Security Flaws". *Scientific Computing*. New Jersey. Retrieved 17 May 2015.

[29] McConnell, Mike (January 2009). *KeyNote Address*. Biometric Consortium Conference. Tampa Convention Center, Tampa, Florida. Retrieved 20 February 2010.

[30] Schneier, Bruce. "The Internet: Anonymous Forever". Retrieved 1 October 2011.

[31] Breckenridge K. (2005). "The Biometric State: The Promise and Peril of Digital Government in the New South Africa". *Journal of Southern African Studies*, 31:2, 267–82

[32] Epstein C. (2007), "Guilty Bodies, Productive Bodies, Destructive Bodies: Crossing the Biometric Borders". *International Political Sociology*, 1:2, 149–64

[33] Pugliese J. (2010), *Biometrics: Bodies, Technologies, Biopolitics*. New York: Routledge

[34] French National Consultative Ethics Committee for Health and Life Sciences (2007), Opinion N° 98, "Biometrics, identifying data and human rights"

[35] Agamben, G. (2008). "No to bio-political tattooing". *Communication and Critical/Cultural Studies*, 5(2), 201–202. Reproduced from Le Monde (10 January 2004).

[36] Agamben G.(1998), *Homo Sacer: Sovereign Power and Bare Life*. Trans. Daniel Heller-Roazen. Stanford: Stanford University Press

[37] Gao, Wei; Ai, Haizhou. *Face Gender Classification on Consumer Images in a Multiethnic Environment*. pp. 169–178. doi:10.1007/978-3-642-01793-3_18.

[38] Walker, Elizabeth (2015). "Biometric Boom: How the private sector Commodifies Human characteristics". *Fordham Intellectual Property, Media & Entertainment Law Journal*.

[39] Browne, Simone (2015). *Dark Matters: On the Surveillance of Blackness*. Duke University Press. p. 116.

[40] Mordini, E; Massari, S. (2008), "Body, Biometrics and Identity" *Bioethics*, 22, 9:488

[41] UNICEF, Birth Registration

[42] Dahan M., Gelb A. (2015) "The Role of Identification in the Post-2015 Development Agenda" – World Bank Working Paper No. 98294 08/2015;

[43] Mordini E, Rebera A (2011) "No Identification Without Representation: Constraints on the Use of Biometric Identification Systems". *Review of Policy Research*, 29, 1: 5–20

[44] Mordini E, Ashton H,(2012), "The Transparent Body – Medical Information, Physical Privacy and Respect for Body Integrity", in Mordini E, Tzovaras D (eds), *Second Generation Biometrics: the Ethical and Social Context*. Springer-Verlag: Berlin

[45] Mordini E, Tzovaras D,(2012), *Second Generation Biometrics: the Ethical and Social Context*. Springer-Verlag: Berlin

[46] Pfleeger, Charles; Pfleeger, Shari (2007). *Security in Computing* (4th ed.). Boston: Pearson Education. p. 220. ISBN 978-0-13-239077-4.

[47] Kent, Jonathan (31 March 2005). "Malaysia car thieves steal finger". *BBC Online*. Kuala Lumpur. Retrieved 11 December 2010.

[48] "Office of Personnel Management data breach". *Wikipedia*. 2017-09-08.

[49] N. K. Ratha, J. H. Connell, and R. M. Bolle, "Enhancing security and privacy in biometrics-based authentication systems", *IBM Systems Journal*, vol. 40, pp. 614–634, 2001.

[50] "Cancelable biometrics – Scholarpedia". www.scholarpedia.org. Retrieved 2015-11-05.

[51] Feng, Y. C.; Yuen, P. C.; Jain, A. K. (2010-03-01). "A Hybrid Approach for Generating Secure and Discriminating Face Template". *IEEE Transactions on Information Forensics and Security*. **5** (1): 103–117. ISSN 1556-6013. doi:10.1109/TIFS.2009.2038760.

[52] S. Tulyakov, F. Farooq, and V. Govindaraju, "Symmetric Hash Functions for Fingerprint Minutiae", *Proc. Int'l Workshop Pattern Recognition for Crime Prevention, Security, and Surveillance*, pp. 30–38, 2005

[53] A. B. J. Teoh, A. Goh, and D. C. L. Ngo, "Random Multispace Quantization as an Analytic Mechanism for BioHashing of Biometric and Random Identity Inputs", *Pattern Analysis and Machine Intelligence, IEEE Transactions on*, vol. 28, pp. 1892–1901, 2006.

[54] M. Savvides, B. V. K. V. Kumar, and P. K. Khosla, "'Corefaces' – Robust Shift-Invariant PCA based Correlation Filter for Illumination Tolerant Face Recognition", presented at IEEE Computer Society Conference on Computer Vision and Pattern Recognition (CVPR'04), 2004.

[55] M. A. Dabbah, W. L. Woo, and S. S. Dlay, "Secure Authentication for Face Recognition", presented at Computational Intelligence in Image and Signal Processing, 2007. CIISP 2007. IEEE Symposium on, 2007.

[56] Ratha, N. K., J. H. Connell, and R. M. Bolle. (2001). "Enhancing security and privacy in biometrics based authentication systems". *IBM Systems Journal* 40(3): 614–634.

[57] Mordini E, Ashton H (2012), "The Transparent Body – Medical Information, Physical Privacy and Respect for Body Integrity'". In Mordini E, Tzovaras D (eds), *Second Generation Biometrics: the Ethical and Social Context*. Berlin: Springer-Verlag, 2057–83

[58] Mordini E (2013) *Biometrics*. In Henk A. M. J. ten Have, Bert Gordijn (eds) *Handbook of Global Bioethics* Berlin: Springer, 341–356

[59] "Testimony of Deputy Assistant Secretary for Policy Kathleen Kraninger, Screening Coordination, and Director Robert A. Mocny, US-VISIT, National Protection and Programs Directorate, before the House Appropriations Committee, Subcommittee on Homeland Security, 'Biometric Identification'". US Department of Homeland Security. March 2009. Retrieved 20 February 2010.

[60] Magnuson, S (January 2009). "Defense department under pressure to share biometric data." *NationalDefenseMagazine.org*. Retrieved 20 February 2010.

[61] Defense Science Board (DSB) (September 2006). "On Defense Biometrics" (PDF). Unclassified Report of the Defense Science Board Task Force. Washington, D.C.: Office of the Under Secretary of Defense For Acquisition, Technology, and Logistics: 84. Retrieved 20 February 2010. |chapter= ignored (help)

[62] web article dated 24 February 2015 in *planet biometrics* entitled "Biometric voter registration launches in Tanzania" accessed 21 January 2016

[63] Gelb, Alan; Julia Clark (2013). *Identification for Development: The Biometrics Revolution*. The Center for Global Development.

[64] "ICTs in Elections Database | International IDEA". *www.idea.int*. Retrieved 2017-07-19.

[65] "If the EMB uses technology to collect voter registration data, is biometric data captured and used during registration? | International IDEA". *www.idea.int*. Retrieved 2017-07-19.

[66] "The Biometric ID Grid: A Country-by-Country Guide : The Corbett Report". *www.corbettreport.com*. Retrieved 2017-07-19.

[67] "Biometric Voter Registration and Voter Identification —". *aceproject.org*. Retrieved 2017-07-19.

1.13 Further reading

- White Paper – What Are Biometrics?

- Biometrics Glossary – Glossary of Biometric Terms based on information derived from the National Science and Technology Council (NSTC) Subcommittee on Biometrics. Published by Fulcrum Biometrics, LLC, July 2013

- Biomtrics Institute Privacy Code, September 2006

- Biometric Vulnerability Assessment Framework, Published by the Biometrics Institute, 2007–2011

- TechCast Article Series, Vivian Chu and Gayathri Rajendran, GWU, Use of Biometrics.

- Delac, K., Grgic, M. (2004). A Survey of Biometric Recognition Methods.

- Biometric Technology Application Manual. Published by the National Biometric Security Project (NBSP), the BTAM is a comprehensive reference manual on biometric technology applications.

- "Fingerprints Pay For School Lunch". (2001). Retrieved 2008-03-02.

- "Germany to phase-in biometric passports from November 2005". (2005). E-Government News. Retrieved 2006-06-11.

- Oezcan, V. (2003). "Germany Weighs Biometric Registration Options for Visa Applicants", Humboldt University Berlin. Retrieved 2006-06-11.

- Ulrich Hottelet: Hidden champion – Biometrics between boom and big brother, German Times, January 2007.

- , The Fundamentals of Digital Forensics in Computer Reactive Security(2) by Kenneth Okereafor, January 2010.

- Paul Benjamin Lowry, Jackson Stephens, Aaron Moyes, Sean Wilson, and Mark Mitchell (2005). "Biometrics, a critical consideration in information security management", in Margherita Pagani, ed. *Encyclopedia of Multimedia Technology and Networks*, Idea Group Inc., pp. 69–75.

- Mordini E., Green M. (eds) (2008), *Identity, Security, and Democracy*, IOS Press NATO Series, Brussels

- Mordini E, Tzovaras D (ads) (2012), *Second Generation Biometrics: the Ethical and Social Context*, Springer, The International Library of Ethics, Law and Technology, Berlin: Springer-Verlag

- Marcelo Luiz Brocardo, Issa Traore, and Isaac Woungang. 2015. Authorship verification of e-mail and tweet messages applied for continuous authentication. J. Comput. Syst. Sci. 81, 8 (December 2015), 1429–1440. DOI=https://dx.doi.org/10.1016/j.jcss.2014.12.019

1.14 External links

- The dictionary definition of biometrics at Wiktionary

Chapter 2

Access control

A sailor allows a driver to enter a military installation.

In the fields of physical security and information security, **access control** (**AC**) is the selective restriction of access to a place or other resource.*[1] The act of *accessing* may mean consuming, entering, or using. Permission to access a resource is called *authorization*.

Locks and login credentials are two analogous mechanisms of access control.

2.1 Physical security

Main article: Physical security

Geographical access control may be enforced by personnel (e.g., border guard, bouncer, ticket checker), or with a device such as a turnstile. There may be fences to avoid circumventing this access control. An alternative of access control in the strict sense (physically controlling access itself) is a system of checking authorized presence, see e.g. Ticket controller (transportation). A variant is exit control, e.g. of a shop (checkout) or a country.

The term access control refers to the practice of restricting entrance to a property, a building, or a room to authorized persons. Physical access control can be achieved by a human (a guard, bouncer, or receptionist), through mechan-

Drop Arm Optical Turnstiles Manufactured by Q-Lane Turnstiles LLc

Underground entrance to the New York City Subway system

ical means such as locks and keys, or through technological means such as access control systems like the mantrap. Within these environments, physical key management may also be employed as a means of further managing and monitoring access to mechanically keyed areas or access to certain small assets.

2.1. PHYSICAL SECURITY

ter or exit, where they are allowed to exit or enter, and when they are allowed to enter or exit. Historically, this was partially accomplished through keys and locks. When a door is locked, only someone with a key can enter through the door, depending on how the lock is configured. Mechanical locks and keys do not allow restriction of the key holder to specific times or dates. Mechanical locks and keys do not provide records of the key used on any specific door, and the keys can be easily copied or transferred to an unauthorized person. When a mechanical key is lost or the key holder is no longer authorized to use the protected area, the locks must be re-keyed.

Electronic access control uses computers to solve the limitations of mechanical locks and keys. A wide range of credentials can be used to replace mechanical keys. The electronic access control system grants access based on the credential presented. When access is granted, the door is unlocked for a predetermined time and the transaction is recorded. When access is refused, the door remains locked and the attempted access is recorded. The system will also monitor the door and alarm if the door is forced open or held open too long after being unlocked.

2.1.1 Access control system operation

When a credential is presented to a reader, the reader sends the credential's information, usually a number, to a control panel, a highly reliable processor. The control panel compares the credential's number to an access control list, grants or denies the presented request, and sends a transaction log to a database. When access is denied based on the access control list, the door remains locked. If there is a match between the credential and the access control list, the control panel operates a relay that in turn unlocks the door. The control panel also ignores a door open signal to prevent an alarm. Often the reader provides feedback, such as a flashing red LED for an access denied and a flashing green LED for an access granted.

The above description illustrates a single factor transaction. Credentials can be passed around, thus subverting the access control list. For example, Alice has access rights to the server room, but Bob does not. Alice either gives Bob her credential, or Bob takes it; he now has access to the server room. To prevent this, two-factor authentication can be used. In a two factor transaction, the presented credential and a second factor are needed for access to be granted; another factor can be a PIN, a second credential, operator intervention, or a biometric input.

There are three types (factors) of authenticating information:[*][2]

- something the user knows, e.g. a password, pass-

Physical security access control with a hand geometry scanner

Example of fob based access control using an ACT reader

Physical access control is a matter of who, where, and when. An access control system determines who is allowed to en-

phrase or PIN

- something the user has, such as smart card or a key fob
- something the user is, such as fingerprint, verified by biometric measurement

Passwords are a common means of verifying a user's identity before access is given to information systems. In addition, a fourth factor of authentication is now recognized: someone you know, whereby another person who knows you can provide a human element of authentication in situations where systems have been set up to allow for such scenarios. For example, a user may have their password, but have forgotten their smart card. In such a scenario, if the user is known to designated cohorts, the cohorts may provide their smart card and password, in combination with the extant factor of the user in question, and thus provide two factors for the user with the missing credential, giving three factors overall to allow access.

2.1.2 Credential

A credential is a physical/tangible object, a piece of knowledge, or a facet of a person's physical being that enables an individual access to a given physical facility or computer-based information system. Typically, credentials can be something a person knows (such as a number or PIN), something they have (such as an access badge), something they are (such as a biometric feature), or some combination of these items. This is known as multi-factor authentication. The typical credential is an access card or key-fob, and newer software can also turn users' smartphones into access devices.*[3]

There are many card technologies including magnetic stripe, bar code, Wiegand, 125 kHz proximity, 26-bit card-swipe, contact smart cards, and contactless smart cards. Also available are key-fobs, which are more compact than ID cards, and attach to a key ring. Biometric technologies include fingerprint, facial recognition, iris recognition, retinal scan, voice, and hand geometry. The built-in biometric technologies found on newer smartphones can also be used as credentials in conjunction with access software running on mobile devices.*[4] In addition to older more traditional card access technologies, newer technologies such as Near field communication (NFC) and Bluetooth low energy also have potential to communicate user credentials to readers for system or building access.*[5]*[6]*[7]

2.1.3 Access control system components

An access control point can be a door, turnstile, parking gate, elevator, or other physical barrier, where granting access can be electronically controlled. Typically, the access point is a door. An electronic access control door can contain several elements. At its most basic, there is a stand-alone electric lock. The lock is unlocked by an operator with a switch. To automate this, operator intervention is replaced by a reader. The reader could be a keypad where a code is entered, it could be a card reader, or it could be a biometric reader. Readers do not usually make an access decision, but send a card number to an access control panel that verifies the number against an access list. To monitor the door position a magnetic door switch can be used. In concept, the door switch is not unlike those on refrigerators or car doors. Generally only entry is controlled, and exit is uncontrolled. In cases where exit is also controlled, a second reader is used on the opposite side of the door. In cases where exit is not controlled, free exit, a device called a request-to-exit (REX) is used. Request-to-exit devices can be a push-button or a motion detector. When the button is pushed, or the motion detector detects motion at the door, the door alarm is temporarily ignored while the door is opened. Exiting a door without having to electrically unlock the door is called mechanical free egress. This is an important safety feature. In cases where the lock must be electrically unlocked on exit, the request-to-exit device also unlocks the door.

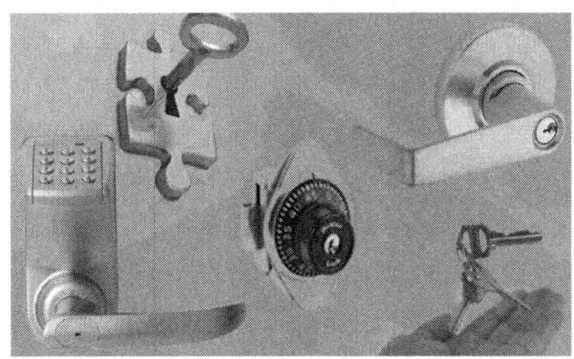

Various control system components

2.1.4 Access control topology

Access control decisions are made by comparing the credential to an access control list. This look-up can be done by a host or server, by an access control panel, or by a reader. The development of access control systems has seen a steady push of the look-up out from a central host to the edge of the system, or the reader. The predominant topology circa 2009 is hub and spoke with a control panel as the hub, and the readers as the spokes. The look-up and control functions are by the control panel. The spokes communicate through a serial connection; usually RS-485. Some manufactures are pushing the decision making to the edge

2.1. PHYSICAL SECURITY

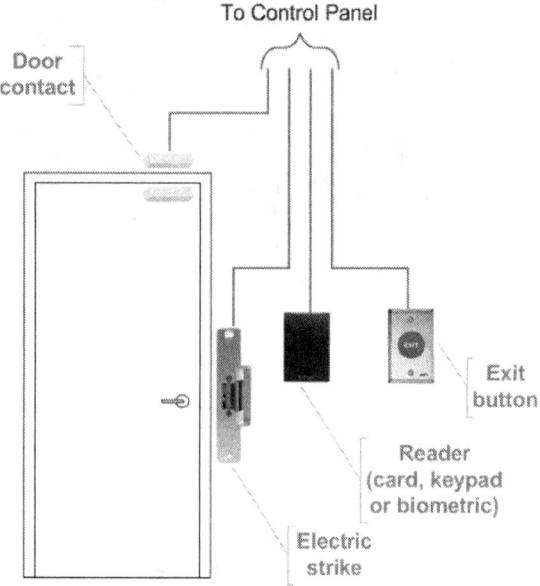

Typical access control door wiring

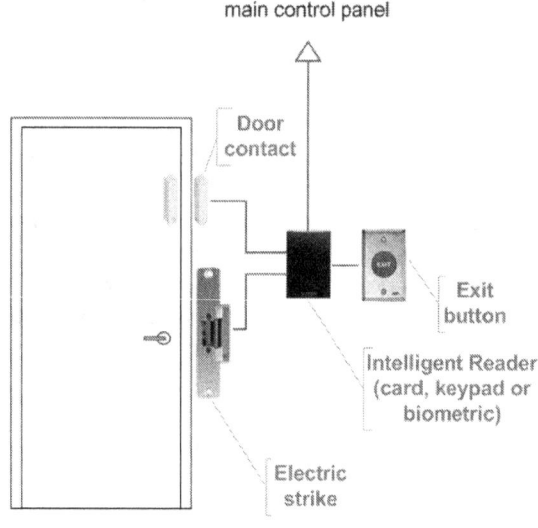

Access control door wiring when using intelligent readers

by placing a controller at the door. The controllers are IP enabled, and connect to a host and database using standard networks.

2.1.5 Types of readers

Access control readers may be classified by the functions they are able to perform:

- Basic (non-intelligent) readers: simply read card number or PIN, and forward it to a control panel. In case of biometric identification, such readers output the ID number of a user. Typically, Wiegand protocol is used for transmitting data to the control panel, but other options such as RS-232, RS-485 and Clock/Data are not uncommon. This is the most popular type of access control readers. Examples of such readers are RF Tiny by RFLOGICS, ProxPoint by HID, and P300 by Farpointe Data.

- Semi-intelligent readers: have all inputs and outputs necessary to control door hardware (lock, door contact, exit button), but do not make any access decisions. When a user presents a card or enters a PIN, the reader sends information to the main controller, and waits for its response. If the connection to the main controller is interrupted, such readers stop working, or function in a degraded mode. Usually semi-intelligent readers are connected to a control panel via an RS-485 bus. Examples of such readers are InfoProx Lite IPL200 by CEM Systems, and AP-510 by Apollo.

- Intelligent readers: have all inputs and outputs necessary to control door hardware; they also have memory and processing power necessary to make access decisions independently. Like semi-intelligent readers, they are connected to a control panel via an RS-485 bus. The control panel sends configuration updates, and retrieves events from the readers. Examples of such readers could be InfoProx IPO200 by CEM Systems, and AP-500 by Apollo. There is also a new generation of intelligent readers referred to as "IP readers". Systems with IP readers usually do not have traditional control panels, and readers communicate directly to a PC that acts as a host.

Some readers may have additional features such as an LCD and function buttons for data collection purposes (i.e. clock-in/clock-out events for attendance reports), camera/speaker/microphone for intercom, and smart card read/write support.

Access control readers may also be classified by their type of identification technology.

2.1.6 Access control system topologies

1. Serial controllers. Controllers are connected to a host PC via a serial RS-485 communication line (or via 20mA current loop in some older systems). External RS-232/485 converters or internal RS-485 cards have to be installed, as standard PCs do not have RS-485 communication ports.

Advantages:

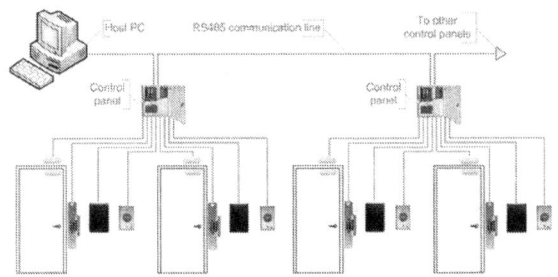

Access control system using serial controllers

- RS-485 standard allows long cable runs, up to 4000 feet (1200 m)

- Relatively short response time. The maximum number of devices on an RS-485 line is limited to 32, which means that the host can frequently request status updates from each device, and display events almost in real time.

- High reliability and security as the communication line is not shared with any other systems.

Disadvantages:

- RS-485 does not allow Star-type wiring unless splitters are used

- RS-485 is not well suited for transferring large amounts of data (i.e. configuration and users). The highest possible throughput is 115.2 kbit/sec, but in most system it is downgraded to 56.2 kbit/sec, or less, to increase reliability.

- RS-485 does not allow the host PC to communicate with several controllers connected to the same port simultaneously. Therefore, in large systems, transfers of configuration, and users to controllers may take a very long time, interfering with normal operations.

- Controllers cannot initiate communication in case of an alarm. The host PC acts as a master on the RS-485 communication line, and controllers have to wait until they are polled.

- Special serial switches are required, in order to build a redundant host PC setup.

- Separate RS-485 lines have to be installed, instead of using an already existing network infrastructure.

- Cable that meets RS-485 standards is significantly more expensive than regular Category 5 UTP network cable.

- Operation of the system is highly dependent on the host PC. In the case that the host PC fails, events from controllers are not retrieved, and functions that require interaction between controllers (i.e. anti-passback) stop working.

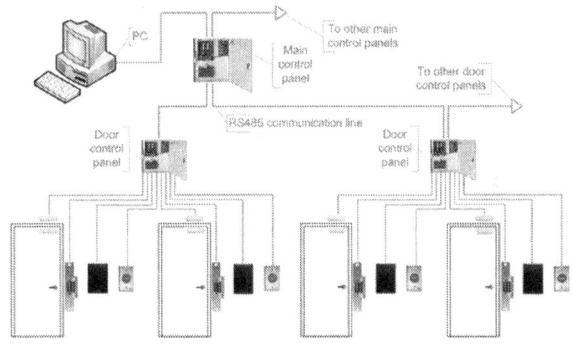

Access control system using serial main and sub-controllers

2. Serial main and sub-controllers. All door hardware is connected to sub-controllers (a.k.a. door controllers or door interfaces). Sub-controllers usually do not make access decisions, and instead forward all requests to the main controllers. Main controllers usually support from 16 to 32 sub-controllers.

Advantages:

- Work load on the host PC is significantly reduced, because it only needs to communicate with a few main controllers.

- The overall cost of the system is lower, as sub-controllers are usually simple and inexpensive devices.

- All other advantages listed in the first paragraph apply.

Disadvantages:

- Operation of the system is highly dependent on main controllers. In case one of the main controllers fails, events from its sub-controllers are not retrieved, and functions that require interaction between sub-controllers (i.e. anti-passback) stop working.

- Some models of sub-controllers (usually lower cost) do not have the memory or processing power to make access decisions independently. If the main controller fails, sub-controllers change to degraded mode in which doors are either completely locked or unlocked, and no events are recorded. Such sub-controllers should be avoided, or used only in areas that do not require high security.

2.1. PHYSICAL SECURITY

- Main controllers tend to be expensive, therefore such a topology is not very well suited for systems with multiple remote locations that have only a few doors.
- All other RS-485-related disadvantages listed in the first paragraph apply.

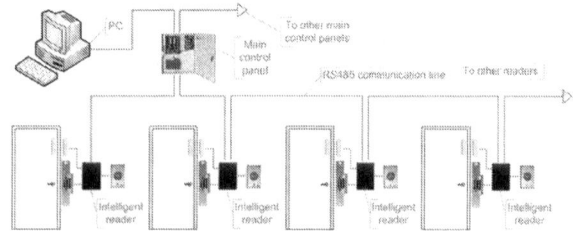

Access control system using serial main controller and intelligent readers

3. Serial main controllers & intelligent readers. All door hardware is connected directly to intelligent or semi-intelligent readers. Readers usually do not make access decisions, and forward all requests to the main controller. Only if the connection to the main controller is unavailable, will the readers use their internal database to make access decisions and record events. Semi-intelligent reader that have no database and cannot function without the main controller should be used only in areas that do not require high security. Main controllers usually support from 16 to 64 readers. All advantages and disadvantages are the same as the ones listed in the second paragraph.

4. Serial controllers with terminal servers. In spite of the rapid development and increasing use of computer networks, access control manufacturers remained conservative, and did not rush to introduce network-enabled products. When pressed for solutions with network connectivity, many chose the option requiring less efforts: addition of a terminal server, a device that converts serial data for transmission via LAN or WAN.

Advantages:

- Allows utilizing the existing network infrastructure for connecting separate segments of the system.
- Provides a convenient solution in cases when the installation of an RS-485 line would be difficult or impossible.

Disadvantages:

- Increases complexity of the system.
- Creates additional work for installers: usually terminal servers have to be configured independently, and not through the interface of the access control software.

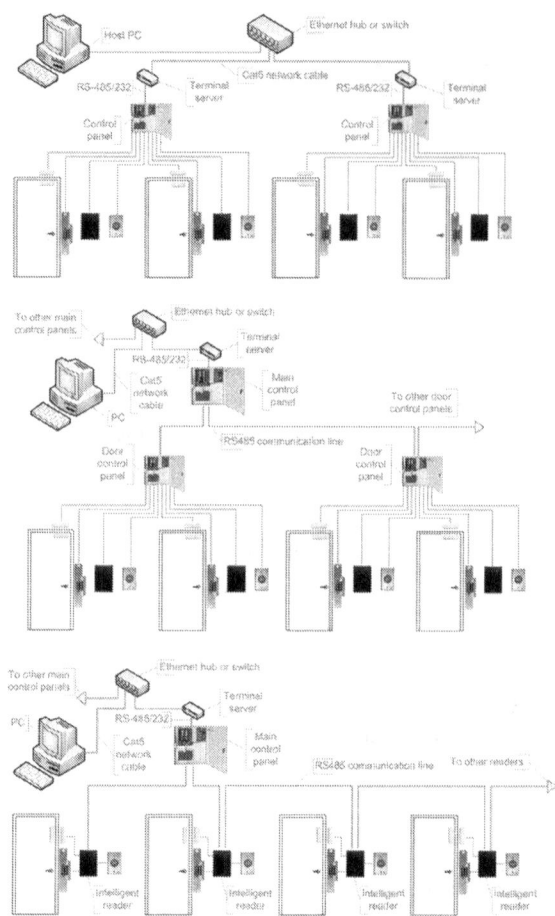

Access control systems using serial controllers and terminal servers

- Serial communication link between the controller and the terminal server acts as a bottleneck: even though the data between the host PC and the terminal server travels at the 10/100/1000Mbit/sec network speed, it must slow down to the serial speed of 112.5 kbit/sec or less. There are also additional delays introduced in the process of conversion between serial and network data.

All the RS-485-related advantages and disadvantages also apply.

5. Network-enabled main controllers. The topology is nearly the same as described in the second and third paragraphs. The same advantages and disadvantages apply, but the on-board network interface offers a couple of valuable improvements. Transmission of configuration and user data to the main controllers is faster, and may be done in parallel. This makes the system more responsive, and does not interrupt normal operations. No special hardware is required in order to achieve redundant host PC setup: in the case that the primary host PC fails, the secondary host PC may start

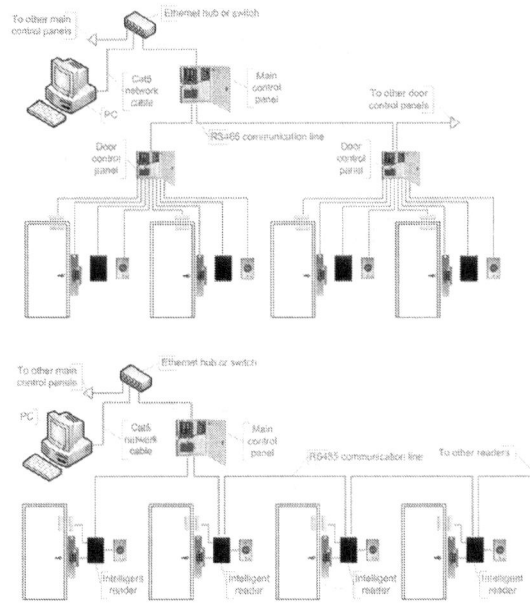

Access control system using network-enabled main controllers

polling network controllers. The disadvantages introduced by terminal servers (listed in the fourth paragraph) are also eliminated.

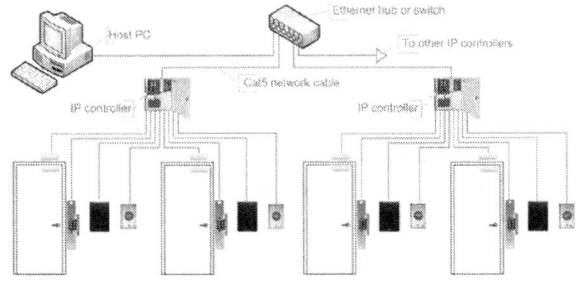

Access control system using IP controllers

6. IP controllers. Controllers are connected to a host PC via Ethernet LAN or WAN.

Advantages:

- An existing network infrastructure is fully utilized, and there is no need to install new communication lines.

- There are no limitations regarding the number of controllers (as the 32 per line in cases of RS-485).

- Special RS-485 installation, termination, grounding and troubleshooting knowledge is not required.

- Communication with the controllers may be done at the full network speed, which is important if transferring a lot of data (databases with thousands of users, possibly including biometric records).

- In case of an alarm, controllers may initiate connection to the host PC. This ability is important in large systems, because it serves to reduce network traffic caused by unnecessary polling.

- Simplifies installation of systems consisting of multiple sites that are separated by large distances. A basic Internet link is sufficient to establish connections to the remote locations.

- Wide selection of standard network equipment is available to provide connectivity in various situations (fiber, wireless, VPN, dual path, PoE)

Disadvantages:

- The system becomes susceptible to network related problems, such as delays in case of heavy traffic and network equipment failures.

- Access controllers and workstations may become accessible to hackers if the network of the organization is not well protected. This threat may be eliminated by physically separating the access control network from the network of the organization. Also it should be noted that most IP controllers utilize either Linux platform or proprietary operating systems, which makes them more difficult to hack. Industry standard data encryption is also used.

- Maximum distance from a hub or a switch to the controller (if using a copper cable) is 100 meters (330 ft).

- Operation of the system is dependent on the host PC. In case the host PC fails, events from controllers are not retrieved and functions that require interaction between controllers (i.e. anti-passback) stop working. Some controllers, however, have a peer-to-peer communication option in order to reduce dependency on the host PC.

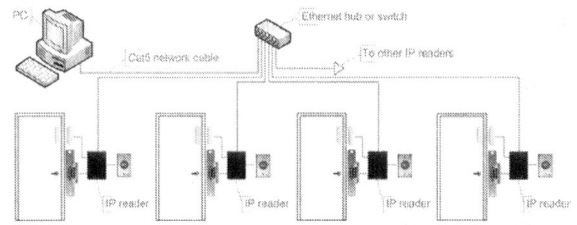

Access control system using IP readers

7. IP readers. Readers are connected to a host PC via Ethernet LAN or WAN.

Advantages:

- Most IP readers are PoE capable. This feature makes it very easy to provide battery backed power to the entire system, including the locks and various types of detectors (if used).

- IP readers eliminate the need for controller enclosures.

- There is no wasted capacity when using IP readers (e.g. a 4-door controller would have 25% of unused capacity if it was controlling only 3 doors).

- IP reader systems scale easily: there is no need to install new main or sub-controllers.

- Failure of one IP reader does not affect any other readers in the system.

Disadvantages:

- In order to be used in high-security areas, IP readers require special input/output modules to eliminate the possibility of intrusion by accessing lock and/or exit button wiring. Not all IP reader manufacturers have such modules available.

- Being more sophisticated than basic readers, IP readers are also more expensive and sensitive, therefore they should not be installed outdoors in areas with harsh weather conditions, or high probability of vandalism, unless specifically designed for exterior installation. A few manufacturers make such models.

The advantages and disadvantages of IP controllers apply to the IP readers as well.

2.1.7 Security risks

The most common security risk of intrusion through an access control system is by simply following a legitimate user through a door, and this is referred to as tailgating. Often the legitimate user will hold the door for the intruder. This risk can be minimized through security awareness training of the user population, or more active means such as turnstiles. In very high security applications this risk is minimized by using a sally port, sometimes called a security vestibule or mantrap, where operator intervention is required presumably to assure valid identification.

The second most common risk is from levering a door open. This is relatively difficult on properly secured doors with strikes or high holding force magnetic locks. Fully implemented access control systems include forced door monitoring alarms. These vary in effectiveness, usually failing from high false positive alarms, poor database configuration, or lack of active intrusion monitoring. Most newer

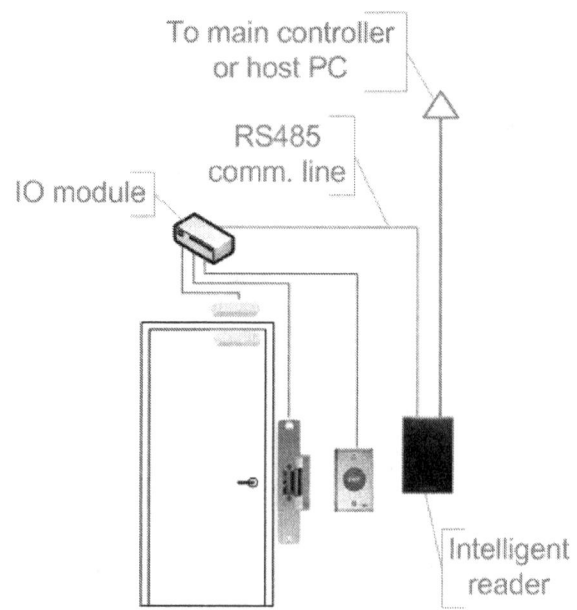

Access control door wiring when using intelligent readers and IO module

access control systems incorporate some type of door prop alarm to inform system administrators of a door left open longer than a specified length of time.

The third most common security risk is natural disasters. In order to mitigate risk from natural disasters, the structure of the building, down to the quality of the network and computer equipment vital. From an organizational perspective, the leadership will need to adopt and implement an All Hazards Plan, or Incident Response Plan. The highlights of any incident plan determined by the National Incident Management System must include Pre-incident planning, during incident actions, disaster recovery, and after action review.*[8]

Similar to levering is crashing through cheap partition walls. In shared tenant spaces the divisional wall is a vulnerability. A vulnerability along the same lines is the breaking of sidelights.

Spoofing locking hardware is fairly simple and more elegant than levering. A strong magnet can operate the solenoid controlling bolts in electric locking hardware. Motor locks, more prevalent in Europe than in the US, are also susceptible to this attack using a doughnut shaped magnet. It is also possible to manipulate the power to the lock either by removing or adding current, although most Access Control systems incorporate battery back-up systems and the locks are almost always located on the secure side of the door.

Access cards themselves have proven vulnerable to sophisticated attacks. Enterprising hackers have built portable readers that capture the card number from a user's proximity card. The hacker simply walks by the user, reads the card, and then presents the number to a reader securing the door. This is possible because card numbers are sent in the clear, no encryption being used. To counter this, dual authentication methods, such as a card plus a PIN should always be used.

Many access control credentials unique serial numbers are programmed in sequential order during manufacturing. Known as a sequential attack, if an intruder has a credential once used in the system they can simply increment or decrement the serial number until they find a credential that is currently authorized in the system. Ordering credentials with random unique serial numbers is recommended to counter this threat.*[9]

Finally, most electric locking hardware still have mechanical keys as a fail-over. Mechanical key locks are vulnerable to bumping.

The need-to-know principle

Further information: Principle of least privilege

The need to know principle can be enforced with user access controls and authorization procedures and its objective is to ensure that only authorized individuals gain access to information or systems necessary to undertake their duties.

2.2 Computer security

Further information: Computer access control

In computer security, general access control includes authorization, authentication, access approval, and audit. A more narrow definition of access control would cover only access approval, whereby the system makes a decision to grant or reject an access request from an already authenticated subject, based on what the subject is authorized to access. Authentication and access control are often combined into a single operation, so that access is approved based on successful authentication, or based on an anonymous access token. Authentication methods and tokens include passwords, biometric scans, physical keys, electronic keys and devices, hidden paths, social barriers, and monitoring by humans and automated systems.

In any access-control model, the entities that can perform actions on the system are called *subjects*, and the entities representing resources to which access may need to be controlled are called *objects* (see also Access Control Matrix). Subjects and objects should both be considered as software entities, rather than as human users: any human users can only have an effect on the system via the software entities that they control.

Although some systems equate subjects with *user IDs*, so that all processes started by a user by default have the same authority, this level of control is not fine-grained enough to satisfy the principle of least privilege, and arguably is responsible for the prevalence of malware in such systems (see computer insecurity).

In some models, for example the object-capability model, any software entity can potentially act as both subject and object.

As of 2014, access-control models tend to fall into one of two classes: those based on capabilities and those based on access control lists (ACLs).

- In a capability-based model, holding an unforgeable reference or *capability* to an object provides access to the object (roughly analogous to how possession of one's house key grants one access to one's house); access is conveyed to another party by transmitting such a capability over a secure channel

- In an ACL-based model, a subject's access to an object depends on whether its identity appears on a list associated with the object (roughly analogous to how a bouncer at a private party would check an ID to see if a name appears on the guest list); access is conveyed by editing the list. (Different ACL systems have a variety of different conventions regarding who or what is responsible for editing the list and how it is edited.)

Both capability-based and ACL-based models have mechanisms to allow access rights to be granted to all members of a *group* of subjects (often the group is itself modeled as a subject).

Access control systems provide the essential services of *authorization*, *identification and authentication (I&A)*, *access approval*, and *accountability* where:

- authorization specifies what a subject can do

- identification and authentication ensure that only legitimate subjects can log on to a system

- access approval grants access during operations, by association of users with the resources that they are allowed to access, based on the authorization policy

- accountability identifies what a subject (or all subjects associated with a user) did

2.3 Access control models

Access to accounts can be enforced through many types of controls.*[10]

1. Attribute-based Access Control (ABAC)
   An access control paradigm whereby access rights are granted to users through the use of policies which evaluate attributes (user attributes, resource attributes and environment conditions)*[11]

2. Discretionary Access Control (DAC)
   In DAC, the data owner determines who can access specific resources. For example, a system administrator may create a hierarchy of files to be accessed based on certain permissions.

3. History-Based Access Control (HBAC)
   Access is granted or declined based on the real-time evaluation of a history of activities of the inquiring party, e.g. behavior, time between requests, content of requests.*[12] For example, the access to a certain service or data source can be granted or declined on the personal behavior, e.g. the request interval exceeds one query per second.

4. Identity-Based Access Control (IBAC)
   Using this network administrators can more effectively manage activity and access based on individual needs.*[13]

5. Mandatory Access Control (MAC)
   In MAC, users do not have much freedom to determine who has access to their files. For example, security clearance of users and classification of data (as confidential, secret or top secret) are used as security labels to define the level of trust.

6. Organization-Based Access control (OrBAC)
   OrBAC model allows the policy designer to define a security policy independently of the implementation*[14]

7. Role-Based Access Control (RBAC)
   RBAC allows access based on the job title. RBAC largely eliminates discretion when providing access to objects. For example, a human resources specialist should not have permissions to create network accounts; this should be a role reserved for network administrators.

8. Rule-Based Access Control (RAC)
   RAC method is largely context based. Example of this would be only allowing students to use the labs during a certain time of day.

9. Responsibility Based Access control
   Information is accessed based on the responsibilities assigned to an actor or a business role*[15]

2.4 Telecommunication

In telecommunication, the term *access control* is defined in U.S. Federal Standard 1037C*[16] with the following meanings:

1. A service feature or technique used to permit or deny use of the components of a communication system.

2. A technique used to define or restrict the rights of individuals or application programs to obtain data from, or place data onto, a storage device.

3. The definition or restriction of the rights of individuals or application programs to obtain data from, or place data into, a storage device.

4. The process of limiting access to the resources of an AIS (Automated Information System) to authorized users, programs, processes, or other systems.

5. That function performed by the resource controller that allocates system resources to satisfy user requests.

This definition depends on several other technical terms from Federal Standard 1037C.

2.5 In object-oriented programming

In object-oriented programming languages, *access control* is a part of the apparatus of achieving encapsulation, one of four fundamentals of object-oriented programming. The goal is to establish a clear separation between interface (visible and accessible parts of the class) and implementation (internal representation and helper methods).

Also known as data hiding, it ensures exclusive data access to class members (both variables and methods) and protects object integrity by preventing corruption by a client programmer/ client classes. Rule of thumb is to use the more restrictive access level for your data, unless there is a compelling reason to expose it. This also helps to reduce interdependencies between classes - leading to lower coupling and fewer regression bugs.*[17]

In object-oriented programming, *access control* is typically implemented using access modifiers in the object or class. Although access modifiers may be syntactically different between languages, they all attempt to achieve the same goal; Define which variables and methods are visible and to whom.

Several programming languages (e.g. Java, C++, C#, Ruby) use the same **public**, **protected** and **private** access modifiers. These are the keywords which allow a programmer to establish access levels to classes and class members (both data and methods). Their exact use in each programming language is varied, depending on the language philosophy, but there are more similarities than differences."[18]

2.5.1 Comparison of use of access modifier keywords in different OOP languages

*[19] *[20]

Note: in Ruby, **private** methods always have **self** as an implicit receiver. Therefore, they can only be used on their current object.

In some languages there are mechanisms to override access modifies to gain access to the private components of an object. One such example is the friend class in C++.

2.5.2 Attribute accessors

Special public member methods - **accessors** (aka **getters**) and **mutator methods** (often called **setters**) are used to control changes to class variables in order to prevent unauthorized access and data corruption.

2.6 Public policy

In public policy, access control to restrict access to systems ("authorization") or to track or monitor behavior within systems ("accountability") is an implementation feature of using trusted systems for security or social control.

2.7 See also

- Alarm device, Alarm management, Security alarm
- Card reader, Common Access Card, Magnetic stripe card, Proximity card, Smart card, Optical turnstile, Access badge
- Castle, Fortification
- Computer security, Logical security, .htaccess, Wiegand effect, XACML, Credential
- Door security, Lock picking, Lock (security device), Electronic lock, Safe, Safe-cracking, Bank vault

- Fingerprint scanner, Photo identification, Biometrics
- Identity management, Identity document, OpenID, IP Controller, IP reader
- Key management, Key cards
- Lock screen
- Physical security information management
- Physical Security Professional
- Prison, Barbed tape, Mantrap
- Security, Security engineering, Security lighting, Security management, Security policy

2.8 References

[1] RFC 4949

[2] Federal Financial Institutions Examination Council (2008). "Authentication in an Internet Banking Environment" (PDF). Retrieved 2009-12-31.

[3] "MicroStrategy's office of the future includes mobile identity and cybersecurity". Washington Post. 2014-04-14. Retrieved 2014-03-30.

[4] "iPhone 5S: A Biometrics Turning Point?". BankInfoSecurity.com. 2013-09-16. Retrieved 2014-03-30.

[5] "NFC access control: cool and coming, but not close". Security Systems News. 2013-09-25. Retrieved 2014-03-30.

[6] "Ditch Those Tacky Key Chains: Easy Access with EC Key". Wireless Design and Development. 2012-06-11. Retrieved 2014-03-31.

[7] "Kisi And KeyMe, Two Smart Phone Apps, Might Make House Keys Obsolete". *The Huffington Post*. The Huffington Post. Retrieved 2 September 2015.

[8] "Incident Command System :: NIMS Online :: Serving the National Incident Management System (NIMS) Community.". 2007-03-18. Archived from the original on March 18, 2007. Retrieved 2016-03-06.

[9] "Smart access control polices for residential & commercial buildings". Retrieved 11 September 2017.

[10] "Cybersecurity: Access Control". 4 February 2014. Retrieved 11 September 2017.

[11] "SP 800-162, Guide to Attribute Based Access Control (ABAC) Definition and Considerations" (PDF). NIST. 2014. Retrieved 2015-12-08.

2.9. EXTERNAL LINKS

[12] Schapranow, Matthieu-P. (2014). *Real-time Security Extensions for EPCglobal Networks*. Springer. ISBN 978-3-642-36342-9.

[13] http://patft.uspto.gov/netacgi/nph-Parser?Sect1=PTO2&Sect2=HITOFF&p=1&u=%2Fnetahtml%2FPTO%2Fsearch-bool.html&r=1&f=G&l=50&co1=AND&d=PTXT&s1=8,984,620.PN.&OS=PN/8,984,620&RS=PN/8,984,620%5B%5D

[14] "OrBAC: Organization Based Access Control - The official OrBAC model website". *orbac.org*. Retrieved 11 September 2017.

[15] http://ceur-ws.org/Vol-599/BUISTAL2010_Paper5.pdf

[16] http://www.its.bldrdoc.gov/fs-1037/other/a.pdf

[17] "What is Data Hiding? - Definition from". Retrieved 11 September 2017.

[18] "Controlling Access to Members of a Class (The Java™ Tutorials > Learning the Java Language > Classes and Objects)". *docs.oracle.com*. Retrieved 11 September 2017.

[19] ealigam@gmail.com, Satish Talim / Original design: Erwin Aligam –. "Ruby Access Control: Ruby Study Notes - Best Ruby Guide, Ruby Tutorial". *rubylearning.com*. Retrieved 11 September 2017.

[20] "Classes (I) - C++ Tutorials". *www.cplusplus.com*. Retrieved 11 September 2017.

- U.S. Federal Standard 1037C
- U.S. MIL-STD-188
- U.S. National Information Systems Security Glossary
- Harris, Shon, All-in-one CISSP Exam Guide, 6th Edition, McGraw Hill Osborne, Emeryville, California, 2012.
- "Integrated Security Systems Design" - Butterworth/Heinenmann - 2007 - Thomas L. Norman, CPP/PSP/CSC Author
- NIST.gov - Computer Security Division - Computer Security Resource Center - ATTRIBUTE BASED ACCESS CONTROL (ABAC) - OVERVIEW

2.9 External links

- eXtensible Access Control Markup Language. An OASIS standard language/model for access control. Also XACML.

Chapter 3

Fingerprint

This article is about human fingerprints. For other uses, see Fingerprint (disambiguation).
"Thumbprint" redirects here. For other uses, see Thumbprint (disambiguation).

A **fingerprint** in its narrow sense is an impression left by the friction ridges of a human finger.[*][1] The recovery of fingerprints from a crime scene is an important method of forensic science. Fingerprints are easily deposited on suitable surfaces (such as glass or metal or polished stone) by the natural secretions of sweat from the eccrine glands that are present in epidermal ridges. These are sometimes referred to as "Chanced Impressions".

In a wider use of the term, fingerprints are the traces of an impression from the friction ridges of any part of a human or other primate hand. A print from the sole of the foot can also leave an impression of friction ridges.

Deliberate impressions of fingerprints may be formed by ink or other substances transferred from the peaks of friction ridges on the skin to a relatively smooth surface such as a fingerprint card.[*][2] Fingerprint records normally contain impressions from the pad on the last joint of fingers and thumbs, although fingerprint cards also typically record portions of lower joint areas of the fingers.

Human fingerprints are detailed, nearly unique, difficult to alter, and durable over the life of an individual, making them suitable as long-term markers of human identity. They may be employed by police or other authorities to identify individuals who wish to conceal their identity, or to identify people who are incapacitated or deceased and thus unable to identify themselves, as in the aftermath of a natural disaster. Fingerprint analysis, in use since the early 20th century, has led to many crimes being solved.[*][3] This means that many criminals consider gloves essential.[*][4][*][5] In 2015, the identification of sex by use of a fingerprint test has been reported.[*][6][*][7]

3.1 Biology

A friction ridge is a raised portion of the epidermis on the digits (fingers and toes), the palm of the hand or the sole of the foot, consisting of one or more connected ridge units of friction ridge skin.[*][1] These are sometimes known as "epidermal ridges" which are caused by the underlying interface between the dermal papillae of the dermis and the interpapillary (rete) pegs of the epidermis. These epidermal ridges serve to amplify vibrations triggered, for example, when fingertips brush across an uneven surface, better transmitting the signals to sensory nerves involved in fine texture perception.[*][8] These ridges may also assist in gripping rough surfaces and may improve surface contact in wet conditions.[*][9]

3.2 Types

Before computerization, manual filing systems were used in large fingerprint repositories. Manual classification systems were based on the general ridge patterns of several or all fingers (such as the presence or absence of circular patterns). This allowed the filing and retrieval of paper records in large collections based on friction ridge patterns alone. The most popular systems used the pattern class of each finger to form a key (a number) to assist lookup in a filing system. Classification systems include the Roscher system, the Juan Vucetich system, and the Henry Classification System. The Roscher system was developed in Germany and implemented in both Germany and Japan, the Vucetich system (developed by a Croatian-born Buenos Aires Police Officer) was developed in Argentina and implemented throughout South America, and the Henry system was developed in India and implemented in most English-speaking countries.[*][10]

In the Henry system of classification, there are three basic fingerprint patterns: loop, whorl, and arch,[*][11] which constitute 60–65%, 30–35%, and 5% of all fingerprints respectively. There are also more complex classification systems that break down patterns even further, into plain arches or tented arches,*[10] and into loops that may be radial or ulnar, depending on the side of the hand toward which the tail points. Ulnar loops start on the pinky-side of the finger, the side closer to the ulna, the lower arm bone. Radial loops start on the thumb-side of the finger, the side closer to the radius. Whorls may also have sub-group classifications including plain whorls, accidental whorls, double loop whorls, peacock's eye, composite, and central pocket loop whorls.*[10]

Other common fingerprint patterns include the tented arch, the plain arch, and the central pocket loop.

The system used by most experts, although complex, is similar to the Henry System of Classification. It consists of five fractions, in which R stands for right, L for left, i for index finger, m for middle finger, t for thumb, r for ring finger and p(pinky) for little finger. The fractions are as follows: $Ri/Rt + Rr/Rm + Lt/Rp + Lm/Li + Lp/Lr$. The numbers assigned to each print are based on whether or not they are whorls. A whorl in the first fraction is given a 16, the second an 8, the third a 4, the fourth a 2, and 0 to the last fraction. Arches and loops are assigned values of 0. Lastly, the numbers in the numerator and denominator are added up, using the scheme:

$$(Ri + Rr + Lt + Lm + Lp)/(Rt + Rm + Rp + Li + Lr)$$

and a 1 is added to both top and bottom, to exclude any possibility of division by zero. For example, if the right ring finger and the left index finger have whorls, the fractions would look like this:

$0/0 + 8/0 + 0/0 + 0/2 + 0/0 + 1/1$, and the calculation: $(0 + 8 + 0 + 0 + 0 + 1)/(0 + 0 + 0 + 2 + 0 + 1) = 9/3 = 3$.

Using this system reduces the number of prints that the print in question needs to be compared to. For example, the above set of prints would only need to be compared to other sets of fingerprints with a value of 3.*[12]

3.3 Dactyloscopy

Fingerprint identification, known as dactyloscopy,*[13] or hand print identification, is the process of comparing two instances of friction ridge skin impressions (see Minutiae), from human fingers or toes, or even the palm of the hand or sole of the foot, to determine whether these impressions could have come from the same individual. The flexibility

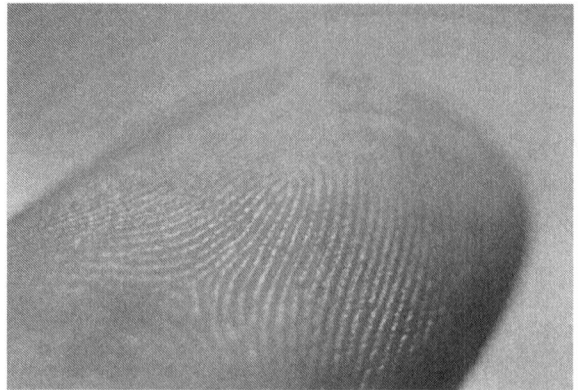

The friction ridges on a finger

of friction ridge skin means that no two finger or palm prints are ever exactly alike in every detail; even two impressions recorded immediately after each other from the same hand may be slightly different. Fingerprint identification, also referred to as individualization, involves an expert, or an expert computer system operating under threshold scoring rules, determining whether two friction ridge impressions are likely to have originated from the same finger or palm (or toe or sole).

An intentional recording of friction ridges is usually made with black printer's ink rolled across a contrasting white background, typically a white card. Friction ridges can also be recorded digitally, usually on a glass plate, using a technique called Live Scan. A "latent print" is the chance recording of friction ridges deposited on the surface of an object or a wall. Latent prints are invisible to the naked eye, whereas "patent prints" or "plastic prints" are viewable with the unaided eye. Latent prints are often fragmentary and require the use of chemical methods, powder, or alternative light sources in order to be made clear. Sometimes an ordinary bright flashlight will make a latent print visible.

When friction ridges come into contact with a surface that will take a print, material that is on the friction ridges such as perspiration, oil, grease, ink or blood, will be transferred to the surface. Factors which affect the quality of friction ridge impressions are numerous. Pliability of the skin, deposition pressure, slippage, the material from which the surface is made, the roughness of the surface and the substance deposited are just some of the various factors which can cause a latent print to appear differently from any known recording of the same friction ridges. Indeed, the conditions surrounding every instance of friction ridge deposition are unique and never duplicated. For these reasons, fingerprint examiners are required to undergo extensive training. The scientific study of fingerprints is called dermatoglyphics.

An image of a fingerprint created by the friction ridge structure

3.4 Types

3.4.1 Exemplar

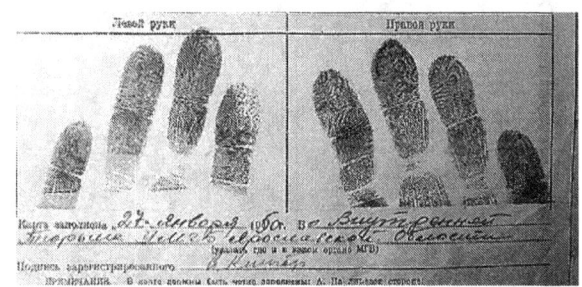

Exemplar prints on paper using ink

Exemplar prints, or known prints, is the name given to fingerprints deliberately collected from a subject, whether for purposes of enrollment in a system or when under arrest for a suspected criminal offense. During criminal arrests, a set of exemplar prints will normally include one print taken from each finger that has been rolled from one edge of the nail to the other, plain (or slap) impressions of each of the four fingers of each hand, and plain impressions of each thumb. Exemplar prints can be collected using live scan or by using ink on paper cards.

3.4.2 Latent

Barely visible latent prints on a knife

Although the word latent means hidden or invisible, in modern usage for forensic science the term latent prints means any chance or accidental impression left by friction ridge skin on a surface, regardless of whether it is visible or invisible at the time of deposition. Electronic, chemical and physical processing techniques permit visualization of invisible latent print residues whether they are from natural sweat on the skin or from a contaminant such as motor oil, blood, ink, paint or some other form of dirt. The different types of fingerprint patterns, such as arch, loop and whorl, will be described below.

Latent prints may exhibit only a small portion of the surface of a finger and this may be smudged, distorted, overlapped by other prints from the same or from different individuals, or any or all of these in combination. For this reason, latent prints usually present an "inevitable source of error in making comparisons", as they generally "contain less clarity, less content, and less undistorted information than a fingerprint taken under controlled conditions, and much, much less detail compared to the actual patterns of ridges and grooves of a finger." *[14]

3.4.3 Patent

Patent prints are chance friction ridge impressions which are obvious to the human eye and which have been caused by the transfer of foreign material from a finger onto a surface. Some obvious examples would be impressions from flour and wet clay. Because they are already visible and have no need of enhancement they are generally photographed rather than being lifted in the way that latent prints are. An attempt to preserve the actual print is always made for later presentation in court, and there are many techniques used to do this. Patent prints can be left on a surface by materials such as ink, dirt, or blood.

3.4.4 Plastic

A plastic print is a friction ridge impression left in a material that retains the shape of the ridge detail. Although very few criminals would be careless enough to leave their prints in a lump of wet clay, this would make a perfect plastic print.*[15] Commonly encountered examples are melted candle wax, putty removed from the perimeter of window panes and thick grease deposits on car parts. Such prints are already visible and need no enhancement, but investigators must not overlook the potential that invisible latent prints deposited by accomplices may also be on such surfaces. After photographically recording such prints, attempts should be made to develop other non-plastic impressions deposited from sweat or other contaminants.

3.4.5 Electronic recording

There has been a newspaper report of a man selling stolen watches sending images of them on a mobile phone, and those images included parts of his hands in enough detail for police to be able to identify fingerprint patterns.*[16]

Recent studies found that the improving cameras with increasing resolution of smartphones might have a high impact on users' security: The back-facing camera of a device can be used to capture an image of the user's index finger, which on smartphones using biometric means of authentication is often used to authenticate a user against the smartphone.*[17]

At the 31st Chaos Communication Congress, hardware hacker starbug presented how DSLRs with high resolution and equipped with a long focus lens can be used to capture images of hands, or more specifically, fingers in order to use them for spoofing.*[18]

3.4.6 Footprints

Main article: Footprint

Friction ridge skin present on the soles of the feet and toes (plantar surfaces) is as unique in its ridge detail as are the fingers and palms (palmar surfaces). When recovered at crime scenes or on items of evidence, sole and toe impressions can be used in the same manner as finger and palm prints to effect identifications. The footprints of infants, along with the thumb or index finger prints of mothers, are still commonly recorded in hospitals to assist in verifying the identity of infants. It is not uncommon for military records of flight personnel to include bare foot inked impressions. Friction ridge skin protected inside flight boots tends to survive the trauma of a plane crash (and accompanying fire) better than fingers.

3.5 Capture and detection

3.5.1 Live scan devices

Fingerprint being scanned

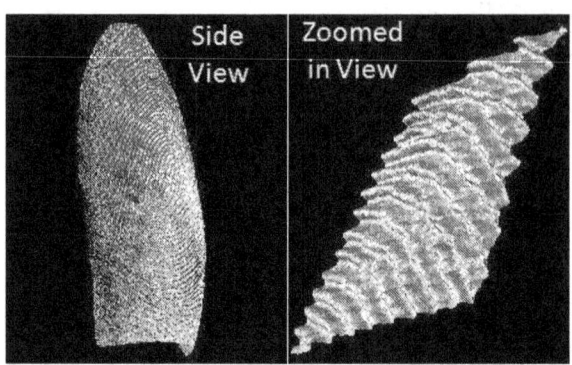

3D fingerprint[19]*

Fingerprint image acquisition is considered to be the most critical step in an automated fingerprint authentication system, as it determines the final fingerprint image quality, which has a drastic effect on the overall system performance. There are different types of fingerprint readers on the market, but the basic idea behind each is to measure the physical difference between ridges and valleys.

All the proposed methods can be grouped into two major families: solid-state fingerprint readers and optical fingerprint readers. The procedure for capturing a fingerprint using a sensor consists of rolling or touching with the finger onto a sensing area, which according to the physical principle in use (optical, ultrasonic, capacitive or thermal) captures the difference between valleys and ridges. When a finger touches or rolls onto a surface, the elastic skin deforms. The quantity and direction of the pressure applied by the user, the skin conditions and the projection of an irregular 3D object (the finger) onto a 2D flat plane introduce distortions, noise and inconsistencies in the captured fingerprint image. These problems result in inconsistent and non-uniform irregularities in the image.[*][20] During each acquisition, therefore, the results of the imaging are different and uncontrollable. The representation of the same fingerprint changes every time the finger is placed on the sensor plate, increasing the complexity of any attempt to match fingerprints, impairing the system performance and consequently, limiting the widespread use of this biometric technology.

In order to overcome these problems, as of 2010, non-contact or touchless 3D fingerprint scanners have been developed.[*][21][*][22] Acquiring detailed 3D information, 3D fingerprint scanners take a digital approach to the analog process of pressing or rolling the finger. By modelling the distance between neighboring points, the fingerprint can be imaged at a resolution high enough to record all the necessary detail.[*][23]

3.5.2 Scanning dead or unconscious people

Placing the hand of a dead or unconscious person on a scanner to gain unauthorized access has become a common plot device. However, a *MythBusters* episode revealed that this doesn't work (at least with the scanners available to the program). But Adam Savage and Jamie Hyneman found a way to convert fingerprints lifted from the hand to a photographic form that the sensor would accept. For obvious reasons, they refuse to reveal the technique.

3.5.3 Latent detection

In the 1930s criminal investigators in the United States first discovered the existence of latent fingerprints on the surfaces of fabrics, most notably on the insides of gloves discarded by perpetrators.[*][24]

Since the late nineteenth century, fingerprint identification methods have been used by police agencies around the world to identify suspected criminals as well as the victims of crime. The basis of the traditional fingerprinting technique is simple. The skin on the palmar surface of the hands and feet forms ridges, so-called papillary ridges, in patterns that are unique to each individual and which do not change

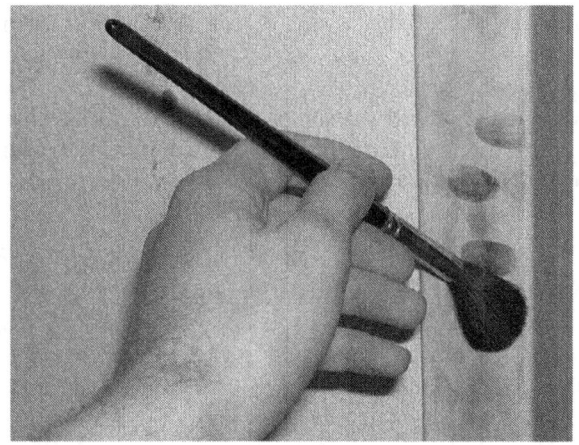
Use of fine powder and brush to reveal latent fingerprints

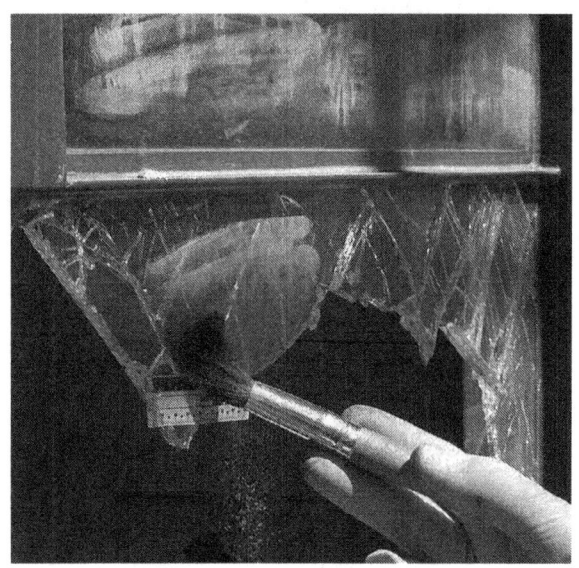
Fingerprints dusting of a burglary scene

over time. Even identical twins (who share their DNA) do not have identical fingerprints. The best way to render latent fingerprints visible, so that they can be photographed, can be complex and may depend, for example, on the type of surfaces on which they have been left. It is generally necessary to use a 'developer', usually a powder or chemical reagent, to produce a high degree of visual contrast between the ridge patterns and the surface on which a fingerprint has been deposited.

Developing agents depend on the presence of organic materials or inorganic salts for their effectiveness, although the water deposited may also take a key role. Fingerprints are typically formed from the aqueous-based secretions of the eccrine glands of the fingers and palms with additional material from sebaceous glands primarily from the forehead. This latter contamination results from the common human behaviors of touching the face and hair. The resulting latent fingerprints consist usually of a substantial proportion of water with small traces of amino acids and chlorides mixed with a fatty, sebaceous component which contains a number of fatty acids and triglycerides. Detection of a small proportion of reactive organic substances such as urea and amino acids is far from easy.

Fingerprints at a crime scene may be detected by simple powders, or by chemicals applied *in situ*. More complex techniques, usually involving chemicals, can be applied in specialist laboratories to appropriate articles removed from a crime scene. With advances in these more sophisticated techniques, some of the more advanced crime scene investigation services from around the world were, as of 2010, reporting that 50% or more of the fingerprints recovered from a crime scene had been identified as a result of laboratory-based techniques.

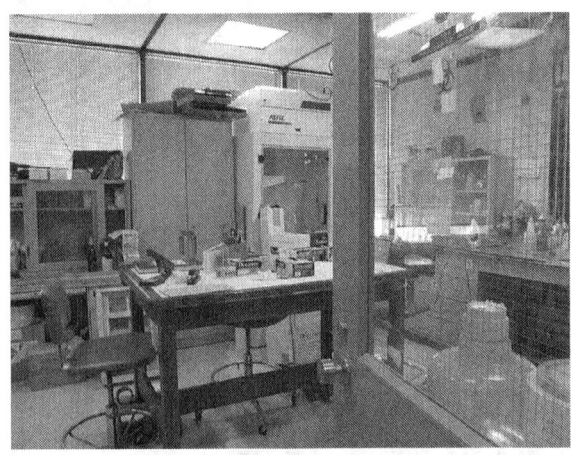

A city fingerprint identification room.

3.5.4 Laboratory techniques

Although there are hundreds of reported techniques for fingerprint detection, many of these are only of academic interest and there are only around 20 really effective methods which are currently in use in the more advanced fingerprint laboratories around the world.

Some of these techniques, such as ninhydrin, diazafluorenone and vacuum metal deposition, show great sensitivity and are used operationally. Some fingerprint reagents are specific, for example ninhydrin or diazafluorenone reacting with amino acids. Others such as ethyl cyanoacrylate polymerisation, work apparently by water-based catalysis and polymer growth. Vacuum metal deposition using gold and zinc has been shown to be non-specific, but can detect fat layers as thin as one molecule.

More mundane methods, such as the application of fine powders, work by adhesion to sebaceous deposits and possibly aqueous deposits in the case of fresh fingerprints. The aqueous component of a fingerprint, whilst initially sometimes making up over 90% of the weight of the fingerprint, can evaporate quite quickly and may have mostly gone after 24 hours. Following work on the use of argon ion lasers for fingerprint detection,*[25] a wide range of fluorescence techniques have been introduced, primarily for the enhancement of chemically developed fingerprints; the inherent fluorescence of some latent fingerprints may also be detected. Fingerprints can for example be visualized in 3D and without chemicals by the use of infrared lasers.*[26]

A comprehensive manual of the operational methods of fingerprint enhancement was last published by the UK Home Office Scientific Development Branch in 2013 and is used widely around the world.*[27]

A technique proposed in 2007 aims to identify an individual's ethnicity, gender, and dietary patterns.*[28]

3.6 Research

The International Fingerprint Research Group (IFRG) which meets biennially, consists of members of the leading fingerprint research groups from Europe, the US, Canada, Australia and Israel and leads the way in the development, assessment and implementation of new techniques for operational fingerprint detection.

One problem for the early twenty-first century is the fact that the organic component of any deposited material is readily destroyed by heat, such as occurs when a gun is fired or a bomb is detonated, when the temperature may reach as high as 500 °C. Encouragingly, however, the non-volatile inorganic component of eccrine secretion has been shown to remain intact even when exposed to temperatures as high as 600 °C.

A technique has been developed that enables fingerprints to be visualised on metallic and electrically conductive surfaces without the need to develop the prints first.*[29] This technique involves the use of an instrument called a scanning Kelvin probe (SKP), which measures the voltage, or electrical potential, at pre-set intervals over the surface of an object on which a fingerprint may have been deposited. These measurements can then be mapped to produce an image of the fingerprint. A higher resolution image can be obtained by increasing the number of points sampled, but at the expense of the time taken for the process. A sampling frequency of 20 points per mm is high enough to visualise a fingerprint in sufficient detail for identification purposes and produces a voltage map in 2–3 hours. As of 2010, this technique had been shown to work effectively on a wide range of forensically important metal surfaces including iron, steel and aluminium. While initial experiments were performed on flat surfaces, the technique has been further developed to cope with irregular or curved surfaces, such as the warped cylindrical surface of fired cartridge cases. Research during 2010 at Swansea University has found that physically removing a fingerprint from a metal surface, for example by rubbing with a tissue, does not necessarily result in the loss of all fingerprint information from that surface. The reason for this is that the differences in potential that are the basis of the visualisation are caused by the interaction of inorganic salts in the fingerprint deposit and the metal surface and begin to occur as soon as the finger comes into contact with the metal, resulting in the formation of metal-ion complexes that cannot easily be removed.

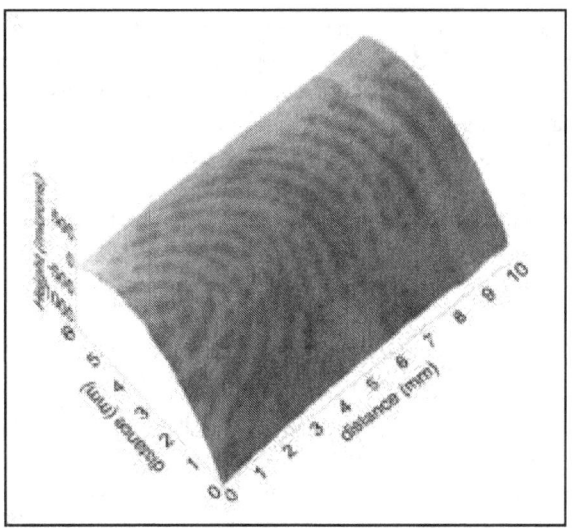

Scanning Kelvin probe scan of the same cartridge case with the fingerprint detected. The Kelvin probe can easily cope with the 3D curvature of the cartridge case, increasing the versatility of the technique.

Cartridge case with an applied fingerprint

Another problem for the early twenty-first century is that during crime scene investigations, a decision has to be made at an early stage whether to attempt to retrieve fingerprints through the use of developers or whether to swab surfaces in an attempt to salvage material for DNA profiling. The two processes are mutually incompatible, as fingerprint developers destroy material that could potentially be used for DNA analysis, and swabbing is likely to make fingerprint identification impossible.

The application of the new scanning Kelvin probe (SKP) fingerprinting technique, which makes no physical contact with the fingerprint and does not require the use of developers, has the potential to allow fingerprints to be recorded whilst still leaving intact material that could subsequently be subjected to DNA analysis. A forensically usable prototype was under development at Swansea University during 2010, in research that was generating significant interest from the British Home Office and a number of different police forces across the UK, as well as internationally. The hope is that this instrument could eventually be manufactured in sufficiently large numbers to be widely used by forensic teams worldwide.*[30]*[31]

3.7 Disappearance of children's latent prints

In 1995, researchers at the Oak Ridge National Laboratory, at the instigation of Detective Art Bohanan of the Knoxville Police Department, discovered that children's fingerprints are considerably more short-lived than adult fingerprints.*[32] The rapid disappearance of children's fingerprints was attributed to a lack of the more waxy oils that become present at the onset of puberty. The lighter fatty acids of children's fingerprints evaporate within a few hours. As of 2010, researchers at Oak Ridge National Laboratory are investigating techniques to capture these lost fingerprints.

3.8 Detection of drug use

The secretions, skin oils and dead cells in a human fingerprint contain residues of various chemicals and their metabolites present in the body. These can be detected and used for forensic purposes. For example, the fingerprints of tobacco smokers contain traces of cotinine, a nicotine metabolite; they also contain traces of nicotine itself. Caution should be used, as its presence may be caused by mere contact of the finger with a tobacco product.

By treating the fingerprint with gold nanoparticles with attached cotinine antibodies, and then subsequently with a fluorescent agent attached to cotinine antibodies, the fingerprint of a smoker becomes fluorescent; non-smokers' fingerprints stay dark.

The same approach, as of 2010, is being tested for use in identifying heavy coffee drinkers, cannabis smokers, and users of various other drugs.*[33]*[34]

In 2008, British researchers developed methods of identifying users of marijuana, cocaine and methadone from their fingerprint residues.*[35]

3.9 United States databases and compression

In the United States, the FBI manages a fingerprint identification system and database called the Integrated Automated Fingerprint Identification System (IAFIS), which currently holds the fingerprints and criminal records of over 51 million criminal record subjects and over 1.5 million civil (non-criminal) fingerprint records. US Visit currently holds a repository of the fingerprints of over 50 million people, primarily in the form of two-finger records. In 2008, US Visit hoped to have changed over to a system recording FBI-standard ten-print records.

Most American law enforcement agencies use Wavelet Scalar Quantization (WSQ), a wavelet-based system for efficient storage of compressed fingerprint images at 500 pixels per inch (ppi). WSQ was developed by the FBI, the Los Alamos National Lab, and the National Institute for Standards and Technology (NIST). For fingerprints recorded at 1000 ppi spatial resolution, law enforcement (including the FBI) uses JPEG 2000 instead of WSQ.

A city fingerprint identification office

3.10 Validity

The validity of forensic fingerprint evidence has been challenged by academics, judges and the media. While fingerprint identification was an improvement on earlier anthropometric systems, the subjective nature of matching, despite a very low error rate, has made this forensic practice controversial.*[36]

Certain specific criticisms are now being accepted by some leaders of the forensic fingerprint community, providing an incentive to improve training and procedures.

3.10.1 Criticism

The words "reliability" and "validity" have specific meanings to the scientific community. Reliability means that successive tests bring the same results. Validity means that these results are judged to accurately reflect the external criteria being measured.

> Although experts are often more comfortable relying on their instincts, this reliance does not always translate into superior predictive ability. For example, in the popular Analysis, Comparison, Evaluation, and Verification (ACE-V) paradigm for fingerprint identification, the verification stage, in which a second examiner confirms the assessment of the original examiner, may increase the consistency of the assessments. But while the verification stage has implications

for the reliability of latent print comparisons, it does not assure their validity.
—Sandy L Zabell, *[37]

The few tests that have been made of the validity of forensic fingerprinting have not been supportive of the method.

> "Despite the absence of objective standards, scientific validation, and adequate statistical studies, a natural question to ask is how well fingerprint examiners actually perform. Proficiency tests do not validate a procedure per se, but they can provide some insight into error rates. In 1995, the Collaborative Testing Service (CTS) administered a proficiency test that, for the first time, was "designed, assembled, and reviewed" by the International Association for Identification (IAI). The results were disappointing. Four suspect cards with prints of all ten fingers were provided together with seven latents. Of 156 people taking the test, only 68 (44%) correctly classified all seven latents. Overall, the tests contained a total of 48 incorrect identifications. David Grieve, the editor of the Journal of Forensic Identification, describes the reaction of the forensic community to the results of the CTS test as ranging from "shock to disbelief", and added:
>
>> 'Errors of this magnitude within a discipline singularly admired and respected for its touted absolute certainty as an identification process have produced chilling and mind-numbing realities. Thirty-four participants, an incredible 22% of those involved, substituted presumed but false certainty for truth. By any measure, this represents a profile of practice that is unacceptable and thus demands positive action by the entire community.'
>
> What is striking about these comments is that they do not come from a critic of the fingerprint community, but from the editor of one of its premier publications."
> —Sandy L Zabell, *[37]

Investigations have been conducted into whether experts can objectively focus on feature information in fingerprints without being misled by extraneous information, such as context.*[38] Fingerprints that have previously been examined and assessed by latent print experts to make a positive identification of suspects have then been re-presented to those same experts in a new context which makes it likely that there will be no match. Within this new context, most of the fingerprint experts made different judgments, thus contradicting their own previous identification decisions.*[38]

Complaints have been made that there have been no published, peer-reviewed studies directly examining the extent to which people can correctly match fingerprints to one another.*[39] Experiments have been carried out using naïve undergraduates to match images of fingerprints. The results of these experiments demonstrate that people can identify fingerprints quite well, and that matching accuracy can vary as a function of both source finger type and image similarity.*[39]

Defense

Fingerprints collected at a crime scene, or on items of evidence from a crime, have been used in forensic science to identify suspects, victims and other persons who touched a surface. Fingerprint identification emerged as an important system within police agencies in the late 19th century, when it replaced anthropometric measurements as a more reliable method for identifying persons having a prior record, often under a false name, in a criminal record repository.*[13] In modern times, researchers can find traces of addictive drugs on just a fingerprint.*[40]

3.10.2 Track record

Fingerprinting has served all governments worldwide during the past 100 years or so to provide identification of criminals. Fingerprints are the fundamental tool in every police agency for the identification of people with a criminal history.*[13] They remain the most commonly gathered forensic evidence worldwide, and in most jurisdictions fingerprint examination is more common than all other forensic examination casework combined. Moreover, it continues to expand, with tens of thousands of people added to fingerprint repositories daily in America alone —far more than other forensic databases.

3.11 Professional certification

Fingerprinting was the basis upon which the first forensic professional organization was formed, the International Association for Identification (IAI), in 1915.*[41] The first professional certification program for forensic scientists was established in 1977, the IAI's Certified Latent Print Examiner program, which issued certificates to those meeting stringent criteria and had the power to revoke certification where an individual's performance warranted it.*[42] Other forensic disciplines have followed suit and established their own certification programs.*[42]

3.12 Errors

3.12.1 Brandon Mayfield and the Madrid bombing

Brandon Mayfield is an Oregon lawyer who was identified as a participant in the 2004 Madrid train bombings based on a fingerprint match by the FBI.*[43] The FBI Latent Print Unit processed a fingerprint collected in Madrid and reported a "100 percent positive" match against one of the 20 fingerprint candidates returned in a search response from their Integrated Automated Fingerprint Identification System. The FBI initially called it an "absolutely incontrovertible match". Subsequently, however, Spanish National Police examiners suggested that the print did not match Mayfield and after two weeks, identified another man whom they claimed the fingerprint did belong to. The FBI acknowledged their error, and a judge released Mayfield, who had spent two weeks in police custody, in May 2004.*[43] In January 2006, a U.S. Justice Department report was released which criticized the FBI for sloppy work but exonerated them of some more serious allegations. The report found that the misidentification had been due to a misapplication of methodology by the examiners involved: Mayfield is an American-born convert*[43] to Islam and his wife is an Egyptian immigrant,*[43] but these are not factors that should have affected fingerprint search technology.

On November 29, 2006, the FBI agreed to pay Brandon Mayfield US$2 million in compensation.*[43] The judicial settlement allowed Mayfield to continue a suit regarding certain other government practices surrounding his arrest and detention. The formal apology stated that the FBI, which erroneously linked him to the 2004 Madrid bombing through a fingerprinting mistake, had taken steps to "ensure that what happened to Mr. Mayfield and the Mayfield family does not happen again." *[43]

3.12.2 René Ramón Sánchez

René Ramón Sánchez, a legal Dominican Republic immigrant to the US, was arrested on July 15, 1995, on a charge of driving while intoxicated. His fingerprints were mistakenly placed on a card containing the name, Social Security number and other data for one Leo Rosario, who was being processed at the same time. Leo Rosario had been arrested for selling cocaine to an undercover police officer. On October 11, 2000, while returning from a visit to relatives in the Dominican Republic, René was misidentified as Leo Rosario at John F. Kennedy International Airport in New York and arrested. Even though he did not match the physical description of Rosario, the erroneously cataloged fingerprints were considered to be more reliable.*[44]

3.12.3 Shirley McKie

Shirley McKie was a police detective in 1997 when she was accused of leaving her thumb print inside a house in Kilmarnock, Scotland, where Marion Ross had been murdered. Although McKie denied having been inside the house, she was arrested in a dawn raid the following year and charged with perjury. The only evidence the prosecution had was this thumb print allegedly found at the murder scene. Two American experts testified on her behalf at her trial in May 1999 and she was found not guilty. The Scottish Criminal Record Office (SCRO) would not admit any error, although Scottish first minister Jack McConnell later said it had been an "honest mistake".

On February 7, 2006, McKie was awarded £750,000 in compensation from the Scottish Executive and the Scottish Criminal Record Office.*[45] Controversy continued to surround the McKie case and the Fingerprint Inquiry into the affair finished taking evidence in November 2009.*[46] The Inquiry Report was published on 11 December 2011.*[47]

3.12.4 Stephan Cowans

Stephan Cowans was convicted of attempted murder in 1997 after he was accused of shooting a police officer whilst fleeing a robbery in Roxbury, Massachusetts. He was implicated in the crime by the testimony of two witnesses, one of whom was the victim. There was also a fingerprint on a glass mug from which the assailant had drunk some water and experts testified that the fingerprint belonged to Cowans. He was found guilty and sent to prison for 35 years. Whilst in prison, Cowans earned money cleaning up biohazards until he could afford to have the evidence against him tested for DNA. The DNA did not match his and he was released. He had already served six years in prison when he was released on January 23, 2004.*[48] Cowans died on October 25, 2007.*[48]

3.12.5 Craig D. Harvey

In April 1993, in the New York State Police Troop C scandal, Craig D. Harvey, a New York State Police trooper, was charged with fabricating evidence. Harvey admitted he and another trooper lifted fingerprints from items the suspect,

John Spencer, touched while in Troop C headquarters during booking. He attached the fingerprints to evidence cards and later claimed that he had pulled the fingerprints from the scene of the murder. The forged evidence was presented during John Spencer's trial and his subsequent conviction resulted in a term of 50 years to life in prison at his sentencing.*[49] Three state troopers were found guilty of fabricating fingerprint evidence and served prison sentences.*[50]

3.13 History

3.13.1 Antiquity and the medieval period

Fingerprints have been found on ancient Babylonian clay tablets, seals, and pottery.*[51]*[52]*[53]*[54] They have also been found on the walls of Egyptian tombs and on Minoan, Greek, and Chinese*[55] pottery, as well as on bricks and tiles from ancient Babylon and Rome. Some of these fingerprints were deposited unintentionally by the potters and masons as a natural consequence of their work, and others were made in the process of adding decoration. However, on some pottery, fingerprints have been impressed so deeply into the clay that they were possibly intended to serve as an identifying mark by the maker.

Fingerprints were used as signatures in ancient Babylon in the second millennium BCE.*[56] In order to protect against forgery, parties to a legal contract would impress their fingerprints into a clay tablet on which the contract had been written.In Ancient India some texts called Naadi were written by a Rishi called Agastya who had a highly developed consciousness, this text is said to predict the past, present and the future lives of all humans from thumb print.the Naadi palm leaves are located based on the thumb impressions (right for men, left for women).*[57] This ancient Indian system of astrology was called Nadi astrology.By 246 BCE, Chinese officials were impressing their fingerprints into the clay seals used to seal documents. With the advent of silk and paper in China, parties to a legal contract impressed their handprints on the document. Sometime before 851 CE, an Arab merchant in China, Abu Zayd Hasan, witnessed Chinese merchants using fingerprints to authenticate loans.*[58] By 702, Japan allowed illiterate petitioners seeking a divorce to "sign" their petitions with a fingerprint.*[59]*[60]

Although ancient peoples probably did not realize that fingerprints could uniquely identify individuals,*[61] references from the age of the Babylonian king Hammurabi (reigned 1792-1750 BCE) indicate that law officials would take the fingerprints of people who had been arrested.*[62] During China's Qin Dynasty, records have shown that officials took hand prints, foot prints as well as finger prints as evidence from a crime scene.*[63] In China, around 300 CE, handprints were used as evidence in a trial for theft. By 650, the Chinese historian Kia Kung-Yen remarked that fingerprints could be used as a means of authentication.*[64] In his *Jami al-Tawarikh* (Universal History), the Persian physician Rashid-al-Din Hamadani (also known as "Rashideddin", 1247–1318) refers to the Chinese practice of identifying people via their fingerprints, commenting: "Experience shows that no two individuals have fingers exactly alike." *[65] In Persia at this time, government documents may have been authenticated with thumbprints.

3.13.2 Europe in the 17th and 18th centuries

In 1665, the Italian physician Marcello Malpighi (1628–1694) briefly mentioned, in his *De externo tactus organo anatomica observatio*, the existence of patterns of ridges and sweat glands on the fingertips.*[66] In 1684, the English physician, botanist, and microscopist Nehemiah Grew (1641–1712) published the first scientific paper to describe the ridge structure of the skin covering the fingers and palms.*[67] In 1685, the Dutch physician Govard Bidloo (1649–1713) published a book on anatomy which also illustrated the ridge structure of the fingers.*[68] A century later, in 1788, the German anatomist Johann Christoph Andreas Mayer (1747–1801) recognized that fingerprints are unique to each individual.*[69]*[70]

3.13.3 Modern era

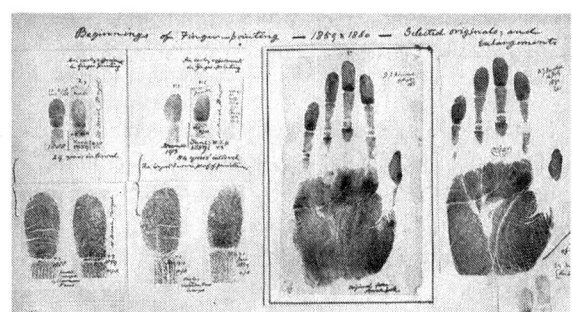

Fingerprints taken by William Herschel 1859/60

Jan Evangelista Purkyně or Purkinje (1787–1869), a Czech physiologist and professor of anatomy at the University of Breslau, published a thesis in 1823 discussing 9 fingerprint patterns, but he did not mention any possibility of using fingerprints to identify people.*[71] In 1840, following the murder of Lord William Russell, a provincial doctor, Robert Blake Overton, wrote to Scotland Yard suggesting checking for fingerprints but the suggestion, though followed up, did not lead to their routine use by the police for another 50 years.*[72] Some years later, the Ger-

3.13. HISTORY

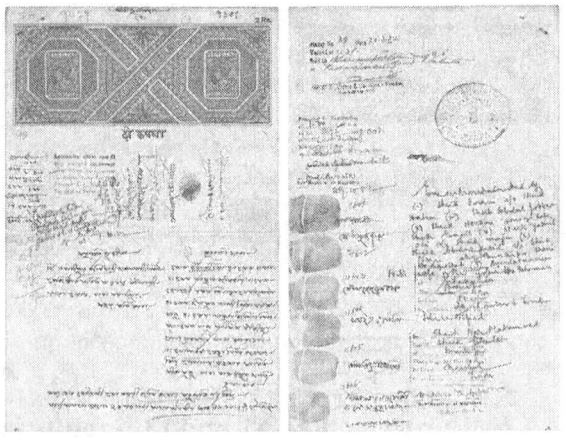

Fingerprints used instead of signatures on an Indian legal document of 1952.

man anatomist Georg von Meissner (1829–1905) studied friction ridges,*[73] and five years after this, in 1858, Sir William James Herschel initiated fingerprinting in India. In 1877 at Hooghly (near Calcutta) he instituted the use of fingerprints on contracts and deeds to prevent the then-rampant repudiation of signatures*[74] and he registered government pensioners' fingerprints to prevent the collection of money by relatives after a pensioner's death.*[75] Herschel also fingerprinted prisoners upon sentencing to prevent various frauds that were attempted in order to avoid serving a prison sentence.

In 1863, Paul-Jean Coulier (1824–1890), professor for chemistry and hygiene at the medical and pharmaceutical school of the Val-de-Grâce military hospital in Paris, discovered that iodine fumes can reveal fingerprints on paper.*[76]

In 1880, Dr. Henry Faulds, a Scottish surgeon in a Tokyo hospital, published his first paper on the subject in the scientific journal *Nature*, discussing the usefulness of fingerprints for identification and proposing a method to record them with printing ink. He also established their first classification and was also the first to identify fingerprints left on a vial.*[77] Returning to the UK in 1886, he offered the concept to the Metropolitan Police in London but it was dismissed at that time.*[78] Faulds wrote to Charles Darwin with a description of his method but, too old and ill to work on it, Darwin gave the information to his cousin, Francis Galton, who was interested in anthropology. Having been thus inspired to study fingerprints for ten years, Galton published a detailed statistical model of fingerprint analysis and identification and encouraged its use in forensic science in his book *Finger Prints*. He had calculated that the chance of a "false positive" (two different individuals having the same fingerprints) was about 1 in 64 billion.*[79]

Juan Vucetich, an Argentine chief police officer, created the first method of recording the fingerprints of individuals on file, associating these fingerprints to the anthropometric system of Alphonse Bertillon, who had created, in 1879, a system to identify individuals by anthropometric photographs and associated quantitative descriptions. In 1892, after studying Galton's pattern types, Vucetich set up the world's first fingerprint bureau. In that same year, Francisca Rojas of Necochea, was found in a house with neck injuries, whilst her two sons were found dead with their throats cut. Rojas accused a neighbour, but despite brutal interrogation, this neighbour would not confess to the crimes. Inspector Alvarez, a colleague of Vucetich, went to the scene and found a bloody thumb mark on a door. When it was compared with Rojas' prints, it was found to be identical with her right thumb. She then confessed to the murder of her sons.

Women clerical employees of the Los Angeles Police Department being fingerprinted and photographed in 1928.

A Fingerprint Bureau was established in Calcutta (Kolkata), India, in 1897, after the Council of the Governor General approved a committee report that fingerprints should be used for the classification of criminal records. Working in the Calcutta Anthropometric Bureau, before it became the first Fingerprint Bureau in the world, were Azizul Haque and Hem Chandra Bose. Haque and Bose were Indian fingerprint experts who have been credited with the primary development of a fingerprint classification system eventually named after their supervisor, Sir Edward Richard Henry.*[80]*[81] The Henry Classification System, co-devised by Haque and Bose, was accepted in England and Wales when the first United Kingdom Fingerprint Bureau was founded in Scotland Yard, the Metropolitan Police headquarters, London, in 1901. Sir Edward Richard Henry subsequently achieved improvements in dactyloscopy.

In the United States, Dr. Henry P. DeForrest used fingerprinting in the New York Civil Service in 1902, and by 1906, New York City Police Department Deputy Commissioner Joseph A. Faurot, an expert in the Bertillon system and a finger print advocate at Police Headquarters, introduced the fingerprinting of criminals to the United States.

The Scheffer case of 1902 is the first case of the identification, arrest and conviction of a murderer based upon fingerprint evidence. Alphonse Bertillon identified the thief and murderer Scheffer, who had previously been arrested and his fingerprints filed some months before, from the fingerprints found on a fractured glass showcase, after a theft in a dentist's apartment where the dentist's employee was found dead. It was able to be proved in court that the fingerprints had been made after the showcase was broken.*[82] A year later, Alphonse Bertillon created a method of getting fingerprints off smooth surfaces and took a further step in the advance of dactyloscopy.

Many criminals wear gloves to avoid leaving fingerprints. However, the gloves themselves can leave prints that are as unique as human fingerprints. After collecting glove prints, law enforcement can match them to gloves that they have collected as evidence or to prints collected at other crime scenes.*[83] In many jurisdictions the act of wearing gloves itself while committing a crime can be prosecuted as an inchoate offense.*[84]

As many offenses are crimes of opportunity, assailants do not always possess gloves when they commit their illegal activities. Thus, assailants have been observed using pulled-down sleeves, pieces of clothing, and other fabrics to handle objects and touch surfaces while committing crimes.*[85]*[86]

3.14 Privacy

3.14.1 Fingerprinting of children

Further information: Biometrics in schools

Various schools have implemented fingerprint locks or made a record of children's fingerprints. In the United Kingdom there have been fingerprint locks in Holland Park School in London,*[87] and children's fingerprints are stored on databases.*[88] There have also been instances in Belgium, at the école Marie-José in Liège,*[89]*[90] in France and in Italy. The non-governmental organization (NGO) Privacy International in 2002 made the cautionary announcement that tens of thousands of UK school children were being fingerprinted by schools, often without the knowledge or consent of their parents.*[91] That same year, the supplier Micro Librarian Systems, which uses a technology similar to that used in US prisons and the German military, estimated that 350 schools throughout Britain were using such systems to replace library cards.*[91] By 2007, it was estimated that 3,500 schools were using such systems.*[92] Under the United Kingdom Data Protection Act, schools in the UK do not have to ask parental consent to allow such practices to take place. Parents opposed to fingerprinting may only bring individual complaints against schools.*[93] In response to a complaint which they are continuing to pursue, in 2010 the European Commission expressed 'significant concerns' over the proportionality and necessity of the practice and the lack of judicial redress, indicating that the practice may break the European Union data protection directive.*[94]

In Belgium, the practice of taking fingerprints from children gave rise to a question in Parliament on February 6, 2007, by Michel de La Motte (Humanist Democratic Centre) to the Education Minister Marie Arena, who replied that it was legal provided that the school did not use them for external purposes, or to survey the private life of children.*[95] At Angers in France, Carqueiranne College in the Var won the Big Brother Award for 2005 and the Commission nationale de l'informatique et des libertés (CNIL), the official organisation in charge of the protection of privacy in France, declared the measures it had introduced "disproportionate."*[96]

In March 2007, the British government was considering fingerprinting all children aged 11 to 15 and adding the prints to a government database as part of a new passport and ID card scheme and disallowing opposition for privacy concerns. All fingerprints taken would be cross-checked against prints from 900,000 unsolved crimes. Shadow Home secretary David Davis called the plan "sinister".*[92] An Early Day Motion which called on the UK Government to conduct a full and open consultation with stakeholders about the use of biometrics in schools, secured the support of 85 Members of Parliament (Early Day Motion 686).*[97] Following the establishment in the United Kingdom of a Conservative and Liberal Democrat coalition government in May 2010, the ID card scheme was scrapped.*[98]

Serious concerns about the security implications of using conventional biometric templates in schools have been raised by a number of leading IT security experts,*[99] one of whom has voiced the opinion that "it is absolutely premature to begin using 'conventional biometrics' in schools".*[100] The vendors of biometric systems claim that their products bring benefits to schools such as improved reading skills, decreased wait times in lunch lines and increased revenues.*[101] They do not cite independent research to support this view. One education specialist wrote in 2007: "I have not been able to find a single piece of published research which suggests that the use of biometrics in schools promotes healthy eating or improves reading skills amongst children... There is absolutely no evidence for such claims".*[102] The Ottawa Police in Canada have advised parents who fear their children may be kidnapped, to fingerprint their children.*[103]

3.15 Other uses

3.15.1 Welfare claimants

It has been alleged that taking the fingerprints of welfare recipients as identification serves as a social stigma that evokes cultural images associated with the processing of criminals.*[104]

3.15.2 Log-in authentication and other locks

Since 2000, electronic fingerprint readers have been introduced for security applications such as log-in authentication for the identification of computer users. However, some less sophisticated devices have been discovered to be vulnerable to quite simple methods of deception, such as fake fingerprints cast in gels. In 2006, fingerprint sensors gained popularity in the notebook PC market. Built-in sensors in laptops, such as ThinkPads, VAIO, HP Pavilion and EliteBook laptops, and others also double as motion detectors for document scrolling, like the scroll wheel.

Following the release of the iPhone 5S model, a group of German hackers announced on September 21, 2013, that they had bypassed Apple's new Touch ID fingerprint sensor by photographing a fingerprint from a glass surface and using that captured image as verification. The spokesman for the group stated: "We hope that this finally puts to rest the illusions people have about fingerprint biometrics. It is plain stupid to use something that you can't change and that you leave everywhere every day as a security token." *[105]

3.15.3 Electronic registration and library access

Fingerprints and, to a lesser extent, iris scans can be used to validate electronic registration, cashless catering, and library access. By 2007, this practice was particularly widespread in UK schools,*[106] and it was also starting to be adopted in some states in the US.

3.16 Absence or mutilation of fingerprints

A very rare medical condition, adermatoglyphia, is characterized by the absence of fingerprints. Affected persons have completely smooth fingertips, palms, toes and soles, but no other medical signs or symptoms.*[107] A 2011 study indicated that adermatoglyphia is caused by the improper expression of the protein SMARCAD1.*[108] The condition has been called *immigration delay disease* by the researchers describing it, because the congenital lack of fingerprints causes delays when affected persons attempt to prove their identity while traveling.*[107] Only five families with this condition have been described as of 2011.*[109]

People with Naegeli–Franceschetti–Jadassohn syndrome and dermatopathia pigmentosa reticularis, which are both forms of ectodermal dysplasia, also have no fingerprints. Both of these rare genetic syndromes produce other signs and symptoms as well, such as thin, brittle hair.

The anti-cancer medication capecitabine may cause the loss of fingerprints.*[110] Swelling of the fingers, such as that caused by bee stings, will in some cases cause the temporary disappearance of fingerprints, though they will return when the swelling recedes.

Since the elasticity of skin decreases with age, many senior citizens have fingerprints that are difficult to capture. The ridges get thicker; the height between the top of the ridge and the bottom of the furrow gets narrow, so there is less prominence.*[111]

Fingerprints can be erased permanently and this can potentially be used by criminals to reduce their chance of conviction. Erasure can be achieved in a variety of ways including simply burning the fingertips, using acids and advanced techniques such as plastic surgery.*[112]*[113]*[114]*[115]*[116] John Dillinger burned his fingers with acid, but prints taken during a previous arrest and upon death still exhibited almost complete relation to one another.*[117]

3.17 Fingerprint recognition

fingerprint authentication refers to the automated method of verifying a match between two human fingerprints. Fingerprints are one of many forms of biometrics used to identify individuals and verify their identity.

The analysis of fingerprints for matching purposes generally requires the comparison of several features of the print pattern. These include patterns, which are aggregate characteristics of ridges, and minutia points, which are unique features found within the patterns.* It is also necessary to know the structure and properties of human skin in order to successfully employ some of the imaging technologies.

3.17.1 Patterns

The three basic patterns of fingerprint ridges are the arch, loop, and whorl:

- arch: The ridges enter from one side of the finger, rise in the center forming an arc, and then exit the other side of the finger.

- loop: The ridges enter from one side of a finger, form a curve, and then exit on that same side.

- whorl: Ridges form circularly around a central point on the finger.

Scientists have found that family members often share the same general fingerprint patterns, leading to the belief that these patterns are inherited.*

3.17.2 Fingerprint processing

Fingerprint processing has three primary functions: enrollment, searching and verification. Among these functions, enrollment which captures fingerprint image from the sensor plays an important role. A reason is that the way people put their fingerprints on a mirror to scan can affect to the result in the searching and verifying process. Regarding to verification function, there are several techniques to match fingerprints such as correlation-based matching, minutiae-based matching, ridge feature-based matching and minutiae-based algorithm. However, the most popular algorithm was minutiae based matching algorithm due to its efficiency and accuracy.

3.17.3 Minutiae features

The major minutia features of fingerprint ridges are ridge ending, bifurcation, and short ridge (or dot). The ridge ending is the point at which a ridge terminates. Bifurcations are points at which a single ridge splits into two ridges. Short ridges (or dots) are ridges which are significantly shorter than the average ridge length on the fingerprint. Minutiae and patterns are very important in the analysis of fingerprints since no two fingers have been shown to be identical.*

3.17.4 Defeats

In 2002 a Japanese cryptographer demonstrated how fingerprint recognition devices can be fooled 4 out of 5 times using a combination of low cunning, cheap kitchen supplies and a digital camera.*[118]

Latent fingerprints from a glass were enhanced with superglue fumes in the form of cyanoacrylate adhesive and photographed. An image editor was then used to improve the contrast and the result printed onto a transparency sheet. The sheet was used to expose a UV sensitive printed-circuit board and etched. The copper imprint were then used for a plastic finger mold. A gelatin found in Gummy bears was molded into a fake finger.*[118] Eleven commercially available fingerprint biometric systems took the fake finger as the real thing. Noted cryptographer Bruce Schneier said "The results are enough to scrap the systems completely, and to send the various fingerprint biometric companies packing." *[118]

3.18 Fingerprint recognition in electronic devices

Two of the first smartphone manufacturers to integrate fingerprint recognition into their phones were Motorola with the Atrix 4G in 2011, and Apple with the iPhone 5S on 10 September 2013. One month after, HTC launched the One Max, which also included fingerprint recognition. In April 2014, Samsung released the Galaxy S5, which integrated a fingerprint sensor on the home button.*[119]

Since December 2015, cheaper smartphones with fingerprint recognition have been released, such as the $100 UMI Fair.*[119] Samsung also recently introduced fingerprint sensors to its mid-range A-series smartphones.

On 25 September 2015 with iPhone 6s, two years after introduction of its first fingerprint scanner in the iPhone 5S, Apple introduced a new generation fingerprint scanner claiming faster response times. In August 2016, OPPO claimed 0,22s response time in its Oppo F1's model.*[120]

Hewlett Packard, Asus, Huawei, Lenovo and Apple are using fingerprint reader in their laptops.*[121]*[122]*[123] Synaptics says the SecurePad sensor is now available for OEMs to start building into their laptops.*[124]

3.19 Fingerprint sensors

A fingerprint sensor is an electronic device used to capture a digital image of the fingerprint pattern. The captured image is called a live scan. This live scan is digitally processed to create a biometric template (a collection of extracted features) which is stored and used for matching. Many technologies have been used including optical, capacitive, RF, thermal, piezoresistive, ultrasonic, piezoelectric, MEMS.*[125] This is an overview of some of the more commonly used fingerprint sensor technologies.

3.19.1 Optical

Optical fingerprint imaging involves capturing a digital image of the print using visible light. This type of sensor is, in essence, a specialized type of digital camera. The top layer of the sensor, where the finger is placed, is known as the touch surface. Beneath this layer is a light-emitting phosphor layer which illuminates the surface of the finger. The light reflected from the finger passes through the phosphor layer to an array of solid state pixels (a charge-coupled device) which captures a visual image of the fingerprint. A scratched or dirty touch surface can cause a bad image of the fingerprint. A disadvantage of this type of sensor is the fact that the imaging capabilities are affected by the quality of skin on the finger. For instance, a dirty or marked finger is difficult to image properly. Also, it is possible for an individual to erode the outer layer of skin on the fingertips to the point where the fingerprint is no longer visible. It can also be easily fooled by an image of a fingerprint if not coupled with a "live finger" detector. However, unlike capacitive sensors, this sensor technology is not susceptible to electrostatic discharge damage.*

Fingerprints can be read from a distance.*[126]

3.19.2 Ultrasonic

Ultrasonic sensors make use of the principles of medical ultrasonography in order to create visual images of the fingerprint. Unlike optical imaging, ultrasonic sensors use very high frequency sound waves to penetrate the epidermal layer of skin. The sound waves are generated using piezoelectric transducers and reflected energy is also measured using piezoelectric materials. Since the dermal skin layer exhibits the same characteristic pattern of the fingerprint, the reflected wave measurements can be used to form an image of the fingerprint. This eliminates the need for clean, undamaged epidermal skin and a clean sensing surface.**[127] LeEco became the first company to introduce this in Smartphone.*[128]

3.19.3 Capacitance

Capacitance sensors use principles associated with capacitance in order to form fingerprint images. In this method of imaging, the sensor array pixels each act as one plate of a parallel-plate capacitor, the dermal layer (which is electrically conductive) acts as the other plate, and the non-conductive epidermal layer acts as a dielectric.

Apple's Touch ID uses a capacitance fingerprint sensor.*[129]

Passive capacitance

A passive capacitance sensor use the principle outlined above to form an image of the fingerprint patterns on the dermal layer of skin. Each sensor pixel is used to measure the capacitance at that point of the array. The capacitance varies between the ridges and valleys of the fingerprint due to the fact that the volume between the dermal layer and sensing element in valleys contains an air gap. The dielectric constant of the epidermis and the area of the sensing element are known values. The measured capacitance values are then used to distinguish between fingerprint ridges and valleys.*

Active capacitance

Active capacitance sensors use a charging cycle to apply a voltage to the skin before measurement takes place. The application of voltage charges the effective capacitor. The electric field between the finger and sensor follows the pattern of the ridges in the dermal skin layer. On the discharge cycle, the voltage across the dermal layer and sensing element is compared against a reference voltage in order to calculate the capacitance. The distance values are then calculated mathematically, and used to form an image of the fingerprint.* Active capacitance sensors measure the ridge patterns of the dermal layer like the ultrasonic method. Again, this eliminates the need for clean, undamaged epidermal skin and a clean sensing surface.*[7]

3.20 Algorithms

Matching algorithms are used to compare previously stored templates of fingerprints against candidate fingerprints for authentication purposes. In order to do this either the original image must be directly compared with the candidate image or certain features must be compared.*

3.20.1 Pre-processing

Pre-processing helped enhancing the quality of an image by filtering and removing unnecessary noises. The minutiae based algorithm only worked effectively in 8-bit gray scale fingerprint image. A reason was that an 8-bit gray fingerprint image was a fundamental base to convert the image to 1-bit image with value 0 for ridges and value 1 for furrows. As a result, the ridges were highlighted with black color while the furrows were highlighted with white color. This process partly removed some noises in an image and helped enhance the edge detection. Furthermore, there are two more steps to improve the best quality for the input image: minutiae extraction and false minutiae removal. The minutiae extraction was carried out by applying ridge thinning algorithm which was to remove redundant pixels of ridges. As a result, the thinned ridges of the fingerprint image are marked with a unique ID so that further operation can be conducted. After the minutiae extraction step, the false minutiae removal was also necessary. The lack of the amount of ink and the cross link among the ridges could cause false minutiae that led to inaccuracy in fingerprint recognition process.

3.20.2 Pattern-based (or image-based) algorithms

Pattern based algorithms compare the basic fingerprint patterns (arch, whorl, and loop) between a previously stored template and a candidate fingerprint. This requires that the images can be aligned in the same orientation. To do this, the algorithm finds a central point in the fingerprint image and centers on that. In a pattern-based algorithm, the template contains the type, size, and orientation of patterns within the aligned fingerprint image. The candidate fingerprint image is graphically compared with the template to determine the degree to which they match.*

3.21 In other species

Some other animals have evolved their own unique prints, especially those whose lifestyle involves climbing or grasping wet objects; these include many primates, such as gorillas and chimpanzees, Australian koalas and aquatic mammal species such as the North American fisher.*[130] According to one study, even with an electron microscope, it can be quite difficult to distinguish between the fingerprints of a koala and a human.*[131] Koalas' independent development of fingerprints is an example of convergent evolution.

3.22 In fiction

3.22.1 Mark Twain

Mark Twain's memoir *Life on the Mississippi* (1883), notable mainly for its account of the author's time on the river, also recounts parts of his later life, and includes tall tales and stories allegedly told to him. Among them is an involved, melodramatic account of a murder in which the killer is identified by a thumbprint.*[132] Twain's novel *Pudd'nhead Wilson*, published in 1893, includes a courtroom drama that turns on fingerprint identification.

3.22.2 Crime fiction

The use of fingerprints in crime fiction has, of course, kept pace with its use in real-life detection. Sir Arthur Conan Doyle wrote a short story about his celebrated sleuth Sherlock Holmes which features a fingerprint: "The Norwood Builder" is a 1903 short story set in 1894 and involves the discovery of a bloody fingerprint which helps Holmes to expose the real criminal and free his client.

The British detective writer R. Austin Freeman's first Thorndyke novel *The Red Thumb-Mark* was published in 1907 and features a bloody fingerprint left on a piece of paper together with a parcel of diamonds inside a safe-box. These become the center of a medico-legal investigation led by Dr. Thorndyke, who defends the accused whose fingerprint matches that on the paper, after the diamonds are stolen.

3.22.3 Film and television

On the television series Bonanza (1959–1973), the first episode with the ethnic Chinese character, Hop Sing, #316 The Mark of Guilt was about fingerprinting and its relationship to Chinese culture. Hop Sing uses his Oriental knowledge of "chops" (unique prints from fingers) to free Little Joe from a murder charge.

The movie *Men in Black*, a popular 1997 science fiction thriller, required Agent J, played by Will Smith, to remove his ten fingerprints by putting his hands on a metal ball, an action deemed necessary by the MIB agency to remove the identity of its agents.

In a 2009 science fiction movie starring Paul Giamatti, *Cold Souls*, a mule who is paid to smuggle souls across borders, wears latex fingerprints to frustrate airport security terminals. She can change her identity by changing her wig, and switching latex fingerprints from the privacy of a restroom, storing extra fingerprints in a ziploc bag, so she can assume an alias that is suitable to her undertaking.

3.23 Other reliable identifiers

Other forms of biometric identification utilizing a physical attribute that is nearly unique to humans include iris recognition, the tongue and DNA profiling, also known as genetic fingerprinting. Forensic dentistry has also been used as an identifier, but bite mark analysis is notable for being unreliable.*[133]

3.24 See also

- Authentication
- Biometrics
- Biometric technology in access control
- Eye vein verification
- Feature extraction
- Fingerprint Verification Competition
- Finger Vein recognition
- Heredity
- Iris recognition
- Medical ultrasonography
- Minutiae
- Piezoelectricity
- Skin
- Touch ID
- Shirley McKie, misidentified fingerprint

3.25 References

[1] "Peer Reviewed Glossary of the Scientific Working Group on Friction Ridge Analysis, Study and Technology (SWG-FAST)" (PDF). Retrieved 2012-09-14.

[2] Olsen, Robert D. Sr (1972). "The Chemical Composition of Palmar Sweat". *Fingerprint and Identification Magazine.* **53** (10).

[3] Hueske, Edward. Firearms and Fingerprints. Facts on File/Infobase Publishing, New York. 2009. ISBN 978-0-8160-5512-8

[4] Horace Cox, ed. (1905). *The Law Times: The Journal and Record: The Law and The Lawyers.* vol. CXIX. London: The Law Times. p. 563.

[5] Hall, Angus. The Crime Busters. Book Sales, United Kingdom/United States. 1989. ISBN 978-1-55521-434-0.

[6] Bhanoo, Sindya N. (20 November 2015). "Science – New Technique Can Identify Gender From a Fingerprint". *New York Times.* Retrieved 21 November 2015.

[7] Huynh, Crystal; Brunelle, Erica; Halámková, Lenka; Agudelo, Juliana; Halámek, Jan (13 October 2015). "Forensic Identification of Gender from Fingerprints". *Analytical Chemistry.* **87** (22): 11531–11536. doi:10.1021/acs.analchem.5b03323. Retrieved 21 November 2015.

[8] "Fake finger reveals the secrets of touch", *Nature,* 29 January 2009, doi:10.1038/news.2009.68

[9] "Fingerprint grip theory rejected". BBC. June 2009. Retrieved March 17, 2010.

[10] Engert, Gerald J. (1964). "International Corner". *Identification News.* **14** (1).

[11] Henry, Edward R., Sir (1900). "Classification and Uses of Finger Prints" (PDF). London: George Rutledge & Sons, Ltd.

[12] Conklin, Barbara Gardner, Robert Gardner, and Dennis Shortelle. Encyclopedia of Forensic Science: a Compendium of Detective Fact and Fiction. Westport, Conn.: Oryx, 2002. Print.

[13] Ashbaugh, David R. "Ridgeology" (PDF). Royal Canadian Mounted Police. Retrieved 2013-10-26.

[14] Zabell, Sandy. "Fingerprint Evidence" (PDF). *Journal of Law and Policy.*

[15] Johnson, P. Lee (1973). "Life of Latents". *Identification News.* **23** (1).

[16] *Manchester Evening News,* June 17, 2010, front page

[17] Fiebig, Tobias; Krissler, Jan; Hänsch, Ronny (August 2014). *Security Impact of High Resolution Smartphone Cameras.* Usenix Association. Retrieved 5 February 2015.

[18] Krissler, Jan. ""Ich sehe, also bin ich du" (Talk at 31C3 conference)" (in German). ccc-tv. Retrieved 5 February 2015.

[19] Kremen, Rachel (September 2009). "Touchless 3-D Fingerprinting: A new system offers better speed and accuracy". Technology Review. Retrieved March 17, 2010.

[20] Ross, A.; Jain, A. (2004). "Estimating fingerprint deformation". Proceedings of the International Conference on Biometric Authentication (ICBA).

[21] Wang, Yongchang; Q. Hao; A. Fatehpuria; D. L. Lau; L. G. Hassebrook (2009). "Data Acquisition and Quality Analysis of 3-Dimensional Fingerprints" (PDF). Florida: IEEE conference on Biometrics, Identity and Security. Retrieved March 17, 2010.

[22] Wang, Yongchang; D. L. Lau; L. G. Hassebrook (2010). "Fit-sphere unwrapping and performance analysis of 3D Fingerprints" (PDF). *Applied Optics.* **49** (4): 592–600. doi:10.1364/ao.49.000592.

[23] Wang, Yongchang; Q. Hao; A. Fatehpuria; L. G. Hassebrook; D. L. Lau (July 2010). "Quality and Matching Performance Analysis of 3D Unraveled Fingerprints" (PDF). **49** (7). Optical Engineering: 077202 (1–10). Retrieved August 16, 2010.

[24] "O'Dougherty Urges All Be Fingerprinted: U.S. Attorney Describes Sciences of Crime Detection to Democrats". The Brooklyn Daily Eagle. March 8, 1938. Retrieved July 1, 2014.

[25] Dalrymple, BE; Duff, JM; Menzel, ER. (1977). "Inherent fingerprint luminescence – detection by laser". *Journal of Forensic Sciences*. **22** (1): 106–115.

[26] Google Patent for Patent DE102014203918B4 concerning Methods and apparatus for detecting the surface structure and texture of a sample, "Fingerabdruck-Scanner" in pvt. Polizei Verkehr Technik. Fachzeitschrift für Polizei- und Verkehrsmanagement, Technik und Ausstattung 2017, 43, "Digitaler Pinsel macht Forensik schneller" in Rhein-Zeitung vom 20. Dezember 2016

[27] Fingerprint Source Book: manual of development techniques, published 26 March 2013 retrieved on 9 February 2017; see also Max M. Houck (Ed.): Forensic Fingerprints, London 2016, p. 21, 50 er seq..

[28] Fleming, Nic. "Fingerprints can reveal race and sex".

[29] Swansea University Archived 2007-09-30 at the Wayback Machine., Materials Research Centre, Professor Neil McMurray and Dr. Geraint Williams.

[30] Ward, Mark (April 2006). "Fingerprints hide lifestyle clues". BBC. Retrieved March 17, 2010.

[31] "Bombers Tracked By New Technique". SkyNews. April 2006. Retrieved March 17, 2010.

[32] "Oak Ridge National Laboratory: The Case of the Vanishing Fingerprint". Ornl.gov. 1995-03-27. Retrieved 2012-09-14.

[33] Paul Marks (May 18, 2007) "New fingerprint analysis identifies smokers", *New Scientist* (on-line version).

[34] Tom Simonite (April 3, 2006) "Fingerprints reveal clues to suspects' habits", *New Scientist* (on-line version).

[35] Everts, Sarah (December 2008). "Fingerprints Reveal Drug Use". *Chemical & Engineering News*. **86** (51): 34.

[36] "Specter, Michael "Do Fingerprints Lie" The New Yorker". Michaelspecter.com. 2002-05-27. Retrieved 2012-09-14.

[37] Zabell, Sandy L., Fingerprint Evidence, Journal of Law and Policy (Brooklyn College Law School) 143-77 (2005)

[38] Dror, I.E., Charlton, D. and Péron, A.E. (2006) "Contextual information renders experts vulnerable to making erroneous identifications", *Forensic Science International*, Vol 156, Iss 1, pp 74-78.

[39] Vokey, J.R., Tangen, J.M. and Cole, S.A, (2009), "On the preliminary psychophysics of fingerprint identification", *The Quarterly Journal of Experimental Psychology*, Vol 62, Iss 5, pp 1023–1040.

[40] Muramoto, S; Forbes, TP; van Asten, AC; Gillen, G. "Test Sample for the Spatially Resolved Quantification of Illicit Drugs on Fingerprints Using Imaging Mass Spectrometry". *Analytical Chemistry*. **87**: 5444–5450. PMID 25915085. doi:10.1021/acs.analchem.5b01060.

[41] "International Association for Identification History, retrieved August 2006". Theiai.org. Retrieved 2012-09-14.

[42] Bonebrake, George J (1978). "Report on the Latent Print Certification Program". *Identification News*. **28** (3).

[43] "U.S. Will Pay $2 Million to Lawyer Wrongly Jailed – New York Times" (article), by Eric Lichtbau, *New York Times*, 2006-11-30, webpage: NYT-061130-settle: on Brandon Mayfield mistaken arrest.

[44] New York Times; May 31, 2004; Can Prints Lie? Yes, Man Finds To His Dismay. In front of the immigration judge, the tall, muscular man began to weep. No, he had patiently tried to explain, he was not Leo Rosario, a drug dealer and a prime candidate for deportation. He was telling the truth. He was René Ramón Sánchez, an auto-body worker and merengue singer ...

[45] "'Relief' over fingerprint verdict". *BBC News*. February 7, 2006.

[46] "The Fingerprint Inquiry Scotland".

[47] "Archived websites". *thefingerprintinquiryscotland.org.uk*. April 29, 2015.

[48] Abel, David (2007-10-26). "Man wrongly convicted in Boston police shooting found dead". The Boston Globe.

[49] "An Officer's Guilt Casts Shadow on Trials". New York Times. March 4, 1993. Retrieved 2007-06-21.

[50] "Police Investigation Supervisor Admits Faking Fingerprints". New York Times. July 30, 1993. Retrieved 2007-06-21.

[51] Laufer, Berthold (1912). "History of the finger-print system". Smithsonian Institution Annual Report. Archived from the original on 2008-10-14. Reprinted in "The Print [newsletter of South California Association of Fingerprint Officers]" (PDF). **16** (2). March–April 2000: 1–13. Archived from the original (PDF) on 2008-10-04.

[52] Ashbaugh, David (1999). *Quantitative-Qualitative Friction Ridge Analysis: An Introduction to Basic and Advanced Ridgeology*. Boca Raton, Florida: CRC Press. pp. 11–19. ISBN 0-8493-7007-8.

[53] Åström, Paul (2007). "The study of ancient fingerprints" (PDF). *Journal of Ancient Fingerprints* (1): 2–3. Archived from the original (PDF) on 2008-10-04.

[54] Åström, Paul; Eriksson, Sven A. (1980). "Fingerprints and Archaeology". *Studies in Mediterranean Archaeology series*. Göteborg, Sweden: Paul Åströms Förlag. **28**.

[55] "Finger prints found on pottery".

[56] "网站地图 _ 广东强富裕投资股份有限公司www.articesbase.com". Articesbase.com. Retrieved 2014-08-02.

3.25. REFERENCES

[57] Kris Dhingra (12 October 2007). "Nadi Astrology – Opening The Leaf To Your Future". Delhi Planet, India. Retrieved 28 June 2012.

[58] Reinaud, Joseph Toussaint (1845). "Relation des voyages faits par les Arabes et les Persans dans l'Inde et a la Chine dans le IX Siecle.". **I**. Paris: Imprimerie royale: 42. quoted in: Laufer (1912)

[59] David R. Ashbaugh, *Quantitative-Qualitative Friction Ridge Analysis: An introduction to basic and advanced ridgeology* (Boca Raton, Florida: CRC Press LLC, 1999), page 19.

[60] Cyril John Polson (1951) "Finger prints and finger printing: an historical study", *Journal of Criminal Law and Criminology*, **41** (4) : 495-517 ; see p. 499. Available on-line at: Northwestern University.
Conclusion: Cyril John Polson (1951) "Finger prints and finger printing: an historical study", *Journal of Criminal Law and Criminology*, **41** (5) : 690-704. Available on-line at: Northwestern University.

[61] Cummins, Harold (1941). "Ancient finger prints in clay". *The Scientific Monthly*. **52**: 389–402. Bibcode:1941SciMo..52..389C. Reprinted in *Journal of Criminal Law and Criminology*, volume 34, 4, page 468–481, November–December 1941

[62] Ashbaugh (1999), page 15.

[63] " 千余學者摸清我國民族膚紋 " 家底" 南北是一家" (in Chinese).

[64] See:

- Ashbaugh (1999), page 17.
- Laufer (1912), pp. 642-643.
- E. Chavannes (1905) "Les livres chinois avant l'invention du papier" (Chinese books before the invention of paper), *Journal asiatique*, 10th series, **5** : 5-75 ; see especially p. 56. From p. 56: "*Kia Kong-yen ···(vers 650) ajoute ici la glose:* "*Les entailles faites sur le côté des ces fiches, c'est comme aujourd'hui les emprientes du doigt*" ···" (Kia Kong-yen ···(about 650) adds here the gloss: "The notches made on the sides of these sheets are like fingerprints today" ···)

[65] Cole, Simon (2001). *Suspect Identities: A history of fingerprinting and criminal identification*. Cambridge, Massachusetts: Harvard University Press. pp. 60–61. ISBN 0-674-00455-8.

[66] Malpighi, Marcello (1665). *De Externo Tactus Organo Anatomica Observatio* [*Anatomical Observations of the External Organs of Touch*]. Naples, Italy: Aegidius Longus. p. 7. Malpighi examined a fingertip ("*extremum digiti*") with a microscope, "*··· & dum attentius inaequales illas rugas quasi in gyrum, vel in spiras ductas contemplor,* ···" (···and while I carefully observed those irregular wrinkles as if formed in a circle or in a spiral ···)

[67] Grew, Nehemiah (1684). "The description and use of the pores in the skin of the hands and feet". *Philosophical Transactions of the Royal Society*. **14**: 566–567. Bibcode:1684RSPT...14..566G. doi:10.1098/rstl.1684.0028.

[68] Bidloo, Govard (1685). *Anatomia Humani Corporis* [*Anatomy of the Human Body*]. Amsterdam, Netherlands. The illustrations of the structure of the skin appear in Plate 4 (T. 4) and the illustrations are explained on the following page, *Quartæ Tabulæ*, in Latin.

[69] Mayer, Johann Christoph Andreas (1788). *Anatomische Kupfertafeln nebst dazu gehörigen Erklärungen* [*Anatomical Illustrations (etchings) with Accompanying Explanations, volume 4*]. Berlin, Prussia: Georg Jacob Decker und Sohn. p. 5. From page 5: "*Zweite Figur. ···Obwohl niemals bey zween Menschen die Lagen der Hautwärzgen übereinkommen,* ···" (Second figure. ···Although the positions of the skin papillae never agree between two people, ···) Available on-line at: University of Heidelberg. See also: illustrations of friction ridges.

[70] "The History of Fingerprints". Onin. February 2010. Retrieved March 17, 2010.

[71] Purkyně, Jan Evangelista (1823). *Commentatio de examine physiologico organi visus et systematis cutanei* [*Commentary on the physiological examination of the visual organ and the skin system*]. Breslau, Prussia: University of Breslau Press. p. 43. From page 43: "*Ego hucusque post observationes innumeras novem potissimum varietates flexurarum inveneram ad quas valleculae tactui inservientes in interna parte extremae digitorum phalangis disponuntur.*" (So far, after innumerable observations, I have found nine main varieties of bends in which are arranged the grooves serving touch in the inward part of the fingertip.) Purkyně then lists and characterizes each fingerprint pattern.
See also: Cummins, Harold; Wright Kennedy, Rebecca (September–October 1940). "Purkinje's observations (1823) on finger prints and other skin features". *The Journal of Criminal Law and Criminology*. **31** (3): 343–356. doi:10.2307/1137436.

[72] Alberge, Dalya (9 December 2012). "Vital clue ignored for 50 years". London: Independent. Retrieved 28 December 2015.

[73] von Meissner, Georg (1853). *Beiträge zur Anatomie und Physiologie der Haut* [*Contributions to the Anatomy and Physiology of the Skin*]. Leipzig, Saxony: Leopold Voss.

[74] Herschel, William J (1916). *The Origin of Finger-Printing* (PDF). Oxford University Press. ISBN 978-1-104-66225-7.

[75] Herschel, William James (November 25, 1880). "Skin furrows of the hand" (PDF). *Nature*. **23** (578): 76. Bibcode:1880Natur..23...76H. doi:10.1038/023076b0.

[76] Coulier, Paul-Jean (1863) "Les vapeurs d'iode employées comme moyen de reconnaitre l'altération des écritures"

(Iodine vapors used as a means of recognizing the alteration of writing), *L'Année scientifique et industrielle*, vol. 8, pages 157-160.

Coulier was trying to develop means of detecting forgeries. He would expose a suspect document to iodine vapor, and the iodine would deposit on the paper, revealing otherwise invisible pen indentations. However, ⋯

> It has happened several times to Mr. Coulier that stains form in places where his fingers had touched the paper. When a finger is applied to the paper without rubbing, iodine stains reproduce with wonderful fidelity the papillae [friction ridges] of the skin, and as they have patterns of infinite variety, just like the lines of the hand, the result is that it is not impossible to recognize, in these traces, the individual who touched the paper. It would suffice to put the fingers of the person in question on a sheet of white paper and then, after exposing the page to iodine vapor, one could obtain in this way prints that could be compared, by means of a loupe [lens] or compass, to the prints that are being identified. [Coulier (1863), p. 159.]

See also: Margot, Pierre and Quinche, Nicolas (March–April 2010) "Coulier, Paul-Jean (1824–1890): A precursor in the history of fingermark detection and their potential use for identifying their source (1863)", *Journal of Forensic Identification*, vol. 60, no. 2, pages 129-134.

[77] Faulds, Henry (October 28, 1880). "On the skin-furrows of the hand" (PDF). *Nature.* **22** (574): 605. Bibcode:1880Natur..22..605F. doi:10.1038/022605a0.

[78] Reid, Donald L. (2003). "Dr. Henry Faulds – Beith Commemorative Society". *Journal of Forensic Identification.* **53** (2). See also this on-line article on Henry Faulds: Tredoux, Gavan (December 2003). "Henry Faulds: the Invention of a Fingerprinter". galton.org.

[79] Galton, Francis (1892). "Finger Prints" (PDF). London: MacMillan and Co. From p. 110: "The result is, that the chance of lineations, constructed by the imagination according to strictly natural forms, which shall be found to resemble those of a single finger print in all of their minutiæ, is less than 1 to $2^{24} \times 2^4 \times 2^8$, or 1 to 2^{36}, or 1 to about sixty-four thousand millions."

[80] Tewari, RK; Ravikumar, KV (2000). "History and development of forensic science in India". *J. Postgrad Med* (46): 303–308.

[81] Sodhi, J.S.; Kaur, asjeed (2005). "The forgotten Indian pioneers of finger print science" (PDF). *Current Science.* **88** (1): 185–191.

[82] Berlière, Jean-Marc (October 16, 1902). "Arrestation du premier assassin confondu par ses empreintes digitales". Célébrations Nationales.

[83] Sawer, Patrick (2008-12-13). "Police use glove prints to catch criminals". Telegraph.co.uk. Retrieved 2012-09-14.

[84] James W.H. McCord and Sandra L. McCord, *Criminal Law and Procedure for the paralegal: a systems approach*, supra, p. 127.

[85] I Hope This Isn't Another Bait Car, Man! on YouTube

[86] Caught on cam: Bait Car Thieves on YouTube

[87] Empreintes digitales pour les enfants d'une école de Londres (in French)

[88] Leave Them Kids Alone Archived 2007-03-23 at the Wayback Machine. (in English)

[89] Empreintes digitales pour sécuriser l'école ? Archived 2007-07-01 at the Wayback Machine. (in French)

[90] "Le lecteur d'empreintes dans les écoles crée la polémique" (in French). 7sur7.be. February 5, 2007. Archived from the original on July 21, 2012.

[91] Fingerprinting of UK school kids causes outcry, *The Register*, July 22, 2002 (in English)

[92] Child fingerprint plan considered, *BBC*, March 4, 2007 (in English)

[93] Schools can fingerprint children without parental consent, *The Register*, September 7, 2006 (in English)

[94] Europe tells Britain to justify itself over fingerprinting children in schools Telegraph, published 2010-12-14, accessed 2011-01-13

[95] Prises d'empreintes digitales dans un établissement scolaire Archived 2007-02-24 at the Wayback Machine., Question d'actualité à la Ministre-Présidente en charge de l'Enseignement obligatoire et de Promotion sociale (in French)

[96] Quand la biométrie s'installe dans les cantines au nez et à la barbe de la Cnil, *Zdnet*, September 9, 2003 (in French)

[97] "EDM 686 – Biometric Data Collection In Schools". UK Parliament. 2007-01-19. Retrieved 2009-11-28.

[98] BBC News Channel, May 27, 2010.

[99] Cavoukian, A and Stoianov, A. 2007. Biometric Encrypton: A Positive-Sum Technology that Achieves Strong Authentication, Security *and* Privacy Archived 2007-06-14 at the Wayback Machine..

[100] Kim Cameron, architect of identity and access in the Connected Systems Division at Microsoft. blog

[101] "Fingerprint Software Eliminates Privacy Concerns and Establishes Success". FindBiometrics. Archived from the original on 2009-03-14. Retrieved 2012-09-14.

[102] 2007. Dr. Sandra Leaton Gray of Homerton College, Cambridge: professional opinion.

[103] Child Print Archived 2007-05-02 at the Wayback Machine. (Ottawa Police Service) (in English)/(in French)

[104] Murray, Harry (March 2000). "Deniable Degradation: The Finger-Imaging of Welfare Recipients". *Sociological Forum*. **15** (1): 39–63. ISSN 0884-8971. doi:10.1023/A:1007594003722.

[105] Stephen Musil (22 September 2013). "Hackers claim to have defeated Apple's Touch ID print sensor". *Cnet*. CBS Interactive Inc. Retrieved 23 September 2013.

[106] "Peers slam school fingerprinting". *BBC News*. March 19, 2007. Retrieved 2 September 2010.

[107] Burger, B.; Fuchs, D.; Sprecher, E.; Itin, P. (May 2011). "The immigration delay disease: adermatoglyphia-inherited absence of epidermal ridges". *J Am Acad Dermatol*. **64** (5): 974–80. PMID 20619487. doi:10.1016/j.jaad.2009.11.013.

[108] "The Mystery of the Missing Fingerprints".

[109] Nousbeck, J; Burger, B; Fuchs-Telem, D; et al. (August 2011). "A Mutation in a Skin-Specific Isoform of SMARCAD1 Causes Autosomal-Dominant Adermatoglyphia". *American Journal of Human Genetics*. **89** (2): 302–307. PMC 3155166. PMID 21820097. doi:10.1016/j.ajhg.2011.07.004.

[110] Wong M, Choo SP, Tan EH (July 2009). "Travel warning with capecitabine". *Annals of Oncology*. **20** (7): 1281. PMID 19470576. doi:10.1093/annonc/mdp278.

[111] Harmon, Katherine (2009-03-29). "Can You Lose Your Fingerprints?". *Scientific American*.

[112] Fingerprint Alteration Biometrics research group, Michigan State University.

[113] "Fingerprint alteration" (PDF). Retrieved 2012-09-14.

[114] "Fingerprint alteration" (PDF). Retrieved 2012-09-14.

[115] "Changing of fingerprints". Scafo.org. Retrieved 2012-09-14.

[116] "Fingerprints, detailed information". Forensic-medecine.info. Retrieved 2012-09-14.

[117] Abel, David (July 21, 2010). "To avoid ID, more [Americans] are mutilating fingerprints". *Boston Globe*.

[118] "Gummi bears defeat fingerprint sensors". 16 May 2002. Retrieved 12 April 2016.

[119] "List of All Fingerprint Scanner Enabled Smartphones". 27 March 2017.

[120] "Oppo F1s". *www.oppo.com*.

[121] "HP Spectre x360 (2017) review: The best just keeps getting better". *PCWorld*. Retrieved 2017-08-16.

[122] "Asus Transformer Pro T304 is a Surface Pro clone that kills it on price". *Digital Trends*. 2017-07-28. Retrieved 2017-08-16.

[123] "Lenovo ThinkPad T570 Review". Retrieved 2017-08-16.

[124] "Coming soon: laptops with fingerprint sensors built into the touchpad". *Engadget*. Retrieved 2017-08-16.

[125] Wasserman, Philip (26 December 2005). "Solid-State Fingerprint Scanners - A Survey of Technologies" (PDF). Archived from the original (PDF) on 17 January 2016. Retrieved 18 October 2015.

[126] *A Fingerprint Scanner That Can Capture Prints From 20 Feet Away* 25 June 2012 Popular Science

[127] Rajawat, Deepak (9 June 2016). "What's So Especial about Ultrasonic fingerprint sensors?".

[128] Rajawat, Deepak (8 June 2016). "LeEco Le Max2 and Le 2 with USB Type".

[129] "About Touch ID security on iPhone and iPad". Retrieved 18 October 2015.

[130] "Animal fingerprints". Retrieved September 2, 2010.

[131] Henneberg, Maciej; Lambert, Kosette M.; Leigh, Chris M. (1997). "Fingerprint homoplasy: koalas and humans". *NaturalSCIENCE.com*. **1** (4). Archived from the original on 2006-11-14.

[132] Mark Twain (Samuel Clemens). "The Project Gutenberg EBook of Life On The Mississippi". Retrieved 24 November 2011.

[133] Evidence From Bite Marks, It Turns Out, Is Not So Elementary. New York Times; January 28, 2007

3.26 Further reading

- Ashbaugh, David R. 1999. *Quantitative-Qualitative Friction Ridge Analysis: An Introduction to Basic and Advanced Ridgeology*. Boca Raton, Florida: CRC Press.

- Beavan, Colin. 2001. *Fingerprints: The Origins of Crime Detection and the Murder Case that Launched Forensic Science*. New York: Hyperion.

- Cowger, James C. 1992. *Friction Ridge Skin: Comparison and Identification of Fingerprints*. Boca Raton, Florida: CRC Press.

- Quinche, Nicolas, and Margot, Pierre. 2010. *Coulier, Paul-Jean (1824–1890): A precursor in the history of fingermark detection and their potential use for identifying their source (1863)*. In *Journal of Forensic Identification* (California), 60 (2), March–April 2010, pp. 129–134.

- Scheibert, J, Leurent, S, Provost, A and Debregeas, G. 2009. *The role of fingerprints in the coding of tactile information probed with a biomimetic sensor.* Science 323: 1503–1506.

3.27 External links

General

- Fingerprint Terminology
- Fingerprint Sourcebook Multi-organization compendium on Fingerprints
- The Fingerprint Society – Society for Fingerprint Experts
- International Association for Identification
- Scientific Working Group on Friction Ridge Analysis, Study and Technology International Working Group on Fingerprints
- Interpol Fingerprint Research
- The Science of Fingerprints FBI Publication
- FBI Fingerprint Guide
- FBI Fingerprinting Video Lesson (4-sec QuickTime video of rolling a single inked finger)
- The History of Fingerprints
- Fingerprinting.com
- Fingerprints and Human Identification
- Fingerprint Processing Guide
- Fingerprint Articles at Crime & Clues
- Galton's *Finger Prints*
- Henry, Faulds, and Herschel's works on fingerprints
- Extensive bibliography So. Calif. Assn. of Fingerprint Officers.

Errors and concerns

- Publications Critical of Fingerprint Identification
- Will West as fable
- Do Fingerprints Lie? The New Yorker (2002)
- Why Experts Make Errors, Itiel E. Dror, David Charlton, Journal of Forensic Identification
- Surgeon jailed for removing fingerprints – Sydney Morning Herald (news article)

Science and statistics

- Fingerprint research and evaluation at the U.S. National Institute of Standards and Technology.
- Fingerprint pattern distribution statistics

Chapter 4

Facial recognition system

"Face recognition" redirects here. For the human cognitive process, see Face perception.

A **face recognition system** is a computer application capable of identifying or verifying a person from a digital image or a video frame from a video source. One of the ways to do this is by comparing selected facial features from the image and a face database.

It is typically used in security systems and can be compared to other biometrics such as fingerprint or eye iris recognition systems.[1] Recently, it has also become popular as a commercial identification and marketing tool.[2]

Swiss European surveillance: face recognition and vehicle make, model, color and license plate reader

Close-up of the infrared illuminator. The light is invisible to the human eye, but creates a day-like environment for the surveillance cameras.

4.1 Techniques for face acquisition

4.1.1 Traditional

Some face recognition algorithms identify facial features by extracting landmarks, or features, from an image of the subject's face. For example, an algorithm may analyze the relative position, size, and/or shape of the eyes, nose, cheekbones, and jaw.[3] These features are then used to search for other images with matching features.[4] Other algorithms normalize a gallery of face images and then compress the face data, only saving the data in the image that is useful for face recognition. A probe image is then compared with the face data.[5] One of the earliest successful systems[6] is based on template matching techniques[7] applied to a set of salient facial features, providing a sort of compressed face representation.

Recognition algorithms can be divided into two main approaches, geometric, which looks at distinguishing features, or photometric, which is a statistical approach that distills an image into values and compares the values with templates to eliminate variances.

Popular recognition algorithms include principal component analysis using eigenfaces, linear discriminant analysis, elastic bunch graph matching using the Fisherface algorithm, the hidden Markov model, the multilinear subspace learning using tensor representation, and the neuronal motivated dynamic link matching.

4.1.2 3-dimensional recognition

A newly emerging trend, claimed to achieve improved accuracy, is three-dimensional face recognition. This technique uses 3D sensors to capture information about the shape of a face. This information is then used to identify distinctive features on the surface of a face, such as the contour of the eye sockets, nose, and chin.*[8]

One advantage of 3D face recognition is that it is not affected by changes in lighting like other techniques. It can also identify a face from a range of viewing angles, including a profile view.*[4]*[8] Three-dimensional data points from a face vastly improve the precision of face recognition. 3D research is enhanced by the development of sophisticated sensors that do a better job of capturing 3D face imagery. The sensors work by projecting structured light onto the face. Up to a dozen or more of these image sensors can be placed on the same CMOS chip—each sensor captures a different part of the spectrum....*[9]

Even a perfect 3D matching technique could be sensitive to expressions. For that goal a group at the Technion applied tools from metric geometry to treat expressions as isometries*[10] A company called Vision Access created a firm solution for 3D face recognition. The company was later acquired by the biometric access company Bioscrypt Inc. which developed a version known as 3D FastPass.

A new method is to introduce a way to capture a 3D picture by using three tracking cameras that point at different angles; one camera will be pointing at the front of the subject, second one to the side, and third one at an angle. All these cameras will work together so it can track a subject's face in real time and be able to face detect and recognize.*[11]

4.1.3 Skin texture analysis

Another emerging trend uses the visual details of the skin, as captured in standard digital or scanned images. This technique, called skin texture analysis, turns the unique lines, patterns, and spots apparent in a person's skin into a mathematical space.*[4]

Tests have shown that with the addition of skin texture analysis, performance in recognizing faces can increase 20 to 25 percent.*[4]*[8]

4.1.4 Thermal cameras

A different form of taking input data for face recognition is by using thermal cameras, by this procedure the cameras will only detect the shape of the head and it will ignore the subject accessories such as glasses, hats, or make up. A problem with using thermal pictures for face recognition is that the databases for face recognition is limited. Diego Socolinsky, and Andrea Selinger (2004) research the use of thermal face recognition in real life, and operation sceneries, and at the same time build a new database of thermal face images. The research uses low-sensitive, low-resolution ferro-electric electrics sensors that are capable of acquire long wave thermal infrared (LWIR). The results show that a fusion of LWIR and regular visual cameras has the greater results in outdoor probes. Indoor results show that visual has a 97.05% accuracy, while LWIR has 93.93%, and the Fusion has 98.40%, however on the outdoor proves visual has 67.06%, LWIR 83.03%, and fusion has 89.02%. The study used 240 subjects over the period of 10 weeks to create the new database. The data was collected on sunny, rainy, and cloudy days.*[12]

4.2 Notable users and deployments

The Australian people and New Zealand Customs Services have an automated border processing system called SmartGate that uses face recognition. The system compares the face of the individual with the image in the e-passport microchip to verify that the holder of the passport is the rightful owner.*[13]

Law enforcement agencies in the United States, including the Los Angeles County Sheriff, use arrest mugshot databases in their forensic investigative work. Law enforcement has been rapidly building a database of photos in recent years.

The U.S. Department of State operates one of the largest face recognition systems in the world with 117 million American adults including mostly law abiding citizens in its database. The photos are typically drawn from driver's license photos.*[14] Although it is still far from completion, it is being put to use in certain cities to give clues as to who was in the photo. The FBI uses the photos as an investigative lead not for positive identification.*[15]

In recent years Maryland has used face recognition by comparing people's faces to their driver's license photos. The system drew controversy when it was used in Baltimore to arrest unruly protesters after the death of Freddie Gray in police custody.*[16] Many other states are using or developing a similar system however some states have laws prohibiting its use.

The FBI has also instituted its Next Generation Identification program to include face recognition, as well as more traditional biometrics like fingerprints and iris scans, which can pull from both criminal and civil databases.*[17]

The Tocumen International Airport in Panama operates an airport-wide surveillance system using hundreds of live face recognition cameras to identify wanted individuals passing through the airport.*[18]

Major Canadian airports will be using a new facial recognition program as part of the Primary Inspection Kiosk program that will compare people's faces to their passports. This program will first come to Ottawa International Airport in early 2017 and to other airports in 2018.*[19]

In 2017, Time & Attendance company ClockedIn released facial recognition as a form of attendance tracking for businesses and organisations looking to have a more automated system of keeping track of hours worked as well as for security and health & safety control.

In May 2017, A man was arrested using an Automatic Facial Recognition (AFR) system mounted on a van operated by the South Wales Police. Ars Technica reported that "this appears to be the first time it[AFR] has led to an arrest".*[20]

Automatic Facial Recognition systems resemble other mobile CCTV systems

4.2.1 Additional uses

In addition to being used for security systems, authorities have found a number of other applications for face recognition systems. While earlier post-9/11 deployments were well publicized trials, more recent deployments are rarely written about due to their covert nature.

At Super Bowl XXXV in January 2001, police in Tampa Bay, Florida used Viisage face recognition software to search for potential criminals and terrorists in attendance at the event. 19 people with minor criminal records were potentially identified.*[21]*[22]

In the 2000 presidential election, the Mexican government employed face recognition software to prevent voter fraud. Some individuals had been registering to vote under several different names, in an attempt to place multiple votes. By comparing new face images to those already in the voter database, authorities were able to reduce duplicate registrations.*[23] Similar technologies are being used in the United States to prevent people from obtaining fake identification cards and driver's licenses.*[24]*[25]

There are also a number of potential uses for face recognition that are currently being developed. For example, the technology could be used as a security measure at ATMs. Instead of using a bank card or personal identification number, the ATM would capture an image of the customer's face, and compare it to the account holder's photo in the bank database to confirm the customer's identity.*[4]

Face recognition systems are used to unlock software on mobile devices. An independently developed Android Marketplace app called Visidon Applock makes use of the phone's built-in camera to take a picture of the user. Face recognition is used to ensure only this person can use certain apps which they choose to secure.*[26]

Face detection and face recognition are integrated into the iPhoto application for Macintosh, to help users organize and caption their collections.*[27]

Because of certain limitations of fingerprint recognition systems, face recognition systems could be used as an alternative way to confirm employee attendance at work for the claimed hours.

Another use could be a portable device to assist people with prosopagnosia in recognizing their acquaintances.

In September 2017, Apple Inc. launched its new flagship device, iPhone X, with facial recognition technology, named Face ID.*[28]

4.3 Advantages and disadvantages

4.3.1 Compared to other technologies

Among the different biometric techniques, face recognition may not be most reliable and efficient. However, one key advantage is that it does not require the cooperation of the test subject to work. Properly designed systems installed in airports, multiplexes, and other public places can identify individuals among the crowd, without passers-by even being aware of the system. Other biometrics like fingerprints, iris scans, and speech recognition cannot perform this kind of mass identification. However, questions have been raised on the effectiveness of face recognition software in cases of railway and airport security.

4.3.2 Weaknesses

Face recognition is far from perfect and struggles to perform under certain conditions. Ralph Gross, a researcher at the Carnegie Mellon Robotics Institute, describes one obstacle related to the viewing angle of the face: "Face recognition has been getting pretty good at full frontal faces and 20 degrees off, but as soon as you go towards profile, there've been problems." *[8]

Current face recognition still often misidentifies people which can sometimes lead to controversy. Google was criticized for racism in its system when a black couple were misidentified as gorillas.*[29] Face recognition software generally doesn't do as well in identifying minorities when most of the subjects used in testing the technology were from the majority group.

Other conditions where face recognition does not work well include poor lighting, sunglasses, hats, scarves, beards, long hair, makeup or other objects partially covering the subject's face, and low resolution images.*[4]

Another serious disadvantage is that many systems are less effective if facial expressions vary. Even a big smile can render the system less effective. For instance: Canada now allows only neutral facial expressions in passport photos.*[30]

There is also inconstancy in the datasets used by researchers. Researchers may use anywhere from several subjects to scores of subjects, and a few hundred images to thousands of images. It is important for researchers to make available the datasets they used to each other, or have at least a standard dataset.*[31]

4.3.3 Effectiveness

Critics of the technology complain that the London Borough of Newham scheme has, as of 2004, never recognized a single criminal, despite several criminals in the system's database living in the Borough and the system having been running for several years. "Not once, as far as the police know, has Newham's automatic face recognition system spotted a live target." *[22]*[32] This information seems to conflict with claims that the system was credited with a 34% reduction in crime (hence why it was rolled out to Birmingham also).*[33] However it can be explained by the notion that when the public is regularly told that they are under constant video surveillance with advanced face recognition technology, this fear alone can reduce the crime rate, whether the face recognition system technically works or does not. This has been the basis for several other face recognition based security systems, where the technology itself does not work particularly well but the user's perception of the technology does.

An experiment in 2002 by the local police department in Tampa, Florida, had similarly disappointing results.*[22]

A system at Boston's Logan Airport was shut down in 2003 after failing to make any matches during a two-year test period.*[34]

As of 2016, facial recognition is still not effective for most applications even though the accuracy has been substantially improved. Although systems are often advertised as having accuracy near 100%, this is misleading as the studies often uses much smaller sample sizes than would be necessary for large scale applications. Because facial recognition is not completely accurate, it creates a list of potential matches. A human operator must then look through these potential matches and studies show the operators pick the correct match out of the list only about half the time. This causes the issue of targeting the wrong suspect.*[15]*[35]

4.3.4 Privacy issues

Civil rights right organizations and privacy campaigners such as the Electronic Frontier Foundation*[36] and the ACLU*[37] express concern that privacy is being compromised by the use of surveillance technologies. Some fear that it could lead to a "total surveillance society," with the government and other authorities having the ability to know the whereabouts and activities of all citizens around the clock. This knowledge has been, is being, and could continue to be deployed to prevent the lawful exercise of rights of citizens to criticize those in office, specific government policies or corporate practices. Many centralized power structures with such surveillance capabilities have abused their privileged access to maintain control of the political and economic apparatus, and to curtail populist reforms.*[38]

Face recognition can be used not just to identify an individual, but also to unearth other personal data associated with an individual – such as other photos featuring the individual, blog posts, social networking profiles, Internet behavior, travel patterns, etc. – all through facial features alone.*[39] Moreover, individuals have limited ability to avoid or thwart face recognition tracking unless they hide their faces. This fundamentally changes the dynamic of day-to-day privacy by enabling any marketer, government agency, or random stranger to secretly collect the identities and associated personal information of any individual captured by the face recognition system.*[39]

Social media web sites such as Facebook have very large numbers of photographs of people, annotated with names. This represents a database which may be abused by governments for face recognition purposes.*[40] Face recognition was used in Russia to harass women allegedly involved in online pornography.*[41] In Russia there is an app 'Find-Face' which can identify faces with about 70% accuracy using the social media app called VK. This app would not be possible in other countries which do not use VK as their social media platform photos are not stored the same way as with VK.*[42]

In July 2012, a hearing was held before the Subcommittee on Privacy, Technology and the Law of the Committee on the Judiciary, United States Senate, to address issues surrounding what face recognition technology means for privacy and civil liberties.*[43]

In 2014, the National Telecommunications and Information Association (NTIA) began a multi-stakeholder process to engage privacy advocates and industry representatives to establish guidelines regarding the use of face recognition technology by private companies.*[44] In June 2015, privacy advocates left the bargaining table over what they felt was an impasse based on the industry representatives being unwilling to agree to consent requirements for the collection of face recognition data.*[45] The NTIA and industry representatives continued without the privacy representatives, and draft rules are expected to be presented in the spring of 2016.*[46]

States have begun enacted legislation to protect citizen's biometric data privacy. Illinois enacted the Biometric Information Privacy Act in 2008.*[47] Facebook's DeepFace has become the subject of several class action lawsuits under the Biometric Information Privacy Act, with claims alleging that Facebook is collecting and storing face recognition data of its users without obtaining informed consent, in direct violation of the Biometric Information Privacy Act.*[48] The most recent case was dismissed in January 2016 because the court lacked jurisdiction.*[49] Therefore, it is still unclear if the Biometric Information Privacy Act will be effective in protecting biometric data privacy rights.

4.4 History

Pioneers of automated face recognition include Woody Bledsoe, Helen Chan Wolf, and Charles Bisson.

During 1964 and 1965, Bledsoe, along with Helen Chan and Charles Bisson, worked on using the computer to recognize human faces (Bledsoe 1966a, 1966b; Bledsoe and Chan 1965). He was proud of this work, but because the funding was provided by an unnamed intelligence agency that did not allow much publicity, little of the work was published. Given a large database of images (in effect, a book of mug shots) and a photograph, the problem was to select from the database a small set of records such that one of the image records matched the photograph. The success of the method could be measured in terms of the ratio of the answer list to the number of records in the database. Bledsoe (1966a) described the following difficulties:

This project was labeled man-machine because the human extracted the coordinates of a set of features from the photographs, which were then used by the computer for recognition. Using a graphics tablet (GRAFACON or RAND TABLET), the operator would extract the coordinates of features such as the center of pupils, the inside corner of eyes, the outside corner of eyes, point of widows peak, and so on. From these coordinates, a list of 20 distances, such as width of mouth and width of eyes, pupil to pupil, were computed. These operators could process about 40 pictures an hour. When building the database, the name of the person in the photograph was associated with the list of computed distances and stored in the computer. In the recognition phase, the set of distances was compared with the corresponding distance for each photograph, yielding a distance between the photograph and the database record. The closest records are returned.

Because it is unlikely that any two pictures would match in head rotation, lean, tilt, and scale (distance from the camera), each set of distances is normalized to represent the face in a frontal orientation. To accomplish this normalization, the program first tries to determine the tilt, the lean, and the rotation. Then, using these angles, the computer undoes the effect of these transformations on the computed distances. To compute these angles, the computer must know the three-dimensional geometry of the head. Because the actual heads were unavailable, Bledsoe (1964) used a standard head derived from measurements on seven heads.

After Bledsoe left PRI in 1966, this work was continued at the Stanford Research Institute, primarily by Peter Hart. In experiments performed on a database of over 2000 photographs, the computer consistently outperformed humans when presented with the same recognition tasks (Bledsoe 1968). Peter Hart (1996) enthusiastically recalled the project with the exclamation, "It really worked!"

By about 1997, the system developed by Christoph von der Malsburg and graduate students of the University of Bochum in Germany and the University of Southern California in the United States outperformed most systems with those of Massachusetts Institute of Technology and the University of Maryland rated next. The Bochum system was developed through funding by the United States Army Research Laboratory. The software was sold as ZN-Face and used by customers such as Deutsche Bank and operators of airports and other busy locations. The software was "robust enough to make identifications from less-than-perfect face views. It can also often see through such impediments to identification as mustaches, beards, changed hair styles and glasses—even sunglasses".*[50]

In about January 2007, image searches were "based on the

text surrounding a photo," for example, if text nearby mentions the image content. Polar Rose technology can guess from a photograph, in about 1.5 seconds, what any individual may look like in three dimensions, and claimed they "will ask users to input the names of people they recognize in photos online" to help build a database. Identix, a company out of Minnesota, has developed the software, FaceIt. FaceIt can pick out someone's face in a crowd and compare it to databases worldwide to recognize and put a name to a face. The software is written to detect multiple features on the human face. It can detect the distance between the eyes, width of the nose, shape of cheekbones, length of jawlines and many more facial features. The software does this by putting the image of the face on a faceprint, a numerical code that represents the human face. Face recognition software used to have to rely on a 2D image with the person almost directly facing the camera. Now, with FaceIt, a 3D image can be compared to a 2D image by choosing 3 specific points off of the 3D image and converting it into a 2D image using a special algorithm that can be scanned through almost all databases. In 2006, the performance of the latest face recognition algorithms were evaluated in the Face Recognition Grand Challenge (FRGC). High-resolution face images, 3-D face scans, and iris images were used in the tests. The results indicated that the new algorithms are 10 times more accurate than the face recognition algorithms of 2002 and 100 times more accurate than those of 1995. Some of the algorithms were able to outperform human participants in recognizing faces and could uniquely identify identical twins.[8] [51]

U.S. Government-sponsored evaluations and challenge problems[52] have helped spur over two orders-of-magnitude in face-recognition system performance. Since 1993, the error rate of automatic face-recognition systems has decreased by a factor of 272. The reduction applies to systems that match people with face images captured in studio or mugshot environments. In Moore's law terms, the error rate decreased by one-half every two years.[9]

Low-resolution images of faces can be enhanced using face hallucination. Further improvements in high resolution, megapixel cameras in the last few years have helped to resolve the issue of insufficient resolution.

4.5 Emotion detection

Facial recognition systems have been used for emotion recognition[53][54] In 2016 Facebook acquired emotion detection startup FacioMetrics.[55][56]

4.6 Anti facial recognition systems

In January 2013 Japanese researchers from the National Institute of Informatics created 'privacy visor' glasses that uses nearly infrared light to make the face underneath it unrecognizable to face recognition software.[57] The latest version uses a titanium frame, light-reflective material and a mask which uses angles and patterns to disrupt facial recognition technology through both absorbing and bouncing back light sources.[58][59][60][61] In December 2016 a form of anti-CCTV and facial recognition sunglasses called 'reflectacles' were invented by a custom-spectacle-craftsmen based in Chicago named Scott Urban.[62] They reflect infrared and, optionally, visible light which makes the users face a white blur to cameras. The project easily surpassed its crowdfunding goal of $28,000 and reflectacles will be commercially available by June 2017.[63] It is conceivable that such technology might be fused with future head-mounted displays such as potential successors of HoloLens.

Another method to protect from facial recognition systems are specific haircuts and make-up patterns that prevent the used algorithms to detect a face.[64]

4.7 See also

- AI effect
- Applications of artificial intelligence
- Automatic number plate recognition
- Biometric technology in access control
- Coke Zero Facial Profiler
- Computer processing of body language
- Computer vision
- Eigenface
- Face detection
- Face ID
- Face perception
- FindFace
- Glasgow Face Matching Test
- Iris recognition
- MALINTENT
- Artificial intelligence for video surveillance

- Multimedia information retrieval
- Multilinear subspace learning
- Pattern recognition, analogy and case-based reasoning
- Retinal scan
- Super recognisers
- Template matching
- Three-dimensional face recognition
- Vein matching
- Gait analysis

Lists

- List of computer vision topics
- List of emerging technologies
- Outline of artificial intelligence

4.8 References

[1] "Face Recognition Applications". Animetrics. Retrieved 2008-06-04.

[2] "Facial Recognition: Who's Tracking You in Public?". *Consumer Reports*. Retrieved 2016-04-05.

[3] "Airport Facial Recognition Passenger Flow Management". *hrsid.com*.

[4] Bonsor, K. "How Facial Recognition Systems Work". Retrieved 2008-06-02.

[5] Smith, Kelly. "Face Recognition" (PDF). Retrieved 2008-06-04.

[6] R. Brunelli and T. Poggio, "Face Recognition: Features versus Templates", IEEE Trans. on PAMI, 1993, (15)10:1042-1052

[7] R. Brunelli, *Template Matching Techniques in Computer Vision: Theory and Practice*, Wiley, ISBN 978-0-470-51706-2, 2009 (*TM book*)

[8] Williams, Mark. "Better Face-Recognition Software". Retrieved 2008-06-02.

[9] Crawford, Mark. "Facial recognition progress report". *SPIE Newsroom*. Retrieved 2011-10-06.

[10] Kimmel, Ron. "Three-dimensional face recognition" (PDF). Retrieved 2005-01-01.

[11] Duhn, S. von; Ko, M. J.; Yin, L.; Hung, T.; Wei, X. (1 September 2007). "Three-View Surveillance Video Based Face Modeling for Recognition". pp. 1–6. doi:10.1109/BCC.2007.4430529 – via IEEE Xplore.

[12] Socolinsky, Diego A.; Selinger, Andrea (1 January 2004). "Thermal Face Recognition in an Operational Scenario". IEEE Computer Society. pp. 1012–1019 – via ACM Digital Library.

[13] http://www.customs.govt.nz/features/smartgate/howsmartgateworks/Pages/default.aspx

[14] FORTUNE. "Here's How Many Adult Faces Are Scanned From Facial Recognition Databases".

[15] "The trouble with facial recognition technology (in the real world)".

[16] Knezevich, Kevin Rector, Alison. "Maryland's use of facial recognition software questioned by researchers, civil liberties advocates".

[17] "Next Generation Identification". *FBI*. Retrieved 2016-04-05.

[18] Vogel, Ben. "Panama puts names to more faces". IHS Jane's Airport Review. Archived from the original on 2014-10-12. Retrieved 2014-10-07. Under the USD11 million contract, a cluster of sixty computers, a fibre optic network, and 150 surveillance cameras were installed in the terminal and at about 30 gates.

[19] "Facial recognition technology is coming to Canadian airports this spring". *CBC News*. Retrieved 2017-03-03.

[20] https://arstechnica.com/tech-policy/2017/06/police-automatic-face-recognition/

[21] Greene, Lisa (2001-02-15). "Face scans match few suspects" (SHTML). *St. Petersburg Times*. Archived from the original on 2014-11-30. Retrieved 2011-06-30. By using Viisage software, police matched 19 people's faces to photos of people arrested in the past for minor pickpocketing, fraud and other charges. They weren't charged with any game-day misdeeds. THIS IS A FARCE

[22] Krause, Mike (2002-01-14). "Is face recognition just high-tech snake oil?". *Enter Stage Right*. ISSN 1488-1756. Archived from the original on 2002-01-24. Retrieved 2011-06-30.

[23] "Mexican Government Adopts FaceIt Face Recognition Technology to Eliminate Duplicate Voter Registrations in Upcoming Presidential Election". Business Wire. 2000-05-11. Retrieved 2008-06-02.

[24] House, David. "Facial recognition at DMV". Oregon Department of Transportation. Archived from the original on 2007-02-05. Retrieved 2007-09-17. Oregon DMV is going to start using "facial recognition" software, a new tool in the prevention of fraud, required by a new state law. The law is designed to prevent someone from obtaining a driver license or ID card under a false name.

[25] Schultz, Zac. "Facial Recognition Technology Helps DMV Prevent Identity Theft". WMTV News, Gray Television. Retrieved 2007-09-17. Madison: ...The Department of Motor Vehicles is using... facial recognition technology [to prevent ID theft]

[26] Visidon. "Visidon AppLock - Android Apps on Google Play". *google.com*. Archived from the original on 2013-12-28.

[27] "iPhoto '09 & iPhoto '11: Improving face recognition results". *apple.com*.

[28] "iPhone 8, iPhone 8 Plus and iPhone X are Launched".

[29] "Google Photos' 'racist' error highlights facial recognition's limits". 1 July 2015.

[30] "Passport Canada - Photos". passportcanada.gc.ca. Archived from the original on 2009-03-01.

[31] Albiol,A., Albiol,A., Oliver,J., Mossi,J.M.(2012). Who is who at different cameras: people re-identification using depth cameras. Computer Vision, IET. Vol 6(5), 378-387.

[32] Meek, James (2002-06-13). "Robo cop". London: UK Guardian newspaper.

[33] "Birmingham City Centre CCTV Installs Visionics' FaceIt". Business Wire. 2008-06-02.

[34] Willing, Richard (2003-09-02). "Airport anti-terror systems flub tests; Face-recognition technology fails to flag 'suspects'" (Abstract). USA Today. Retrieved 2007-09-17.

[35] White, David; Dunn, James D.; Schmid, Alexandra C.; Kemp, Richard I. (14 October 2015). "Error Rates in Users of Automatic Face Recognition Software". *PLOS ONE*. **10** (10): e0139827. PMC 4605725. PMID 26465631. doi:10.1371/journal.pone.0139827 – via PLoS Journals.

[36] "EFF Sues FBI For Access to Facial-Recognition Records". *Electronic Frontier Foundation*.

[37] "Q&A On Face-Recognition". *American Civil Liberties Union*.

[38] "Civil Liberties & Facial Recognition Software". About.com, The New York Times Company. pp. *pp.* 2. Archived from the original on 2006-03-01. Retrieved 2007-09-17. A few examples which have already arisen from surveillance video are: using license plates to blackmail gay married people, stalking women, tracking estranged spouses...

[39] Harley Geiger (2011-12-06). "Facial Recognition and Privacy". Center for Democracy & Technology. Retrieved 2012-01-10.

[40] Martin Koste (2013-10-28). "A Look Into Facebook's Potential To Recognize Anybody's Face". NPR. Archived from the original on 2013-11-01. Retrieved 2013-12-25.

[41] "Facial Recognition is getting really accurate, and we have not prepared". 11 October 2016.

[42] "This creepy facial recognition app is taking Russia by storm". 18 May 2016.

[43] What Facial Recognition Technology Means for Privacy and Civil Liberties: Hearing before the Subcommittee on Privacy, Technology and the Law of the Committee on the Judiciary, United States Senate, One Hundred Twelfth Congress, Second Session, July 18, 2012

[44] "Privacy Multistakeholder Process: Facial Recognition Technology". *National Telecommunications and Information Association*. Retrieved April 5, 2016.

[45] McCabe, David. "Facial recognition talks break down as privacy advocates withdraw". *TheHill*. Retrieved 2016-04-05.

[46] Weaver, Dustin. "Business eyes facial recognition guidelines". *TheHill*. Retrieved 2016-04-05.

[47] "740 ILCS 14/ Biometric Information Privacy Act.". *www.ilga.gov*. Archived from the original on 2016-04-16. Retrieved 2016-04-05.

[48] "Facebook Keeps Getting Sued Over Face-Recognition Software, And Privacy Groups Say We Should Be Paying More Attention". *International Business Times*. Retrieved 2016-04-05.

[49] Herra, Dana. "Judge tosses Illinois privacy law class action vs Facebook over photo tagging; California cases still pending". *cookcountyrecord.com*. Retrieved 2016-04-05.

[50] "Mugspot Can Find A Face In The Crowd -- Face-Recognition Software Prepares To Go To Work In The Streets". ScienceDaily. 12 November 1997. Retrieved 2007-11-06.

[51] R. Kimmel and G. Sapiro (2003-04-30). "The Mathematics of Face Recognition". SIAM News. Retrieved 2003-04-30.

[52] "Face Homepage". *nist.gov*.

[53] "Emotion detector: Facial expression recognition to improve learning, gaming". Science Daily. Retrieved 4 January 2017.

[54] "Facial Recognition Market - Global Forecast to 2021". Digital Journal. Retrieved 4 January 2017.

[55] Constine, Josh. "Like by smiling? Facebook acquires emotion detection startup FacioMetrics". TechCrunch. Retrieved 4 January 2017.

[56] "Facebook acquires FacioMetrics to add 'fun effects' to photos and videos". VentureBeat. Retrieved 4 January 2017.

[57] "These Goofy-Looking Glasses Could Make You Invisible to Facial Recognition Technology". *Slate*. 18 January 2013. Retrieved 22 January 2013.

[58] Hongo, Jun. "Eyeglasses with Face Un-Recognition Function to Debut in Japan". Wall Street Journal. Retrieved 9 February 2017.

[59] Osborne, Charlie. "Privacy visor which blocks facial recognition software set for public release". ZDNet. Retrieved 9 February 2017.

[60] Stone, Maddie. "These Glasses Block Facial Recognition Technology". Gizmodo. Retrieved 9 February 2017.

[61] "How Japan's Privacy Visor fools face-recognition cameras". PCWorld. Retrieved 9 February 2017.

[62] http://www.magneticmag.com/2016/12/be-seen-and-unseen-reflectacles-are-the-sunglasses-of-the-future/

[63] "Reflectacles - Reflective Eyewear and Sunglasses".

[64] Harvey, Adam. "CV Dazzle: Camouflage from Face Detection". cvdazzle.com. Retrieved 2017-09-15.

4.9 Further reading

- What are Biometrics? White Paper Published by Aware, Inc., January 2014

- Farokhi, Sajad; Shamsuddin, Siti Mariyam; Flusser, Jan; Sheikh, U.U; Khansari, Mohammad; Jafari-Khouzani, Kourosh (2014). "Near infrared face recognition by combining Zernike moments and undecimated discrete wavelet transform". *Digital Signal Processing*. **31** (1). doi:10.1016/j.dsp.2014.04.008.

- "The Face Detection Algorithm Set to Revolutionize Image Search" (Feb. 2015), *MIT Technology Review*

- Garvie, Clare; Bedoya, Alvaro; Frankle, Jonathan (October 18, 2016). *Perpetual Line Up: Unregulated Police Face Recognition in America*. Center on Privacy & Technology at Georgetown Law. Retrieved October 22, 2016.

  - "Facial Recognition Software 'Sounds Like Science Fiction,' but May Affect Half of Americans". *As It Happens*. Canadian Broadcasting Corporation. October 20, 2016. Retrieved October 22, 2016. Interview with Alvaro Bedoya, executive director of the Center on Privacy & Technology at Georgetown Law and co-author of *Perpetual Line Up: Unregulated Police Face Recognition in America*.

- A Photometric Stereo Approach to Face Recognition". University of the West of England. http://www1.uwe.ac.uk/et/mvl/projects/facerecognition.aspx

4.10 External links

- Media related to Face recognition system at Wikimedia Commons

Chapter 5

DNA

For a non-technical introduction to the topic, see Introduction to genetics. For other uses, see DNA (disambiguation).

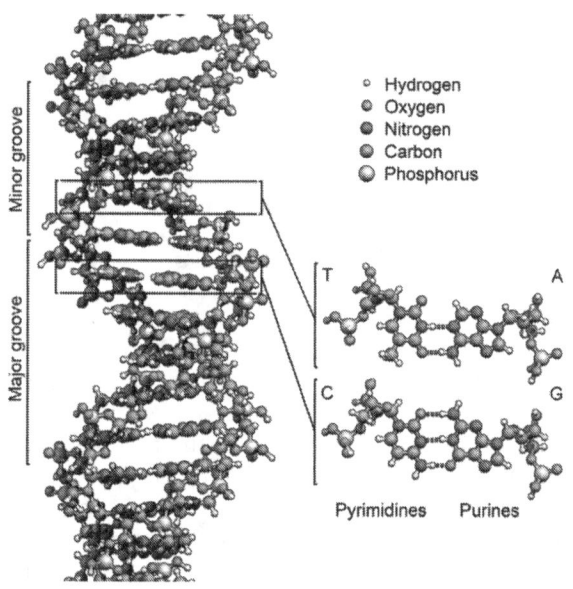

The structure of the DNA double helix. The atoms in the structure are colour-coded by element and the detailed structures of two base pairs are shown in the bottom right.

Deoxyribonucleic acid (/diˈɒksiˌraɪboʊsnjəˌkliːɪk, -ˌkleɪɪk/;[1] **DNA**) is a molecule that carries the genetic instructions used in the growth, development, functioning and reproduction of all known living organisms and many viruses. DNA and ribonucleic acid (RNA) are nucleic acids; alongside proteins, lipids and complex carbohydrates (polysaccharides), they are one of the four major types of macromolecules that are essential for all known forms of life. Most DNA molecules consist of two biopolymer strands coiled around each other to form a double helix.

The two DNA strands are called polynucleotides since they are composed of simpler monomer units called nucleotides.[2][3] Each nucleotide is composed of one

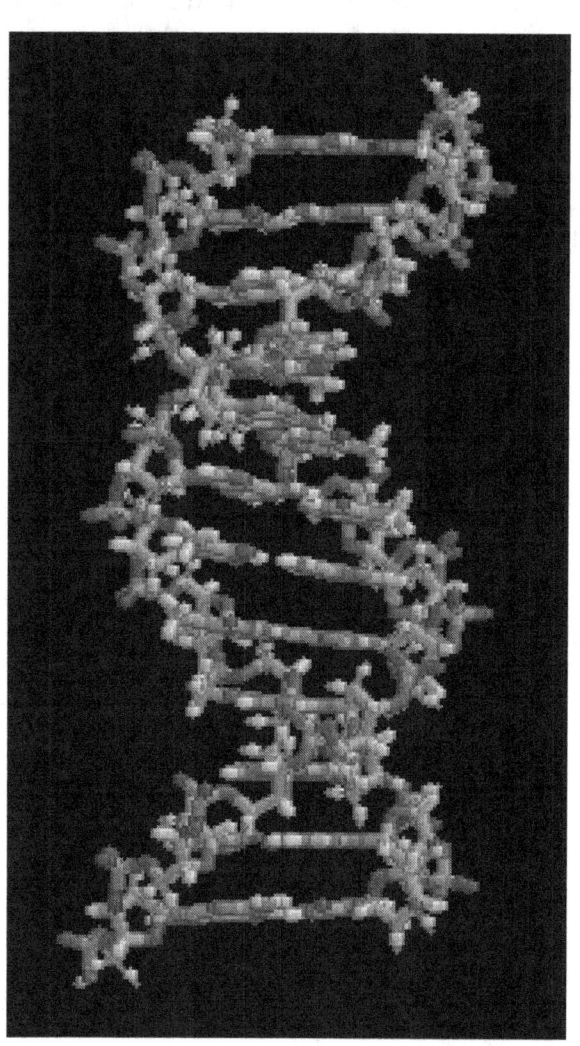

The structure of part of a DNA double helix

of four nitrogen-containing nucleobases —cytosine (C), guanine (G), adenine (A) or thymine (T) —a sugar called deoxyribose and a phosphate group. The nucleotides are joined to one another in a chain by covalent bonds between the sugar of one nucleotide and the phosphate of the next, resulting in an alternating sugar-phosphate backbone.

The nitrogenous bases of the two separate polynucleotide strands are bound together, according to base pairing rules (A with T and C with G), with hydrogen bonds to make double-stranded DNA. The total amount of related DNA base pairs on Earth is estimated at 5.0×10^{37} and weighs 50 billion tonnes.*[4] In comparison, the total mass of the biosphere has been estimated to be as much as 4 trillion tons of carbon (TtC).*[5]

DNA stores biological information. The DNA backbone is resistant to cleavage, and both strands of the double-stranded structure store the same biological information. This information is replicated as and when the two strands separate. A large part of DNA (more than 98% for humans) is non-coding, meaning that these sections do not serve as patterns for protein sequences.

The two strands of DNA run in opposite directions to each other and are thus antiparallel. Attached to each sugar is one of four types of nucleobases (informally, *bases*). It is the sequence of these four nucleobases along the backbone that encodes biological information. RNA strands are created using DNA strands as a template in a process called transcription. Under the genetic code, these RNA strands are translated to specify the sequence of amino acids within proteins in a process called translation.

Within eukaryotic cells DNA is organized into long structures called chromosomes. During cell division these chromosomes are duplicated in the process of DNA replication, providing each cell its own complete set of chromosomes. Eukaryotic organisms (animals, plants, fungi and protists) store most of their DNA inside the cell nucleus and some of their DNA in organelles, such as mitochondria or chloroplasts.*[6] In contrast prokaryotes (bacteria and archaea) store their DNA only in the cytoplasm. Within the eukaryotic chromosomes, chromatin proteins such as histones compact and organize DNA. These compact structures guide the interactions between DNA and other proteins, helping control which parts of the DNA are transcribed.

DNA was first isolated by Friedrich Miescher in 1869. Its molecular structure was first identified by James Watson and Francis Crick at the Cavendish Laboratory within the University of Cambridge in 1953, whose model-building efforts were guided by X-ray diffraction data acquired by Raymond Gosling, who was a post-graduate student of Rosalind Franklin. DNA is used by researchers as a molecular tool to explore physical laws and theories, such as the ergodic theorem and the theory of elasticity. The unique material properties of DNA have made it an attractive molecule for material scientists and engineers interested in micro- and nano-fabrication. Among notable advances in this field are DNA origami and DNA-based hybrid materials.*[7]

5.1 Properties

Chemical structure of DNA; hydrogen bonds shown as dotted lines

DNA is a long polymer made from repeating units called nucleotides.*[8]*[9] The structure of DNA is dynamic along its length, being capable of coiling into tight loops, and other shapes.*[10] In all species it is composed of two helical chains, bound to each other by hydrogen bonds. Both chains are coiled round the same axis, and have the same pitch of 34 ångströms (3.4 nanometres). The pair of chains has a radius of 10 ångströms (1.0 nanometre).*[11] According to another study, when measured in a different solution, the DNA chain measured 22 to 26 ångströms wide (2.2 to 2.6 nanometres), and one nucleotide unit measured 3.3 Å (0.33 nm) long.*[12] Although each individual nucleotide repeating unit is very small, DNA polymers can be very large molecules containing millions to hundreds of millions of nucleotides. For instance, the DNA in the largest human chromosome, chromosome number 1, consists of approximately 220 million base pairs*[13] and would be 85 mm long if straightened.

In living organisms, DNA does not usually exist as a single molecule, but instead as a pair of molecules that are held tightly together.*[14]*[15] These two long strands entwine like vines, in the shape of a double helix. The nucleotide contains both a segment of the backbone of the molecule (which holds the chain together) and a nucleobase (which interacts with the other DNA strand in the helix). A nucleobase linked to a sugar is called a nucleoside and a base linked to a sugar and one or more phosphate groups is called a nucleotide. A polymer comprising multiple linked nucleotides (as in DNA) is called a polynucleotide.*[16]

The backbone of the DNA strand is made from alternating phosphate and sugar residues.*[17] The sugar in DNA is 2-deoxyribose, which is a pentose (five-carbon) sugar. The sugars are joined together by phosphate groups that form phosphodiester bonds between the third and fifth carbon atoms of adjacent sugar rings. These asymmetric bonds mean a strand of DNA has a direction. In a double helix, the direction of the nucleotides in one strand is opposite to their direction in the other strand: the strands are *antiparallel*. The asymmetric ends of DNA strands are said to have a directionality of *five prime* (5′) and *three prime* (3′), with the 5′ end having a terminal phosphate group and the 3′ end a terminal hydroxyl group. One major difference between DNA and RNA is the sugar, with the 2-deoxyribose in DNA being replaced by the alternative pentose sugar ribose in RNA.*[15]

The DNA double helix is stabilized primarily by two forces: hydrogen bonds between nucleotides and base-stacking interactions among aromatic nucleobases.*[19] In the aqueous environment of the cell, the conjugated π bonds of nucleotide bases align perpendicular to the axis of the DNA molecule, minimizing their interaction with the solvation shell. The four bases found in DNA are adenine (A), cytosine (C), guanine (G) and thymine (T). These four bases are attached to the sugar-phosphate to form the complete nucleotide, as shown for adenosine monophosphate. Adenine pairs with thymine and guanine pairs with cytosine. It was represented by A-T base pairs and G-C base pairs.*[20]*[21]

5.1.1 Nucleobase classification

The nucleobases are classified into two types: the purines, A and G, being fused five- and six-membered heterocyclic compounds, and the pyrimidines, the six-membered rings C and T.*[15] A fifth pyrimidine nucleobase, uracil (U), usually takes the place of thymine in RNA and differs from thymine by lacking a methyl group on its ring. In addition to RNA and DNA, many artificial nucleic acid analogues have been created to study the properties of nucleic acids, or for use in biotechnology.*[22]

5.1.2 Non canonical bases

Uracil is not usually found in DNA, occurring only as a breakdown product of cytosine. However, in several bacteriophages, *Bacillus subtilis* bacteriophages PBS1 and PBS2 and *Yersinia* bacteriophage piR1-37, thymine has been replaced by uracil.*[23] Another phage - Staphylococcal phage S6 - has been identified with a genome where

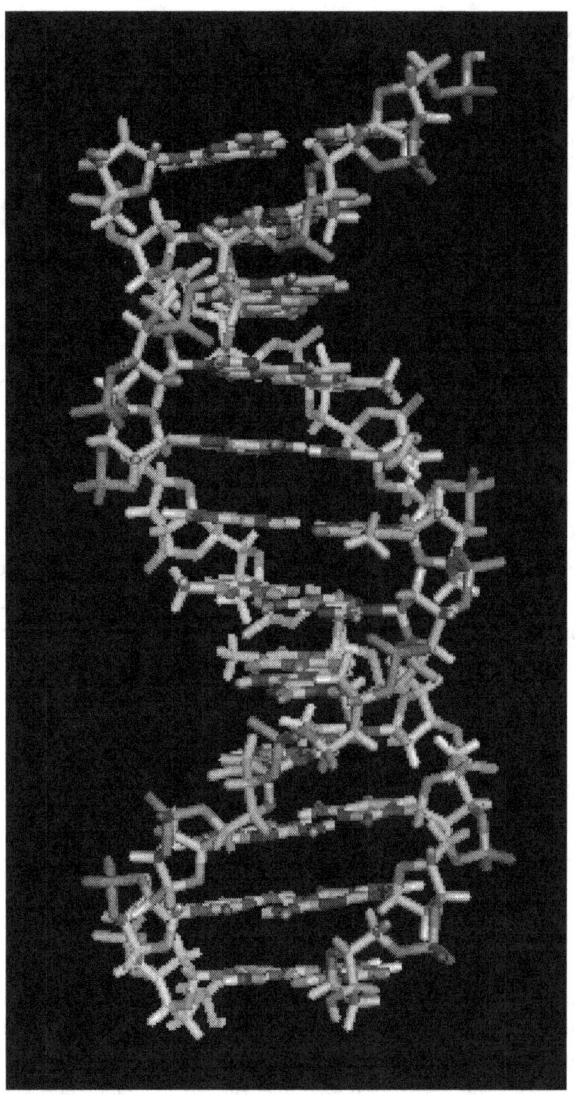

A section of DNA. The bases lie horizontally between the two spiraling strands.[18] (animated version).*

thymine has been replaced by uracil.*[24]

5-hydroxymethyldeoxyuracil (*hm5dU) is also known to replace thymidine in several genomes including the *Bacillus* phages SPO1, ϕe, SP8, H1, 2C and SP82. Another modified uracil - 5-dihydroxypentauracil – has also been described.*[25]

Base J (beta-d-glucopyranosyloxymethyluracil), a modified form of uracil, is also found in several organisms: the flagellates *Diplonema* and *Euglena*, and all the kinetoplastid genera.*[26] Biosynthesis of J occurs in two steps: in the first step, a specific thymidine in DNA is converted into hydroxymethyldeoxyuridine; in the second, HOMedU is glycosylated to form J.*[27] Proteins that bind specifically to this base have been identified.*[28]*[29]*[30] These proteins appear to be distant relatives of the Tet1 onco-

5.1. PROPERTIES

gene that is involved in the pathogenesis of acute myeloid leukemia.*[31] J appears to act as a termination signal for RNA polymerase II.*[32]*[33]

In 1976 a bacteriophage - S-2L - which infects species of the genus *Synechocystis* was found to have all the adenosine bases within its genome replaced by 2,6-diaminopurine.*[34] In 2016 deoxyarchaeosine was found to be present in the genomes of several bacteria and the *Escherichia* phage 9g.*[35]

Modified bases also occur in DNA. The first of these recognised was 5-methylcytosine which was found in the genome of *Mycobacterium tuberculosis* in 1925.*[36] The complete replacement of cytosine by 5-glycosylhydroxymethylcytosine in T even phages (T2, T4 and T6) was observed in 1953*[37] In the genomes of Xanthomonas oryzae bacteriophage Xp12 and halovirus FH the full complement of cystosine has been replaced by 5-methylcytosine.*[38]*[39] 6N-methyadenine was discovered to be present in DNA in 1955.*[40] N6-carbamoyl-methyladenine was described in 1975.*[41] 7-methylguanine was described in 1976.*[42] N4-methylcytosine in DNA was described in 1983.*[43] In 1985 5-hydroxycytosine was found in the genomes of the Rhizobium phages RL38JI and N17.*[44] α-putrescinylthymine occurs in both the genomes of the *Delftia* phage ΦW-14 and the *Bacillus* phage SP10.*[45] α-glutamythymidine is found in the Bacillus phage SP01 and 5-dihydroxypentyluracil is found in the Bacillus phage SP15.

The reason for the presence of these non canonical bases in DNA is not known. It seems likely that at least part of the reason for their presence in bacterial viruses (phages) is to avoid the restriction enzymes present in bacteria. This enzyme system acts at least in part as a molecular immune system protecting bacteria from infection by viruses.

This does not appear to be the entire story. Four modifications to the cytosine residues in human DNA have been reported.*[46] These modifications are the addition of methyl (CH_3)-, hydroxymethyl (CH_2OH)-, formyl (CHO)- and carboxyl (COOH)- groups. These modifications are through to have regulatory functions.

5.1.3 Listing of non canonical bases found in DNA

Seventeen non canonical bases are known to occur in DNA. Most of these are modifications of the canonical bases plus uracil.

- Modified **Adenosine**
  - N6-carbamoyl-methyladenine
  - N6-methyadenine
- Modified **Guanine**
  - 7-Methylguanine
- Modified **Cytosine**
  - N4-Methylcytosine
  - 5-Carboxylcytosine
  - 5-Formylcytosine
  - 5-Glycosylhydroxymethylcytosine
  - 5-Hydroxycytosine
  - 5-Methylcytosine
- Modified **Thymidine**
  - α-Glutamythymidine
  - α-Putrescinylthymine
- **Uracil** and modifications
  - Base J
  - Uracil
  - 5-Dihydroxypentauracil
  - 5-Hydroxymethyldeoxyuracil
- Others
  - Deoxyarchaeosine
  - 2,6-Diaminopurine

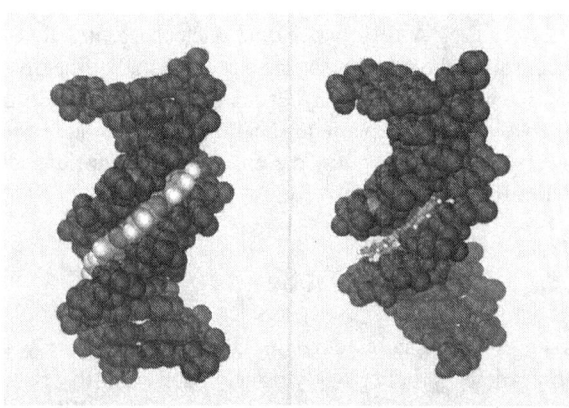

DNA major and minor grooves. The latter is a binding site for the Hoechst stain dye 33258.

5.1.4 Grooves

Twin helical strands form the DNA backbone. Another double helix may be found tracing the spaces, or grooves, between the strands. These voids are adjacent to the base pairs and may provide a binding site. As the strands are not symmetrically located with respect to each other, the grooves are unequally sized. One groove, the major groove, is 22 Å wide and the other, the minor groove, is 12 Å wide.*[47] The width of the major groove means that the edges of the bases are more accessible in the major groove than in the minor groove. As a result, proteins such as transcription factors that can bind to specific sequences in double-stranded DNA usually make contact with the sides of the bases exposed in the major groove.*[48] This situation varies in unusual conformations of DNA within the cell *(see below)*, but the major and minor grooves are always named to reflect the differences in size that would be seen if the DNA is twisted back into the ordinary B form.

5.1.5 Base pairing

Further information: Base pair

In a DNA double helix, each type of nucleobase on one strand bonds with just one type of nucleobase on the other strand. This is called complementary base pairing. Here, purines form hydrogen bonds to pyrimidines, with adenine bonding only to thymine in two hydrogen bonds, and cytosine bonding only to guanine in three hydrogen bonds. This arrangement of two nucleotides binding together across the double helix is called a Watson-Crick base pair. Another type of base pairing is Hoogsteen base pairing where two hydrogen bonds form between guanine and cytosine.*[49] As hydrogen bonds are not covalent, they can be broken and rejoined relatively easily. The two strands of DNA in a double helix can thus be pulled apart like a zipper, either by a mechanical force or high temperature.*[50] As a result of this base pair complementarity, all the information in the double-stranded sequence of a DNA helix is duplicated on each strand, which is vital in DNA replication. This reversible and specific interaction between complementary base pairs is critical for all the functions of DNA in living organisms.*[9]

Top, a **GC** base pair with three hydrogen bonds. Bottom, an **AT** base pair with two hydrogen bonds. Non-covalent hydrogen bonds between the pairs are shown as dashed lines.

The two types of base pairs form different numbers of hydrogen bonds, AT forming two hydrogen bonds, and GC forming three hydrogen bonds (see figures, right). DNA with high GC-content is more stable than DNA with low GC-content.

As noted above, most DNA molecules are actually two polymer strands, bound together in a helical fashion by non-covalent bonds; this double-stranded (**dsDNA**) structure is maintained largely by the intrastrand base stacking interactions, which are strongest for G,C stacks. The two strands can come apart – a process known as melting – to form two single-stranded DNA (**ssDNA**) molecules. Melting occurs at high temperature, low salt and high pH (low pH also melts DNA, but since DNA is unstable due to acid depurination, low pH is rarely used).

The stability of the dsDNA form depends not only on the GC-content (% G,C basepairs) but also on sequence (since stacking is sequence specific) and also length (longer molecules are more stable). The stability can be measured in various ways; a common way is the "melting temperature", which is the temperature at which 50% of the ds molecules are converted to ss molecules; melting temperature is dependent on ionic strength and the concentration of DNA. As a result, it is both the percentage of GC base pairs and the overall length of a DNA double helix that determines the strength of the association between the two strands of DNA. Long DNA helices with a high GC-content have stronger-interacting strands, while short helices with high AT content have weaker-interacting strands.*[51] In biology, parts of the DNA double helix that need to separate easily, such as the TATAAT Pribnow box in some promoters, tend to have a high AT content, making the strands easier to pull apart.*[52]

In the laboratory, the strength of this interaction can be measured by finding the temperature necessary to break the hydrogen bonds, their melting temperature (also called T_m value). When all the base pairs in a DNA double helix melt, the strands separate and exist in solution as two entirely independent molecules. These single-stranded DNA molecules have no single common shape, but some conformations are more stable than others.*[53]

5.1.6 Sense and antisense

Further information: Sense (molecular biology)

A DNA sequence is called "sense" if its sequence is the same as that of a messenger RNA copy that is translated into protein.*[54] The sequence on the opposite strand is called the "antisense" sequence. Both sense and antisense sequences can exist on different parts of the same strand of DNA (i.e. both strands can contain both sense and antisense sequences). In both prokaryotes and eukaryotes, antisense RNA sequences are produced, but the functions of

5.1. PROPERTIES

these RNAs are not entirely clear.*[55] One proposal is that antisense RNAs are involved in regulating gene expression through RNA-RNA base pairing.*[56]

A few DNA sequences in prokaryotes and eukaryotes, and more in plasmids and viruses, blur the distinction between sense and antisense strands by having overlapping genes.*[57] In these cases, some DNA sequences do double duty, encoding one protein when read along one strand, and a second protein when read in the opposite direction along the other strand. In bacteria, this overlap may be involved in the regulation of gene transcription,*[58] while in viruses, overlapping genes increase the amount of information that can be encoded within the small viral genome.*[59]

5.1.7 Supercoiling

Further information: DNA supercoil

DNA can be twisted like a rope in a process called DNA supercoiling. With DNA in its "relaxed" state, a strand usually circles the axis of the double helix once every 10.4 base pairs, but if the DNA is twisted the strands become more tightly or more loosely wound.*[60] If the DNA is twisted in the direction of the helix, this is positive supercoiling, and the bases are held more tightly together. If they are twisted in the opposite direction, this is negative supercoiling, and the bases come apart more easily. In nature, most DNA has slight negative supercoiling that is introduced by enzymes called topoisomerases.*[61] These enzymes are also needed to relieve the twisting stresses introduced into DNA strands during processes such as transcription and DNA replication.*[62]

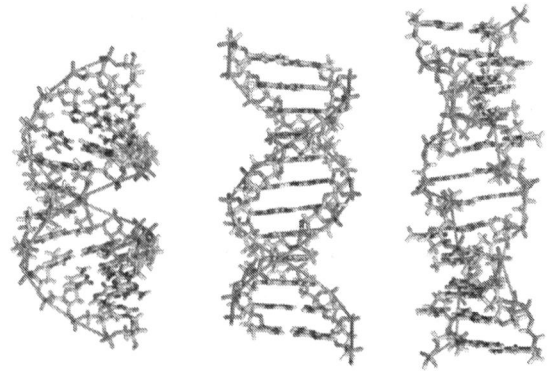

From left to right, the structures of A, B and Z DNA

5.1.8 Alternative DNA structures

Further information: Molecular Structure of Nucleic Acids: A Structure for Deoxyribose Nucleic Acid, Molecular models of DNA, and DNA structure

DNA exists in many possible conformations that include A-DNA, B-DNA, and Z-DNA forms, although, only B-DNA and Z-DNA have been directly observed in functional organisms.*[17] The conformation that DNA adopts depends on the hydration level, DNA sequence, the amount and direction of supercoiling, chemical modifications of the bases, the type and concentration of metal ions, and the presence of polyamines in solution.*[63]

The first published reports of A-DNA X-ray diffraction patterns—and also B-DNA—used analyses based on Patterson transforms that provided only a limited amount of structural information for oriented fibers of DNA.*[64]*[65] An alternative analysis was then proposed by Wilkins *et al.*, in 1953, for the *in vivo* B-DNA X-ray diffraction-scattering patterns of highly hydrated DNA fibers in terms of squares of Bessel functions.*[66] In the same journal, James Watson and Francis Crick presented their molecular modeling analysis of the DNA X-ray diffraction patterns to suggest that the structure was a double-helix.*[11]

Although the *B-DNA form* is most common under the conditions found in cells,*[67] it is not a well-defined conformation but a family of related DNA conformations*[68] that occur at the high hydration levels present in living cells. Their corresponding X-ray diffraction and scattering patterns are characteristic of molecular paracrystals with a significant degree of disorder.*[69]*[70]

Compared to B-DNA, the A-DNA form is a wider right-handed spiral, with a shallow, wide minor groove and a narrower, deeper major groove. The A form occurs under non-physiological conditions in partly dehydrated samples of DNA, while in the cell it may be produced in hybrid pairings of DNA and RNA strands, and in enzyme-DNA complexes.*[71]*[72] Segments of DNA where the bases have been chemically modified by methylation may undergo a larger change in conformation and adopt the Z form. Here, the strands turn about the helical axis in a left-handed spiral, the opposite of the more common B form.*[73] These unusual structures can be recognized by specific Z-DNA binding proteins and may be involved in the regulation of transcription.*[74]

5.1.9 Alternative DNA chemistry

For many years exobiologists have proposed the existence of a shadow biosphere, a postulated microbial biosphere of Earth that uses radically different biochemical and molecular processes than currently known life. One of the proposals was the existence of lifeforms that use arsenic instead of phosphorus in DNA. A report in

2010 of the possibility in the bacterium GFAJ-1, was announced,*[75]*[75]*[76] though the research was disputed,*[76]*[77] and evidence suggests the bacterium actively prevents the incorporation of arsenic into the DNA backbone and other biomolecules.*[78]

5.1.10 Quadruplex structures

Further information: G-quadruplex

At the ends of the linear chromosomes are specialized regions of DNA called telomeres. The main function of these regions is to allow the cell to replicate chromosome ends using the enzyme telomerase, as the enzymes that normally replicate DNA cannot copy the extreme 3′ ends of chromosomes.*[79] These specialized chromosome caps also help protect the DNA ends, and stop the DNA repair systems in the cell from treating them as damage to be corrected.*[80] In human cells, telomeres are usually lengths of single-stranded DNA containing several thousand repeats of a simple TTAGGG sequence.*[81]

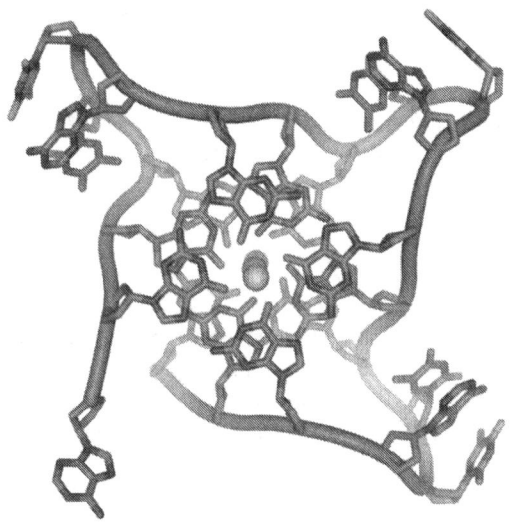

DNA quadruplex formed by telomere repeats. The looped conformation of the DNA backbone is very different from the typical DNA helix. The green spheres in the center represent potassium ions.*[82]

These guanine-rich sequences may stabilize chromosome ends by forming structures of stacked sets of four-base units, rather than the usual base pairs found in other DNA molecules. Here, four guanine bases form a flat plate and these flat four-base units then stack on top of each other, to form a stable G-quadruplex structure.*[83] These structures are stabilized by hydrogen bonding between the edges of the bases and chelation of a metal ion in the centre of each four-base unit.*[84] Other structures can also be formed, with the central set of four bases coming from either a single strand folded around the bases, or several different parallel strands, each contributing one base to the central structure.

In addition to these stacked structures, telomeres also form large loop structures called telomere loops, or T-loops. Here, the single-stranded DNA curls around in a long circle stabilized by telomere-binding proteins.*[85] At the very end of the T-loop, the single-stranded telomere DNA is held onto a region of double-stranded DNA by the telomere strand disrupting the double-helical DNA and base pairing to one of the two strands. This triple-stranded structure is called a displacement loop or D-loop.*[83]

Branched DNA can form networks containing multiple branches.

5.1.11 Branched DNA

Further information: Branched DNA and DNA nanotechnology

In DNA, fraying occurs when non-complementary regions exist at the end of an otherwise complementary double-strand of DNA. However, branched DNA can occur if a third strand of DNA is introduced and contains adjoining regions able to hybridize with the frayed regions of the pre-existing double-strand. Although the simplest example of branched DNA involves only three strands of DNA, complexes involving additional strands and multiple branches are also possible.*[86] Branched DNA can be used in nanotechnology to construct geometric shapes, see the section on uses in technology below.

5.2 Chemical modifications and altered DNA packaging

Structure of cytosine with and without the 5-methyl group. Deamination converts 5-methylcytosine into thymine.

5.2.1 Base modifications and DNA packaging

Further information: DNA methylation and Chromatin remodeling

The expression of genes is influenced by how the DNA is packaged in chromosomes, in a structure called chromatin.

Base modifications can be involved in packaging, with regions that have low or no gene expression usually containing high levels of methylation of cytosine bases. DNA packaging and its influence on gene expression can also occur by covalent modifications of the histone protein core around which DNA is wrapped in the chromatin structure or else by remodeling carried out by chromatin remodeling complexes (see Chromatin remodeling). There is, further, crosstalk between DNA methylation and histone modification, so they can coordinately affect chromatin and gene expression.*[87]

For one example, cytosine methylation produces 5-methylcytosine, which is important for X-inactivation of chromosomes.*[88] The average level of methylation varies between organisms – the worm *Caenorhabditis elegans* lacks cytosine methylation, while vertebrates have higher levels, with up to 1% of their DNA containing 5-methylcytosine.*[89] Despite the importance of 5-methylcytosine, it can deaminate to leave a thymine base, so methylated cytosines are particularly prone to mutations.*[90] Other base modifications include adenine methylation in bacteria, the presence of 5-hydroxymethylcytosine in the brain,*[91] and the glycosylation of uracil to produce the "J-base" in kinetoplastids.*[92]*[93]

5.2.2 Damage

Further information: DNA damage (naturally occurring), Mutation, and DNA damage theory of aging

DNA can be damaged by many sorts of mutagens, which change the DNA sequence. Mutagens include oxidizing agents, alkylating agents and also high-energy electromagnetic radiation such as ultraviolet light and X-rays. The type of DNA damage produced depends on the type of mutagen. For example, UV light can damage DNA by producing thymine dimers, which are cross-links between pyrimidine bases.*[95] On the other hand, oxidants such as free radicals or hydrogen peroxide produce multiple forms of damage, including base modifications, particularly of guanosine, and double-strand breaks.*[96] A typical human cell contains about 150,000 bases that have suffered oxidative damage.*[97] Of these oxidative lesions, the most dangerous are double-strand breaks, as these are difficult to repair and can produce point mutations, insertions, deletions from the DNA sequence, and chromosomal translocations.*[98] These mutations can cause cancer. Because of inherent limits in the DNA repair mechanisms, if humans lived long enough, they would all eventually develop cancer.*[99]*[100] DNA damages that are naturally occurring, due to normal cellular processes that produce reactive oxygen species, the hydrolytic activities of cellular water, etc., also occur frequently. Although most of these damages are repaired, in any cell some DNA damage may remain despite the action of repair processes. These remaining DNA damages accumulate with age in mammalian postmitotic tissues. This accumulation appears to be an important underlying cause of aging.*[101]*[102]*[103]

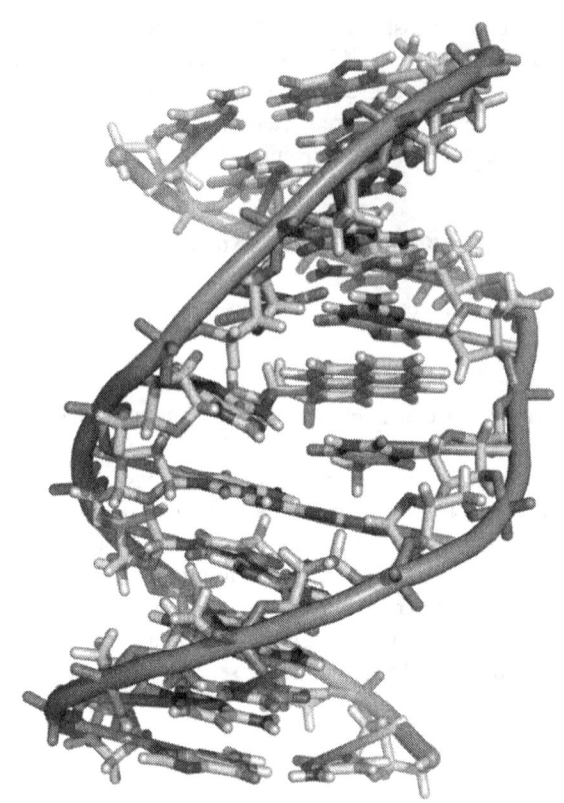

A covalent adduct between a metabolically activated form of benzo[a]pyrene, the major mutagen in tobacco smoke, and DNA[94]

Many mutagens fit into the space between two adjacent base pairs, this is called *intercalation*. Most intercalators are aromatic and planar molecules; examples include ethidium bromide, acridines, daunomycin, and doxorubicin. For an intercalator to fit between base pairs, the bases must separate, distorting the DNA strands by unwinding of the double helix. This inhibits both transcription and DNA replication, causing toxicity and mutations.*[104] As a result, DNA intercalators may be carcinogens, and in the case of thalidomide, a teratogen.*[105] Others such as benzo[a]pyrene diol epoxide and aflatoxin form DNA adducts that induce errors in replication.*[106] Nevertheless, due to their ability to inhibit DNA transcription and replication, other similar toxins are also used in chemotherapy to inhibit rapidly growing cancer cells.*[107]

5.3 Biological functions

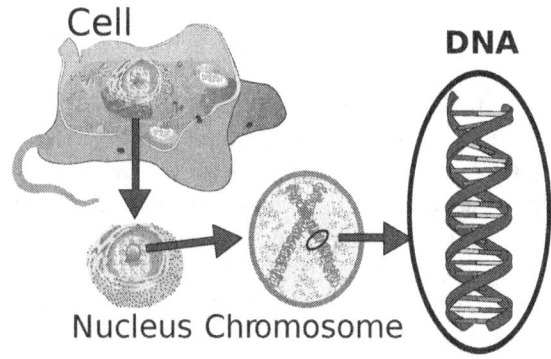

Location of eukaryote nuclear DNA within the chromosomes.

DNA usually occurs as linear chromosomes in eukaryotes, and circular chromosomes in prokaryotes. The set of chromosomes in a cell makes up its genome; the human genome has approximately 3 billion base pairs of DNA arranged into 46 chromosomes.[108] The information carried by DNA is held in the sequence of pieces of DNA called genes. Transmission of genetic information in genes is achieved via complementary base pairing. For example, in transcription, when a cell uses the information in a gene, the DNA sequence is copied into a complementary RNA sequence through the attraction between the DNA and the correct RNA nucleotides. Usually, this RNA copy is then used to make a matching protein sequence in a process called translation, which depends on the same interaction between RNA nucleotides. In alternative fashion, a cell may simply copy its genetic information in a process called DNA replication. The details of these functions are covered in other articles; here the focus is on the interactions between DNA and other molecules that mediate the function of the genome.

5.3.1 Genes and genomes

Further information: Cell nucleus, Chromatin, Chromosome, Gene, and Noncoding DNA

Genomic DNA is tightly and orderly packed in the process called DNA condensation, to fit the small available volumes of the cell. In eukaryotes, DNA is located in the cell nucleus, with small amounts in mitochondria and chloroplasts. In prokaryotes, the DNA is held within an irregularly shaped body in the cytoplasm called the nucleoid.[109] The genetic information in a genome is held within genes, and the complete set of this information in an organism is called its genotype. A gene is a unit of heredity and is a region of DNA that influences a particular characteristic in an organism. Genes contain an open reading frame that can be transcribed, and regulatory sequences such as promoters and enhancers, which control transcription of the open reading frame.

In many species, only a small fraction of the total sequence of the genome encodes protein. For example, only about 1.5% of the human genome consists of protein-coding exons, with over 50% of human DNA consisting of non-coding repetitive sequences.[110] The reasons for the presence of so much noncoding DNA in eukaryotic genomes and the extraordinary differences in genome size, or *C-value*, among species, represent a long-standing puzzle known as the "C-value enigma".[111] However, some DNA sequences that do not code protein may still encode functional non-coding RNA molecules, which are involved in the regulation of gene expression.[112]

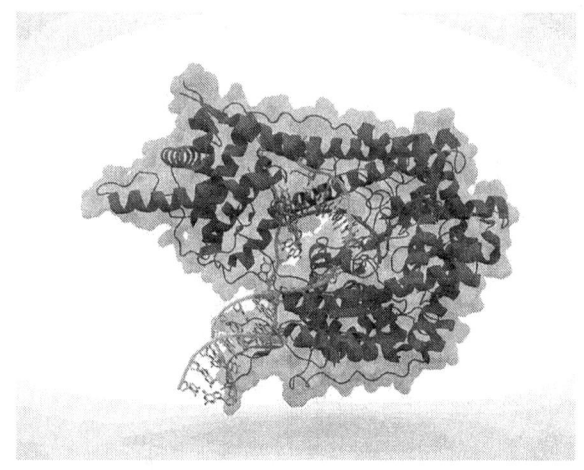

T7 RNA polymerase (blue) producing an mRNA (green) from a DNA template (orange).[113]

Some noncoding DNA sequences play structural roles in chromosomes. Telomeres and centromeres typically contain few genes but are important for the function and stability of chromosomes.[80][114] An abundant form of non-coding DNA in humans are pseudogenes, which are copies of genes that have been disabled by mutation.[115] These sequences are usually just molecular fossils, although they can occasionally serve as raw genetic material for the creation of new genes through the process of gene duplication and divergence.[116]

5.3.2 Transcription and translation

Further information: Genetic code, Transcription (genetics), and Protein biosynthesis

A gene is a sequence of DNA that contains genetic information and can influence the phenotype of an organism.

Within a gene, the sequence of bases along a DNA strand defines a messenger RNA sequence, which then defines one or more protein sequences. The relationship between the nucleotide sequences of genes and the amino-acid sequences of proteins is determined by the rules of translation, known collectively as the genetic code. The genetic code consists of three-letter 'words' called *codons* formed from a sequence of three nucleotides (e.g. ACT, CAG, TTT).

In transcription, the codons of a gene are copied into messenger RNA by RNA polymerase. This RNA copy is then decoded by a ribosome that reads the RNA sequence by base-pairing the messenger RNA to transfer RNA, which carries amino acids. Since there are 4 bases in 3-letter combinations, there are 64 possible codons (4^3 combinations). These encode the twenty standard amino acids, giving most amino acids more than one possible codon. There are also three 'stop' or 'nonsense' codons signifying the end of the coding region; these are the TAA, TGA, and TAG codons.

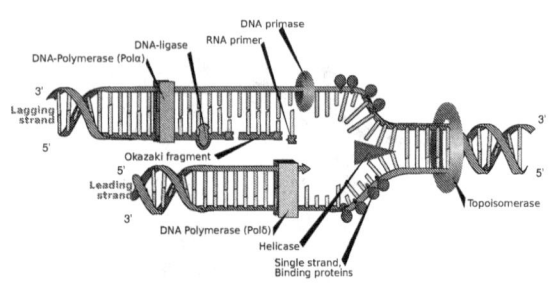

DNA replication. The double helix is unwound by a helicase and topoisomerase. Next, one DNA polymerase produces the leading strand copy. Another DNA polymerase binds to the lagging strand. This enzyme makes discontinuous segments (called Okazaki fragments) before DNA ligase joins them together.

5.3.3 Replication

Further information: DNA replication

Cell division is essential for an organism to grow, but, when a cell divides, it must replicate the DNA in its genome so that the two daughter cells have the same genetic information as their parent. The double-stranded structure of DNA provides a simple mechanism for DNA replication. Here, the two strands are separated and then each strand's complementary DNA sequence is recreated by an enzyme called DNA polymerase. This enzyme makes the complementary strand by finding the correct base through complementary base pairing and bonding it onto the original strand. As DNA polymerases can only extend a DNA strand in a 5′ to 3′ direction, different mechanisms are used to copy the antiparallel strands of the double helix.[117] In this way, the base on the old strand dictates which base appears on the new strand, and the cell ends up with a perfect copy of its DNA.

5.3.4 Extracellular nucleic acids

Naked extracellular DNA (eDNA), most of it released by cell death, is nearly ubiquitous in the environment. Its concentration in soil may be as high as 2 μg/L, and its concentration in natural aquatic environments may be as high at 88 μg/L.*[118] Various possible functions have been proposed for eDNA: it may be involved in horizontal gene transfer;*[119] it may provide nutrients;*[120] and it may act as a buffer to recruit or titrate ions or antibiotics.*[121] Extracellular DNA acts as a functional extracellular matrix component in the biofilms of several bacterial species. It may act as a recognition factor to regulate the attachment and dispersal of specific cell types in the biofilm;*[122] it may contribute to biofilm formation;*[123] and it may contribute to the biofilm's physical strength and resistance to biological stress.*[124]

Cell-free fetal DNA is found in the blood of the mother, and can be sequenced to determine a great deal of information about the developing fetus.*[125]

5.4 Interactions with proteins

All the functions of DNA depend on interactions with proteins. These protein interactions can be non-specific, or the protein can bind specifically to a single DNA sequence. Enzymes can also bind to DNA and of these, the polymerases that copy the DNA base sequence in transcription and DNA replication are particularly important.

5.4.1 DNA-binding proteins

Further information: DNA-binding protein
Interaction of DNA (in orange) with histones (in blue). These proteins' basic amino acids bind to the acidic phosphate groups on DNA.

Structural proteins that bind DNA are well-understood examples of non-specific DNA-protein interactions. Within chromosomes, DNA is held in complexes with structural proteins. These proteins organize the DNA into a compact structure called chromatin. In eukaryotes, this structure involves DNA binding to a complex of small basic proteins called histones, while in prokaryotes multiple types of proteins are involved.*[126]*[127] The histones form a disk-shaped complex called a nucleosome, which contains two complete turns of double-stranded

DNA wrapped around its surface. These non-specific interactions are formed through basic residues in the histones, making ionic bonds to the acidic sugar-phosphate backbone of the DNA, and are thus largely independent of the base sequence.*[128] Chemical modifications of these basic amino acid residues include methylation, phosphorylation, and acetylation.*[129] These chemical changes alter the strength of the interaction between the DNA and the histones, making the DNA more or less accessible to transcription factors and changing the rate of transcription.*[130] Other non-specific DNA-binding proteins in chromatin include the high-mobility group proteins, which bind to bent or distorted DNA.*[131] These proteins are important in bending arrays of nucleosomes and arranging them into the larger structures that make up chromosomes.*[132]

A distinct group of DNA-binding proteins is the DNA-binding proteins that specifically bind single-stranded DNA. In humans, replication protein A is the best-understood member of this family and is used in processes where the double helix is separated, including DNA replication, recombination, and DNA repair.*[133] These binding proteins seem to stabilize single-stranded DNA and protect it from forming stem-loops or being degraded by nucleases.

In contrast, other proteins have evolved to bind to particular DNA sequences. The most intensively studied of these are the various transcription factors, which are proteins that regulate transcription. Each transcription factor binds to one particular set of DNA sequences and activates or inhibits the transcription of genes that have these sequences close to their promoters. The transcription factors do this in two ways. Firstly, they can bind the RNA polymerase responsible for transcription, either directly or through other mediator proteins; this locates the polymerase at the promoter and allows it to begin transcription.*[135] Alternatively, transcription factors can bind enzymes that modify the histones at the promoter. This changes the accessibility of the DNA template to the polymerase.*[136]

As these DNA targets can occur throughout an organism's genome, changes in the activity of one type of transcription factor can affect thousands of genes.*[137] Consequently, these proteins are often the targets of the signal transduction processes that control responses to environmental changes or cellular differentiation and development. The specificity of these transcription factors' interactions with DNA come from the proteins making multiple contacts to the edges of the DNA bases, allowing them to "read" the DNA sequence. Most of these base-interactions are made in the major groove, where the bases are most accessible.*[48]

5.4.2 DNA-modifying enzymes

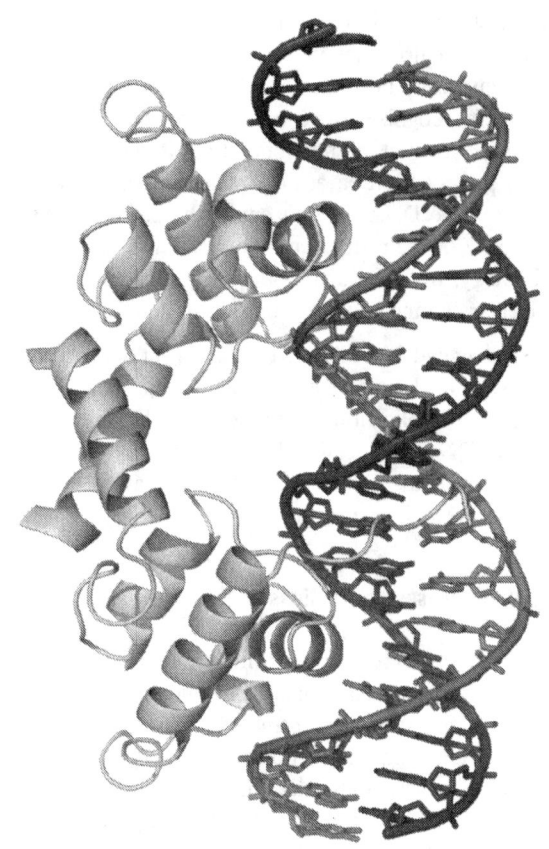

The lambda repressor helix-turn-helix transcription factor bound to its DNA target[134]

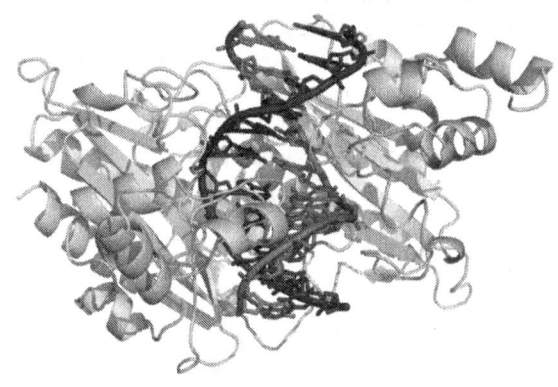

The restriction enzyme EcoRV (green) in a complex with its substrate DNA[138]

Nucleases and ligases

Nucleases are enzymes that cut DNA strands by catalyzing the hydrolysis of the phosphodiester bonds. Nucleases that hydrolyse nucleotides from the ends of DNA strands are called exonucleases, while endonucleases cut within strands.

The most frequently used nucleases in molecular biology are the restriction endonucleases, which cut DNA at specific sequences. For instance, the EcoRV enzyme shown to the left recognizes the 6-base sequence 5'-GATATC-3' and makes a cut at the horizontal line. In nature, these enzymes protect bacteria against phage infection by digesting the phage DNA when it enters the bacterial cell, acting as part of the restriction modification system.*[139] In technology, these sequence-specific nucleases are used in molecular cloning and DNA fingerprinting.

Enzymes called DNA ligases can rejoin cut or broken DNA strands.*[140] Ligases are particularly important in lagging strand DNA replication, as they join together the short segments of DNA produced at the replication fork into a complete copy of the DNA template. They are also used in DNA repair and genetic recombination.*[140]

Topoisomerases and helicases

Topoisomerases are enzymes with both nuclease and ligase activity. These proteins change the amount of supercoiling in DNA. Some of these enzymes work by cutting the DNA helix and allowing one section to rotate, thereby reducing its level of supercoiling; the enzyme then seals the DNA break.*[61] Other types of these enzymes are capable of cutting one DNA helix and then passing a second strand of DNA through this break, before rejoining the helix.*[141] Topoisomerases are required for many processes involving DNA, such as DNA replication and transcription.*[62]

Helicases are proteins that are a type of molecular motor. They use the chemical energy in nucleoside triphosphates, predominantly adenosine triphosphate (ATP), to break hydrogen bonds between bases and unwind the DNA double helix into single strands.*[142] These enzymes are essential for most processes where enzymes need to access the DNA bases.

Polymerases

Polymerases are enzymes that synthesize polynucleotide chains from nucleoside triphosphates. The sequence of their products is created based on existing polynucleotide chains—which are called *templates*. These enzymes function by repeatedly adding a nucleotide to the 3' hydroxyl group at the end of the growing polynucleotide chain. As a consequence, all polymerases work in a 5' to 3' direction.*[143] In the active site of these enzymes, the incoming nucleoside triphosphate base-pairs to the template: this allows polymerases to accurately synthesize the complementary strand of their template. Polymerases are classified according to the type of template that they use.

In DNA replication, DNA-dependent DNA polymerases make copies of DNA polynucleotide chains. To preserve biological information, it is essential that the sequence of bases in each copy are precisely complementary to the sequence of bases in the template strand. Many DNA polymerases have a proofreading activity. Here, the polymerase recognizes the occasional mistakes in the synthesis reaction by the lack of base pairing between the mismatched nucleotides. If a mismatch is detected, a 3' to 5' exonuclease activity is activated and the incorrect base removed.*[144] In most organisms, DNA polymerases function in a large complex called the replisome that contains multiple accessory subunits, such as the DNA clamp or helicases.*[145]

RNA-dependent DNA polymerases are a specialized class of polymerases that copy the sequence of an RNA strand into DNA. They include reverse transcriptase, which is a viral enzyme involved in the infection of cells by retroviruses, and telomerase, which is required for the replication of telomeres.*[79]*[146] For example, HIV reverse transcriptase is an enzyme for AIDS virus replication.*[146] Telomerase is an unusual polymerase because it contains its own RNA template as part of its structure. It synthesizes telomeres at the ends of chromosomes. Telomeres prevent fusion of the ends of neighboring chromosomes and protect chromosome ends from damage.*[80]

Transcription is carried out by a DNA-dependent RNA polymerase that copies the sequence of a DNA strand into RNA. To begin transcribing a gene, the RNA polymerase binds to a sequence of DNA called a promoter and separates the DNA strands. It then copies the gene sequence into a messenger RNA transcript until it reaches a region of DNA called the terminator, where it halts and detaches from the DNA. As with human DNA-dependent DNA polymerases, RNA polymerase II, the enzyme that transcribes most of the genes in the human genome, operates as part of a large protein complex with multiple regulatory and accessory subunits.*[147]

5.5 Genetic recombination

Structure of the Holliday junction intermediate in genetic recombination. The four separate DNA strands are coloured red, blue, green and yellow.*[148]
Further information: Genetic recombination

A DNA helix usually does not interact with other segments of DNA, and in human cells, the different chromosomes even occupy separate areas in the nucleus called "chromosome territories".*[149] This physical separation of different chromosomes is important for the ability of DNA to function as a stable repository for information, as one of the few times chromosomes interact is in chromosomal crossover which occurs during sexual reproduction, when genetic recombination occurs. Chromoso-

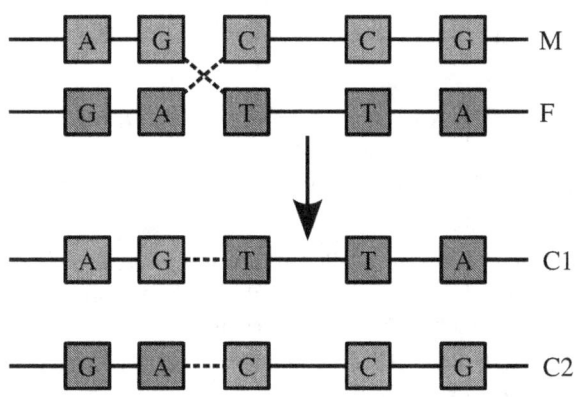

Recombination involves the breaking and rejoining of two chromosomes (M and F) to produce two rearranged chromosomes (C1 and C2).

mal crossover is when two DNA helices break, swap a section and then rejoin.

Recombination allows chromosomes to exchange genetic information and produces new combinations of genes, which increases the efficiency of natural selection and can be important in the rapid evolution of new proteins.[*][150] Genetic recombination can also be involved in DNA repair, particularly in the cell's response to double-strand breaks.[*][151]

The most common form of chromosomal crossover is homologous recombination, where the two chromosomes involved share very similar sequences. Non-homologous recombination can be damaging to cells, as it can produce chromosomal translocations and genetic abnormalities. The recombination reaction is catalyzed by enzymes known as recombinases, such as RAD51.[*][152] The first step in recombination is a double-stranded break caused by either an endonuclease or damage to the DNA.[*][153] A series of steps catalyzed in part by the recombinase then leads to joining of the two helices by at least one Holliday junction, in which a segment of a single strand in each helix is annealed to the complementary strand in the other helix. The Holliday junction is a tetrahedral junction structure that can be moved along the pair of chromosomes, swapping one strand for another. The recombination reaction is then halted by cleavage of the junction and re-ligation of the released DNA.[*][154] Only strands of like polarity exchange DNA during recombination. There are two types of cleavage: east-west cleavage and north-south cleavage. The north-south cleavage nicks both strands of DNA, while the east-west cleavage has one strand of DNA intact. The formation of a Holliday junction during recombination makes it possible for genetic diversity, genes to exchange on chromosomes, and expression of wild-type viral genomes.

5.6 Evolution

Further information: RNA world hypothesis

DNA contains the genetic information that allows all modern living things to function, grow and reproduce. However, it is unclear how long in the 4-billion-year history of life DNA has performed this function, as it has been proposed that the earliest forms of life may have used RNA as their genetic material.[*][155][*][156] RNA may have acted as the central part of early cell metabolism as it can both transmit genetic information and carry out catalysis as part of ribozymes.[*][157] This ancient RNA world where nucleic acid would have been used for both catalysis and genetics may have influenced the evolution of the current genetic code based on four nucleotide bases. This would occur, since the number of different bases in such an organism is a trade-off between a small number of bases increasing replication accuracy and a large number of bases increasing the catalytic efficiency of ribozymes.[*][158] However, there is no direct evidence of ancient genetic systems, as recovery of DNA from most fossils is impossible because DNA survives in the environment for less than one million years, and slowly degrades into short fragments in solution.[*][159] Claims for older DNA have been made, most notably a report of the isolation of a viable bacterium from a salt crystal 250 million years old,[*][160] but these claims are controversial.[*][161][*][162]

Building blocks of DNA (adenine, guanine, and related organic molecules) may have been formed extraterrestrially in outer space.[*][163][*][164][*][165] Complex DNA and RNA organic compounds of life, including uracil, cytosine, and thymine, have also been formed in the laboratory under conditions mimicking those found in outer space, using starting chemicals, such as pyrimidine, found in meteorites. Pyrimidine, like polycyclic aromatic hydrocarbons (PAHs), the most carbon-rich chemical found in the universe, may have been formed in red giants or in interstellar cosmic dust and gas clouds.[*][166]

5.7 Uses in technology

5.7.1 Genetic engineering

Further information: Molecular biology, Nucleic acid methods, and Genetic engineering

Methods have been developed to purify DNA from organisms, such as phenol-chloroform extraction, and to manipulate it in the laboratory, such as restriction digests and the polymerase chain reaction. Modern biology and biochemistry make intensive use of these techniques in recombinant DNA technology. Recombinant DNA is a manmade DNA sequence that has been assembled from other DNA sequences. They can be transformed into organisms in the form of plasmids or in the appropriate format, by using a viral vector.[*][167] The genetically modified organisms produced can be used to produce products such as recombinant proteins, used in medical research,[*][168] or be grown in agriculture.[*][169][*][170]

5.7.2 DNA profiling

Further information: DNA profiling

Forensic scientists can use DNA in blood, semen, skin, saliva or hair found at a crime scene to identify a matching DNA of an individual, such as a perpetrator. This process is formally termed DNA profiling, but may also be called "genetic fingerprinting". In DNA profiling, the lengths of variable sections of repetitive DNA, such as short tandem repeats and minisatellites, are compared between people. This method is usually an extremely reliable technique for identifying a matching DNA.[*][171] However, identification can be complicated if the scene is contaminated with DNA from several people.[*][172] DNA profiling was developed in 1984 by British geneticist Sir Alec Jeffreys,[*][173] and first used in forensic science to convict Colin Pitchfork in the 1988 Enderby murders case.[*][174]

The development of forensic science and the ability to now obtain genetic matching on minute samples of blood, skin, saliva, or hair has led to re-examining many cases. Evidence can now be uncovered that was scientifically impossible at the time of the original examination. Combined with the removal of the double jeopardy law in some places, this can allow cases to be reopened where prior trials have failed to produce sufficient evidence to convince a jury. People charged with serious crimes may be required to provide a sample of DNA for matching purposes. The most obvious defense to DNA matches obtained forensically is to claim that cross-contamination of evidence has occurred. This has resulted in meticulous strict handling procedures with new cases of serious crime. DNA profiling is also used successfully to positively identify victims of mass casualty incidents,[*][175] bodies or body parts in serious accidents, and individual victims in mass war graves, via matching to family members.

DNA profiling is also used in DNA paternity testing to determine if someone is the biological parent or grandparent of a child with the probability of parentage is typically 99.99% when the alleged parent is biologically related to the child. Normal DNA sequencing methods happen after birth, but there are new methods to test paternity while a mother is still pregnant.[*][176]

5.7.3 DNA enzymes or catalytic DNA

Further information: Deoxyribozyme

Deoxyribozymes, also called DNAzymes or catalytic DNA, are first discovered in 1994.[*][177] They are mostly single stranded DNA sequences isolated from a large pool of random DNA sequences through a combinatorial approach called in vitro selection or systematic evolution of ligands by exponential enrichment (SELEX). DNAzymes catalyze variety of chemical reactions including RNA-DNA cleavage, RNA-DNA ligation, amino acids phosphorylation-dephosphorylation, carbon-carbon bond formation, and etc. DNAzymes can enhance catalytic rate of chemical reactions up to 100,000,000,000-fold over the uncatalyzed reaction.[*][178] The most extensively studied class of DNAzymes is RNA-cleaving types which have been used to detect different metal ions and designing therapeutic agents. Several metal-specific DNAzymes have been reported including the GR-5 DNAzyme (lead-specific),[*][177] the CA1-3 DNAzymes (copper-specific),[*][179] the 39E DNAzyme (uranyl-specific) and the NaA43 DNAzyme (sodium-specific).[*][180] The NaA43 DNAzyme, which is reported to be more than 10,000-fold selective for sodium over other metal ions, was used to make a real-time sodium sensor in living cells.

5.7.4 Bioinformatics

Further information: Bioinformatics

Bioinformatics involves the development of techniques to store, data mine, search and manipulate biological data, including DNA nucleic acid sequence data. These have led to widely applied advances in computer science, especially string searching algorithms, machine learning, and database theory.[*][181] String searching or matching algorithms, which find an occurrence of a sequence of letters inside a larger sequence of letters, were developed to search for specific sequences of nucleotides.[*][182] The DNA sequence may be aligned with other DNA sequences to identify homologous sequences and locate the specific mutations that make them distinct. These techniques, especially multiple sequence alignment, are used in studying phylogenetic relationships and protein function.[*][183] Data sets representing entire genomes' worth of DNA sequences, such as those produced by the Human Genome Project, are difficult to use without the annotations that identify the locations of genes and regulatory elements on each chromosome. Regions of DNA sequence that have the characteristic patterns associated with protein- or RNA-coding genes can be identified by gene finding algorithms, which allow researchers to predict the presence of particular gene products and their possible functions in an organism even before they have been isolated experimentally.[*][184] Entire genomes may also be compared, which can shed light on the evolutionary history of particular organism and permit the examination of complex evolutionary events.

5.7.5 DNA nanotechnology

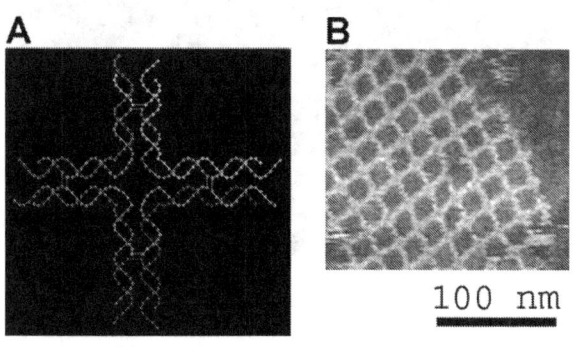

The DNA structure at left (schematic shown) will self-assemble into the structure visualized by atomic force microscopy at right. DNA nanotechnology is the field that seeks to design nanoscale structures using the molecular recognition properties of DNA molecules. Image from Strong, 2004.

Further information: DNA nanotechnology

DNA nanotechnology uses the unique molecular recognition properties of DNA and other nucleic acids to create self-assembling branched DNA complexes with useful properties.[*][185] DNA is thus used as a structural material rather than as a carrier of biological information. This has led to the creation of two-dimensional periodic lattices (both tile-based and using the *DNA origami* method) and three-dimensional structures in the shapes of polyhedra.[*][186] Nanomechanical devices and algorithmic self-assembly have also been demonstrated,[*][187] and these DNA structures have been used to template the arrangement of other molecules such as gold nanoparticles and streptavidin proteins.[*][188]

5.7.6 History and anthropology

Further information: Phylogenetics and Genetic genealogy

Because DNA collects mutations over time, which are then inherited, it contains historical information, and, by comparing DNA sequences, geneticists can infer the evolutionary history of organisms, their phylogeny.[*][189] This field of phylogenetics is a powerful tool in evolutionary biology. If DNA sequences within a species are compared, population geneticists can learn the history of particular populations. This can be used in studies ranging from ecological genetics to anthropology; For example, DNA evidence is being used to try to identify the Ten Lost Tribes of Israel.[*][190][*][191]

5.7.7 Information storage

Main article: DNA digital data storage

In a paper published in *Nature* in January 2013, scientists from the European Bioinformatics Institute and Agilent Technologies proposed a mechanism to use DNA's ability to code information as a means of digital data storage. The group was able to encode 739 kilobytes of data into DNA code, synthesize the actual DNA, then sequence the DNA and decode the information back to its original form, with a reported 100% accuracy. The encoded information consisted of text files and audio files. A prior experiment was published in August 2012. It was conducted by researchers at Harvard University, where the text of a 54,000-word book was encoded in DNA.[*][192][*][193]

Moreover, in living cells, the storage can be turned active by enzymes. Light-gated protein domains fused to DNA processing enzymes are suitable for that task *in vitro*.[*][194][*][195] Fluorescent exonucleases can transmit the output according to the nucleotide they have read.[*][196]

5.8 History of DNA research

Further information: History of molecular biology
  DNA was first isolated by the Swiss physician Friedrich Miescher who, in 1869, discovered a microscopic substance in the pus of discarded surgical bandages. As it resided in the nuclei of cells, he called it "nuclein".[*][197][*][198] In 1878, Albrecht Kossel isolated the non-protein component of "nuclein", nucleic acid, and later isolated its five primary nucleobases.[*][199][*][200] In 1919, Phoebus Levene identified the base, sugar, and phosphate nucleotide unit.[*][201] Levene suggested that DNA consisted of a string of nucleotide units linked together through the phosphate groups. Levene thought the chain was short and the bases repeated in a fixed order. In 1937, William Astbury produced the first X-ray diffraction patterns that showed that DNA had a regular structure.[*][202]

In 1927, Nikolai Koltsov proposed that inherited traits would be inherited via a "giant hereditary molecule"

5.8. HISTORY OF DNA RESEARCH

James Watson and Francis Crick (right), co-originators of the double-helix model, with Maclyn McCarty (left).

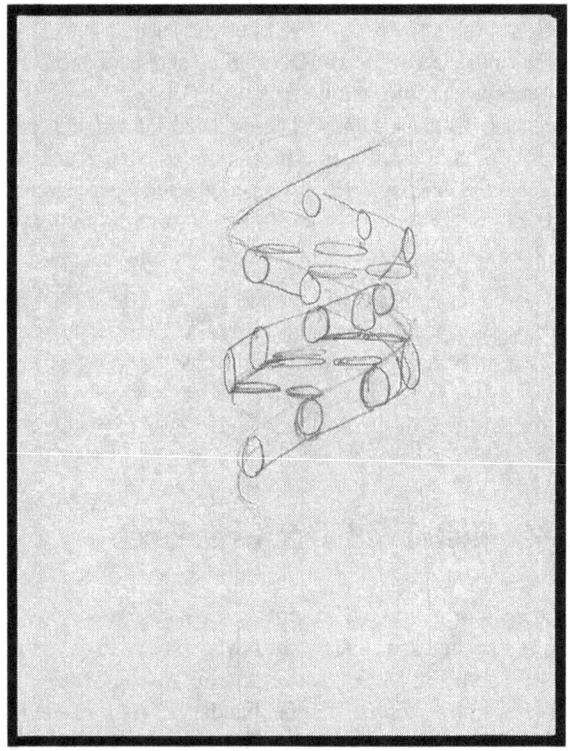

Pencil sketch of the DNA double helix by Francis Crick in 1953

made up of "two mirror strands that would replicate in a semi-conservative fashion using each strand as a template".[203][204] In 1928, Frederick Griffith in his experiment discovered that traits of the "smooth" form of *Pneumococcus* could be transferred to the "rough" form of the same bacteria by mixing killed "smooth" bacteria with the live "rough" form.[205][206] This system provided the first clear suggestion that DNA carries genetic information—the Avery–MacLeod–McCarty experiment—when Oswald Avery, along with coworkers Colin MacLeod and Maclyn McCarty, identified DNA as the transforming principle in 1943.[207] DNA's role in heredity was confirmed in 1952 when Alfred Hershey and Martha Chase in the Hershey–Chase experiment showed that DNA is the genetic material of the T2 phage.[208]

A blue plaque outside The Eagle pub commemorating Crick and Watson

Late in 1951, Francis Crick started working with James Watson at the Cavendish Laboratory within the University of Cambridge. In 1953, Watson and Crick suggested what is now accepted as the first correct double-helix model of DNA structure in the journal *Nature*.[11] Their double-helix, molecular model of DNA was then based on one X-ray diffraction image (labeled as "Photo 51")[209] taken by Rosalind Franklin and Raymond Gosling in May 1952, and the information that the DNA bases are paired. On 28 February 1953 Crick interrupted patrons' lunchtime at The Eagle pub in Cambridge to announce that he and Watson had "discovered the secret of life".[210]

Experimental evidence supporting the Watson and Crick model was published in a series of five articles in the same issue of *Nature*.[211] Of these, Franklin and Gosling's paper was the first publication of their own X-ray diffraction data and original analysis method that partly supported the Watson and Crick model;[65][212] this issue also contained an article on DNA structure by Maurice Wilkins and two of his colleagues, whose analysis and *in vivo* B-DNA X-ray patterns also supported the presence *in vivo* of the double-helical DNA configurations as proposed by Crick and Watson for their double-helix molecular model of DNA in the prior two pages of *Nature*.[66] In 1962, after Franklin's death, Watson, Crick, and Wilkins jointly received the Nobel Prize in Physiology or Medicine.[213] Nobel Prizes are awarded only to living recipients. A debate continues about who should receive credit for the discovery.[214]

In an influential presentation in 1957, Crick laid out the central dogma of molecular biology, which foretold the relationship between DNA, RNA, and proteins, and articulated the "adaptor hypothesis".*[215] Final confirmation of the replication mechanism that was implied by the double-helical structure followed in 1958 through the Meselson–Stahl experiment.*[216] Further work by Crick and coworkers showed that the genetic code was based on non-overlapping triplets of bases, called codons, allowing Har Gobind Khorana, Robert W. Holley, and Marshall Warren Nirenberg to decipher the genetic code.*[217] These findings represent the birth of molecular biology.*[218]

5.9 See also

- Autosome
- Crystallography
- DNA-encoded chemical library
- DNA microarray
- DNA sequencing
- Macromolecule
- Genetic disorder
- Genetic genealogy
- Haplotype
- Comparison of nucleic acid simulation software
- Meiosis
- Mitochondrial DNA
- Nuclear DNA
- Nucleic acid double helix
- Nucleic acid notation
- Nucleic acid sequence
- Pangenesis
- Phosphoramidite
- Ribosomal DNA
- Southern blot
- X-ray scattering techniques
- Xeno nucleic acid
- RNA
- Deoxyribozyme

5.10 References

[1] "deoxyribonucleic acid". *Merriam-Webster Dictionary*.

[2] Alberts B, Johnson A, Lewis J, Raff M, Roberts K, Walter P (2014). *Molecular Biology of the Cell* (6th ed.). Garland. p. Chapter 4: DNA, Chromosomes and Genomes. ISBN 9780815344322.

[3] Purcell A. "DNA". *Basic Biology*.

[4] Nuwer R (18 July 2015). "Counting All the DNA on Earth". *The New York Times*. New York: The New York Times Company. ISSN 0362-4331. Retrieved 2015-07-18.

[5] "The Biosphere: Diversity of Life". *Aspen Global Change Institute*. Basalt, CO. Retrieved 2015-07-19.

[6] Russell P (2001). *iGenetics*. New York: Benjamin Cummings. ISBN 0-8053-4553-1.

[7] Mashaghi A, Katan A (2013). "A physicist's view of DNA". *De Physicus*. **24e** (3): 59–61. Bibcode:2013arXiv1311.2545M. arXiv:1311.2545v1.

[8] Saenger W (1984). *Principles of Nucleic Acid Structure*. New York: Springer-Verlag. ISBN 0-387-90762-9.

[9] Alberts B, Johnson A, Lewis J, Raff M, Roberts K, Peter W (2002). *Molecular Biology of the Cell* (Fourth ed.). New York and London: Garland Science. ISBN 0-8153-3218-1. OCLC 145080076.

[10] Irobalieva RN, Fogg JM, Catanese DJ, Catanese DJ, Sutthibutpong T, Chen M, Barker AK, Ludtke SJ, Harris SA, Schmid MF, Chiu W, Zechiedrich L (October 2015). "Structural diversity of supercoiled DNA". *Nature Communications*. **6**: 8440. PMC 4608029. PMID 26455586. doi:10.1038/ncomms9440.

[11] Watson JD, Crick FH (April 1953). "Molecular structure of nucleic acids; a structure for deoxyribose nucleic acid" (PDF). *Nature*. **171** (4356): 737–8. Bibcode:1953Natur.171..737W. PMID 13054692. doi:10.1038/171737a0.

[12] Mandelkern M, Elias JG, Eden D, Crothers DM (October 1981). "The dimensions of DNA in solution". *Journal of Molecular Biology*. **152** (1): 153–61. PMID 7338906. doi:10.1016/0022-2836(81)90099-1.

[13] Gregory SG, Barlow KF, McLay KE, Kaul R, Swarbreck D, Dunham A, et al. (May 2006). "The DNA sequence and biological annotation of human chromosome 1". *Nature*. **441** (7091): 315–21. Bibcode:2006Natur.441..315G. PMID 16710414. doi:10.1038/nature04727.

[14] Watson JD, Crick FH (April 1953). "Molecular structure of nucleic acids; a structure for deoxyribose nucleic acid" (PDF). *Nature*. **171** (4356): 737–8. Bibcode:1953Natur.171..737W. PMID 13054692. doi:10.1038/171737a0.

5.10. REFERENCES

[15] Berg J., Tymoczko J. and Stryer L. (2002) *Biochemistry*. W. H. Freeman and Company ISBN 0-7167-4955-6

[16] Abbreviations and Symbols for Nucleic Acids, Polynucleotides and their Constituents IUPAC-IUB Commission on Biochemical Nomenclature (CBN). Retrieved 3 January 2006.

[17] Ghosh A, Bansal M (April 2003). "A glossary of DNA structures from A to Z". *Acta Crystallographica. Section D, Biological Crystallography*. **59** (Pt 4): 620–6. PMID 12657780. doi:10.1107/S0907444903003251.

[18] Created from PDB 1D65

[19] Yakovchuk P, Protozanova E, Frank-Kamenetskii MD (2006). "Base-stacking and base-pairing contributions into thermal stability of the DNA double helix". *Nucleic Acids Research*. **34** (2): 564–74. PMC 1360284. PMID 16449200. doi:10.1093/nar/gkj454.

[20] Burton E. Tropp – "*Molecular Biology*" - Jones and Barlett Learning, ISBN 978-0-7637-8663-2

[21] "Watson-Crick Structure of DNA – 1953". *Steven Carr*. Memorial University of Newfoundland. Retrieved 13 July 2016.

[22] Verma S, Eckstein F (1998). "Modified oligonucleotides: synthesis and strategy for users". *Annual Review of Biochemistry*. **67**: 99–134. PMID 9759484. doi:10.1146/annurev.biochem.67.1.99.

[23] Kiljunen S, Hakala K, Pinta E, Huttunen S, Pluta P, Gador A, Lönnberg H, Skurnik M (December 2005). "Yersiniophage phiR1-37 is a tailed bacteriophage having a 270 kb DNA genome with thymidine replaced by deoxyuridine". *Microbiology*. **151** (Pt 12): 4093–102. PMID 16339954. doi:10.1099/mic.0.28265-0.

[24] Uchiyama J, Takemura-Uchiyama I, Sakaguchi Y, Gamoh K, Kato S, Daibata M, Ujihara T, Misawa N, Matsuzaki S (September 2014). "Intragenus generalized transduction in Staphylococcus spp. by a novel giant phage". *The ISME Journal*. **8** (9): 1949–52. PMC 4139722. PMID 24599069. doi:10.1038/ismej.2014.29.

[25] Casella E, Markewych O, Dosmar M, Heman W (1978) Production and expression of dTMP-enriched DNA of bacteriophage SP15. J Virology 28 (3) 753–766

[26] Simpson L (March 1998). "A base called J". *Proceedings of the National Academy of Sciences of the United States of America*. **95** (5): 2037–8. Bibcode:1998PNAS...95.2037S. PMC 33841. PMID 9482833. doi:10.1073/pnas.95.5.2037.

[27] Borst P, Sabatini R (2008). "Base J: discovery, biosynthesis, and possible functions". *Annual Review of Microbiology*. **62**: 235–51. PMID 18729733. doi:10.1146/annurev.micro.62.081307.162750.

[28] Cross M, Kieft R, Sabatini R, Wilm M, de Kort M, van der Marel GA, van Boom JH, van Leeuwen F, Borst P (November 1999). "The modified base J is the target for a novel DNA-binding protein in kinetoplastid protozoans". *The EMBO Journal*. **18** (22): 6573–81. PMC 1171720. PMID 10562569. doi:10.1093/emboj/18.22.6573.

[29] DiPaolo C, Kieft R, Cross M, Sabatini R (February 2005). "Regulation of trypanosome DNA glycosylation by a SWI2/SNF2-like protein". *Molecular Cell*. **17** (3): 441–51. PMID 15694344. doi:10.1016/j.molcel.2004.12.022.

[30] Vainio S, Genest PA, ter Riet B, van Luenen H, Borst P (April 2009). "Evidence that J-binding protein 2 is a thymidine hydroxylase catalyzing the first step in the biosynthesis of DNA base J". *Molecular and Biochemical Parasitology*. **164** (2): 157–61. PMID 19114062. doi:10.1016/j.molbiopara.2008.12.001.

[31] Iyer LM, Tahiliani M, Rao A, Aravind L (June 2009). "Prediction of novel families of enzymes involved in oxidative and other complex modifications of bases in nucleic acids". *Cell Cycle*. **8** (11): 1698–710. PMC 2995806. PMID 19411852. doi:10.4161/cc.8.11.8580.

[32] van Luenen HG, Farris C, Jan S, Genest PA, Tripathi P, Velds A, Kerkhoven RM, Nieuwland M, Haydock A, Ramasamy G, Vainio S, Heidebrecht T, Perrakis A, Pagie L, van Steensel B, Myler PJ, Borst P (August 2012). "Glucosylated hydroxymethyluracil, DNA base J, prevents transcriptional readthrough in Leishmania". *Cell*. **150** (5): 909–21. PMC 3684241. PMID 22939620. doi:10.1016/j.cell.2012.07.030.

[33] Hazelbaker DZ, Buratowski S (November 2012). "Transcription: base J blocks the way". *Current Biology*. **22** (22): R960–2. PMC 3648658. PMID 23174300. doi:10.1016/j.cub.2012.10.010.

[34] Khudyakov I, Ya, Kirnos MD, Alexandrushkina NI, Vanyushin BF (1976) Cyanophage S-2L contains DNA with 2,6-diaminopurine substituted for adenine. Virology 88 (1) 8–18

[35] Thiaville JJ, Kellner SM, Yuan Y, Hutinet G, Thiaville PC, Jumpathong W; Mohapatra S, Brochier-Armanet C, Letarov AV, Hillebrand R, Malik CK, Rizzo CJ, Dedon PC, de Crecy-Lagard V (2016) Novel genomic island modifies DNA with 7-deazaguanine derivatives. Proc Nat Acad Sci USA 113 (11) 1452–1459

[36] Johnson TB, Coghill RD (1925) Pyrimidines. CIII. The discovery of 5-methylcytosine in tuberculinic acid, the nucleic acid of the tubercle bacillus. Journal of the American Chemical Society 47: 2838–2844

[37] Wyatt GR, Cohen SS (1953) Bases of the nucleic acids of some bacterial and animal viruses: the occurrence of 5-hydroxymethylcytosine. Biochem J 55: 774–782

[38] Kuo T-T, Huang T-C, Teng M-H (1968) 5-Methylcytosine replacing cytosine in the deoxyribonucleic acid of a bacteriophage for *Xanthomonas oryzae*. J Mol Biol 34 (2) 373–375

[39] Vogelsang-Wenke H, Oesterhelt D (1988) (1988) Isolation of a halobacterial phage with a fully cytosine-methylated genome. Mol Gen Gen 211 (3) 407–414

[40] Dunn DB, Smith JD (1955) Occurrence of a new base in the deoxyribonucleic acid of a strain of *Bacterium coli*. Nature 175 (4451) 336–337

[41] Allet B, Bukhari AI (1975) Analysis of bacteriophage Mu and λ-Mu hybrid DNAs by specific endonucleases. J Mol Biol 92 (4) 529–540

[42] Nikolskaya II, Lopatina NG, Debov SS (1976) Methylated guanine derivative as a minor base in the DNA of phage DDVI *Shigella disenteriae*. Biochim Biophys Acta Nucleic Acids Protein Synth 435 (2) 206–210

[43] Janulaitis A, Klimasauskas S, Petrusyte M, Butkus, V (1983) Cytosine modification in DNA by *BcnI* methylase yields N4-methylcytosine. FEBS Letters 161 (1) 131–134

[44] Swinton D, Hattman S, Benzinger R, Buchanan-Wollaston V, Beringer J (1985) Replacement of the deoxycytidine residues in Rhizobium bacteriophage RL38JI DNA. FEBS Letters 184 (2) 294–298

[45] Maltman KL, Neuhard J, Warren RAJ (1981) 5-[(Hydroxymethyl)-O-pyrophosphoryl]uracil, an intermediate in the biosynthesis of α-putrescinylthymine in deoxyribonucleic acid of bacteriophage ΦW-14. Biochemistry 20 (12) 3586–3591

[46] Carell T, Kurz MQ, Müller M, Rossa M, Spada F (2017) Non-canonical bases in the genome: The regulatory information layer in DNA. Angew Chem Int Ed Engl

[47] Wing R, Drew H, Takano T, Broka C, Tanaka S, Itakura K, Dickerson RE (October 1980). "Crystal structure analysis of a complete turn of B-DNA". *Nature*. **287** (5784): 755–8. Bibcode:1980Natur.287..755W. PMID 7432492. doi:10.1038/287755a0.

[48] Pabo CO, Sauer RT (1984). "Protein-DNA recognition". *Annual Review of Biochemistry*. **53**: 293–321. PMID 6236744. doi:10.1146/annurev.bi.53.070184.001453.

[49] Nikolova EN, Zhou H, Gottardo FL, Alvey HS, Kimsey IJ, Al-Hashimi HM (2013). "A historical account of Hoogsteen base-pairs in duplex DNA". *Biopolymers*. **99** (12): 955–68. PMC 3844552 ⊚. PMID 23818176. doi:10.1002/bip.22334.

[50] Clausen-Schaumann H, Rief M, Tolksdorf C, Gaub HE (April 2000). "Mechanical stability of single DNA molecules". *Biophysical Journal*. **78** (4): 1997–2007. Bibcode:2000BpJ....78.1997C. PMC 1300792 ⊚. PMID 10733978. doi:10.1016/S0006-3495(00)76747-6.

[51] Chalikian TV, Völker J, Plum GE, Breslauer KJ (July 1999). "A more unified picture for the thermodynamics of nucleic acid duplex melting: a characterization by calorimetric and volumetric techniques". *Proceedings of the National Academy of Sciences of the United States of America*. **96** (14): 7853–8. Bibcode:1999PNAS...96.7853C. PMC 22151 ⊚. PMID 10393911. doi:10.1073/pnas.96.14.7853.

[52] deHaseth PL, Helmann JD (June 1995). "Open complex formation by Escherichia coli RNA polymerase: the mechanism of polymerase-induced strand separation of double helical DNA". *Molecular Microbiology*. **16** (5): 817–24. PMID 7476180. doi:10.1111/j.1365-2958.1995.tb02309.x.

[53] Isaksson J, Acharya S, Barman J, Cheruku P, Chattopadhyaya J (December 2004). "Single-stranded adenine-rich DNA and RNA retain structural characteristics of their respective double-stranded conformations and show directional differences in stacking pattern". *Biochemistry*. **43** (51): 15996–6010. PMID 15609994. doi:10.1021/bi048221v.

[54] Designation of the two strands of DNA JCBN/NC-IUB Newsletter 1989. Retrieved 7 May 2008

[55] Hüttenhofer A, Schattner P, Polacek N (May 2005). "Non-coding RNAs: hope or hype?". *Trends in Genetics*. **21** (5): 289–97. PMID 15851066. doi:10.1016/j.tig.2005.03.007.

[56] Munroe SH (November 2004). "Diversity of antisense regulation in eukaryotes: multiple mechanisms, emerging patterns". *Journal of Cellular Biochemistry*. **93** (4): 664–71. PMID 15389973. doi:10.1002/jcb.20252.

[57] Makalowska I, Lin CF, Makalowski W (February 2005). "Overlapping genes in vertebrate genomes". *Computational Biology and Chemistry*. **29** (1): 1–12. PMID 15680581. doi:10.1016/j.compbiolchem.2004.12.006.

[58] Johnson ZI, Chisholm SW (November 2004). "Properties of overlapping genes are conserved across microbial genomes". *Genome Research*. **14** (11): 2268–72. PMC 525685 ⊚. PMID 15520290. doi:10.1101/gr.2433104.

[59] Lamb RA, Horvath CM (August 1991). "Diversity of coding strategies in influenza viruses". *Trends in Genetics*. **7** (8): 261–6. PMID 1771674. doi:10.1016/0168-9525(91)90326-L.

[60] Benham CJ, Mielke SP (2005). "DNA mechanics". *Annual Review of Biomedical Engineering*. **7**: 21–53. PMID 16004565. doi:10.1146/annurev.bioeng.6.062403.132016.

[61] Champoux JJ (2001). "DNA topoisomerases: structure, function, and mechanism". *Annual Review of Biochemistry*. **70**: 369–413. PMID 11395412. doi:10.1146/annurev.biochem.70.1.369.

[62] Wang JC (June 2002). "Cellular roles of DNA topoisomerases: a molecular perspective". *Nature Reviews. Molecular Cell Biology*. **3** (6): 430–40. PMID 12042765. doi:10.1038/nrm831.

5.10. REFERENCES

[63] Basu HS, Feuerstein BG, Zarling DA, Shafer RH, Marton LJ (October 1988). "Recognition of Z-RNA and Z-DNA determinants by polyamines in solution: experimental and theoretical studies". *Journal of Biomolecular Structure & Dynamics*. **6** (2): 299–309. PMID 2482766. doi:10.1080/07391102.1988.10507714.

[64] Franklin RE, Gosling RG (6 March 1953). "The Structure of Sodium Thymonucleate Fibres I. The Influence of Water Content" (PDF). *Acta Crystallogr*. **6** (8–9): 673–7. doi:10.1107/S0365110X53001939.
Franklin RE, Gosling RG (1953). "The structure of sodium thymonucleate fibres. II. The cylindrically symmetrical Patterson function". *Acta Crystallogr*. **6** (8–9): 678–85. doi:10.1107/S0365110X53001940.

[65] Franklin RE, Gosling RG (April 1953). "Molecular configuration in sodium thymonucleate" (PDF). *Nature*. **171** (4356): 740–1. Bibcode:1953Natur.171..740F. PMID 13054694. doi:10.1038/171740a0.

[66] Wilkins MH, Stokes AR, Wilson HR (April 1953). "Molecular structure of deoxypentose nucleic acids" (PDF). *Nature*. **171** (4356): 738–40. Bibcode:1953Natur.171..738W. PMID 13054693. doi:10.1038/171738a0.

[67] Leslie AG, Arnott S, Chandrasekaran R, Ratliff RL (October 1980). "Polymorphism of DNA double helices". *Journal of Molecular Biology*. **143** (1): 49–72. PMID 7441761. doi:10.1016/0022-2836(80)90124-2.

[68] Baianu, I.C. (1980). "Structural Order and Partial Disorder in Biological systems". *Bull. Math. Biol*. **42** (4): 137–141. doi:10.1007/BF02462372. http://cogprints.org/3822/

[69] Hosemann R., Bagchi R.N., *Direct analysis of diffraction by matter*, North-Holland Publs., Amsterdam – New York, 1962.

[70] Baianu, I.C. (1978). "X-ray scattering by partially disordered membrane systems". *Acta Crystallogr A*. **34** (5): 751–753. Bibcode:1978AcCrA..34..751B. doi:10.1107/S0567739478001540.

[71] Wahl MC, Sundaralingam M (1997). "Crystal structures of A-DNA duplexes". *Biopolymers*. **44** (1): 45–63. PMID 9097733. doi:10.1002/(SICI)1097-0282(1997)44:1<45::AID-BIP4>3.0.CO;2-#.

[72] Lu XJ, Shakked Z, Olson WK (July 2000). "A-form conformational motifs in ligand-bound DNA structures". *Journal of Molecular Biology*. **300** (4): 819–40. PMID 10891271. doi:10.1006/jmbi.2000.3690.

[73] Rothenburg S, Koch-Nolte F, Haag F (December 2001). "DNA methylation and Z-DNA formation as mediators of quantitative differences in the expression of alleles". *Immunological Reviews*. **184**: 286–98. PMID 12086319. doi:10.1034/j.1600-065x.2001.1840125.x.

[74] Oh DB, Kim YG, Rich A (December 2002). "Z-DNA-binding proteins can act as potent effectors of gene expression in vivo". *Proceedings of the National Academy of Sciences of the United States of America*. **99** (26): 16666–71. Bibcode:2002PNAS...9916666O. PMC 139201. PMID 12486233. doi:10.1073/pnas.262672699.

[75] Palmer J (2 December 2010). "Arsenic-loving bacteria may help in hunt for alien life". *BBC News*. Retrieved 2 December 2010.

[76] Bortman, Henry (2 December 2010). "Arsenic-Eating Bacteria Opens New Possibilities for Alien Life". *Space.com*. Retrieved 2 December 2010.

[77] Katsnelson A (2 December 2010). "Arsenic-eating microbe may redefine chemistry of life". *Nature News*. doi:10.1038/news.2010.645.

[78] Cressey D (3 October 2012). "'Arsenic-life' Bacterium Prefers Phosphorus after all". *Nature News*. doi:10.1038/nature.2012.11520.

[79] Greider CW, Blackburn EH (December 1985). "Identification of a specific telomere terminal transferase activity in Tetrahymena extracts". *Cell*. **43** (2 Pt 1): 405–13. PMID 3907856. doi:10.1016/0092-8674(85)90170-9.

[80] Nugent CI, Lundblad V (April 1998). "The telomerase reverse transcriptase: components and regulation". *Genes & Development*. **12** (8): 1073–85. PMID 9553037. doi:10.1101/gad.12.8.1073.

[81] Wright WE, Tesmer VM, Huffman KE, Levene SD, Shay JW (November 1997). "Normal human chromosomes have long G-rich telomeric overhangs at one end". *Genes & Development*. **11** (21): 2801–9. PMC 316649. PMID 9353250. doi:10.1101/gad.11.21.2801.

[82] Created from

[83] Burge S, Parkinson GN, Hazel P, Todd AK, Neidle S (2006). "Quadruplex DNA: sequence, topology and structure". *Nucleic Acids Research*. **34** (19): 5402–15. PMC 1636468. PMID 17012276. doi:10.1093/nar/gkl655.

[84] Parkinson GN, Lee MP, Neidle S (June 2002). "Crystal structure of parallel quadruplexes from human telomeric DNA". *Nature*. **417** (6891): 876–80. Bibcode:2002Natur.417..876P. PMID 12050675. doi:10.1038/nature755.

[85] Griffith JD, Comeau L, Rosenfield S, Stansel RM, Bianchi A, Moss H, de Lange T (May 1999). "Mammalian telomeres end in a large duplex loop". *Cell*. **97** (4): 503–14. PMID 10338214. doi:10.1016/S0092-8674(00)80760-6.

[86] Seeman NC (November 2005). "DNA enables nanoscale control of the structure of matter". *Quarterly Reviews of Biophysics*. **38** (4): 363–71. PMC 3478329. PMID 16515737. doi:10.1017/S0033583505004087.

[87] Hu Q, Rosenfeld MG (2012). "Epigenetic regulation of human embryonic stem cells". *Frontiers in Genetics*. **3**: 238. PMC 3488762. PMID 23133442. doi:10.3389/fgene.2012.00238.

[88] Klose RJ, Bird AP (February 2006). "Genomic DNA methylation: the mark and its mediators". *Trends in Biochemical Sciences*. **31** (2): 89–97. PMID 16403636. doi:10.1016/j.tibs.2005.12.008.

[89] Bird A (January 2002). "DNA methylation patterns and epigenetic memory". *Genes & Development*. **16** (1): 6–21. PMID 11782440. doi:10.1101/gad.947102.

[90] Walsh CP, Xu GL (2006). "Cytosine methylation and DNA repair". *Current Topics in Microbiology and Immunology*. Current Topics in Microbiology and Immunology. **301**: 283–315. ISBN 3-540-29114-8. PMID 16570853. doi:10.1007/3-540-31390-7_11.

[91] Kriaucionis S, Heintz N (May 2009). "The nuclear DNA base 5-hydroxymethylcytosine is present in Purkinje neurons and the brain". *Science*. **324** (5929): 929–30. Bibcode:2009Sci...324..929K. PMC 3263819. PMID 19372393. doi:10.1126/science.1169786.

[92] Ratel D, Ravanat JL, Berger F, Wion D (March 2006). "N6-methyladenine: the other methylated base of DNA". *BioEssays*. **28** (3): 309–15. PMC 2754416. PMID 16479578. doi:10.1002/bies.20342.

[93] Gommers-Ampt JH, Van Leeuwen F, de Beer AL, Vliegenthart JF, Dizdaroglu M, Kowalak JA, Crain PF, Borst P (December 1993). "beta-D-glucosyl-hydroxymethyluracil: a novel modified base present in the DNA of the parasitic protozoan T. brucei". *Cell*. **75** (6): 1129–36. PMID 8261512. doi:10.1016/0092-8674(93)90322-H.

[94] Created from PDB 1JDG

[95] Douki T, Reynaud-Angelin A, Cadet J, Sage E (August 2003). "Bipyrimidine photoproducts rather than oxidative lesions are the main type of DNA damage involved in the genotoxic effect of solar UVA radiation". *Biochemistry*. **42** (30): 9221–6. PMID 12885257. doi:10.1021/bi034593c.

[96] Cadet J, Delatour T, Douki T, Gasparutto D, Pouget JP, Ravanat JL, Sauvaigo S (March 1999). "Hydroxyl radicals and DNA base damage". *Mutation Research*. **424** (1–2): 9–21. PMID 10064846. doi:10.1016/S0027-5107(99)00004-4.

[97] Beckman KB, Ames BN (August 1997). "Oxidative decay of DNA". *The Journal of Biological Chemistry*. **272** (32): 19633–6. PMID 9289489. doi:10.1074/jbc.272.32.19633.

[98] Valerie K, Povirk LF (September 2003). "Regulation and mechanisms of mammalian double-strand break repair". *Oncogene*. **22** (37): 5792–812. PMID 12947387. doi:10.1038/sj.onc.1206679.

[99] Johnson G (28 December 2010). "Unearthing Prehistoric Tumors, and Debate". *The New York Times*. If we lived long enough, sooner or later we all would get cancer.

[100] Alberts B, Johnson A, Lewis J, et al. (2002). "The Preventable Causes of Cancer". *Molecular biology of the cell* (4th ed.). New York: Garland Science. ISBN 0-8153-4072-9. A certain irreducible background incidence of cancer is to be expected regardless of circumstances: mutations can never be absolutely avoided, because they are an inescapable consequence of fundamental limitations on the accuracy of DNA replication, as discussed in Chapter 5. If a human could live long enough, it is inevitable that at least one of his or her cells would eventually accumulate a set of mutations sufficient for cancer to develop.

[101] Bernstein H, Payne CM, Bernstein C, Garewal H, Dvorak K (2008). "Cancer and aging as consequences of un-repaired DNA damage". In Kimura H, Suzuki A. *New Research on DNA Damage*. New York: Nova Science Publishers. pp. 1–47. ISBN 978-1-60456-581-2.

[102] Hoeijmakers JH (October 2009). "DNA damage, aging, and cancer". *The New England Journal of Medicine*. **361** (15): 1475–85. PMID 19812404. doi:10.1056/NEJMra0804615.

[103] Freitas AA, de Magalhães JP (2011). "A review and appraisal of the DNA damage theory of ageing". *Mutation Research*. **728** (1–2): 12–22. PMID 21600302. doi:10.1016/j.mrrev.2011.05.001.

[104] Ferguson LR, Denny WA (September 1991). "The genetic toxicology of acridines". *Mutation Research*. **258** (2): 123–60. PMID 1881402. doi:10.1016/0165-1110(91)90006-H.

[105] Stephens TD, Bunde CJ, Fillmore BJ (June 2000). "Mechanism of action in thalidomide teratogenesis". *Biochemical Pharmacology*. **59** (12): 1489–99. PMID 10799645. doi:10.1016/S0006-2952(99)00388-3.

[106] Jeffrey AM (1985). "DNA modification by chemical carcinogens". *Pharmacology & Therapeutics*. **28** (2): 237–72. PMID 3936066. doi:10.1016/0163-7258(85)90013-0.

[107] Braña MF, Cacho M, Gradillas A, de Pascual-Teresa B, Ramos A (November 2001). "Intercalators as anticancer drugs". *Current Pharmaceutical Design*. **7** (17): 1745–80. PMID 11562309. doi:10.2174/1381612013397113.

[108] Venter JC, Adams MD, Myers EW, Li PW, Mural RJ, Sutton GG, et al. (February 2001). "The sequence of the human genome". *Science*. **291** (5507): 1304–51. Bibcode:2001Sci...291.1304V. PMID 11181995. doi:10.1126/science.1058040.

[109] Thanbichler M, Wang SC, Shapiro L (October 2005). "The bacterial nucleoid: a highly organized and dynamic structure". *Journal of Cellular Biochemistry*. **96** (3): 506–21. PMID 15988757. doi:10.1002/jcb.20519.

[110] Wolfsberg TG, McEntyre J, Schuler GD (February 2001). "Guide to the draft human genome". *Nature*. **409** (6822): 824–6. Bibcode:2001Natur.409..824W. PMID 11236998. doi:10.1038/35057000.

5.10. REFERENCES

[111] Gregory TR (January 2005). "The C-value enigma in plants and animals: a review of parallels and an appeal for partnership". *Annals of Botany*. **95** (1): 133–46. PMID 15596463. doi:10.1093/aob/mci009.

[112] Birney E, Stamatoyannopoulos JA, Dutta A, Guigó R, Gingeras TR, Margulies EH, et al. (June 2007). "Identification and analysis of functional elements in 1% of the human genome by the ENCODE pilot project". *Nature*. **447** (7146): 799–816. Bibcode:2007Natur.447..799B. PMC 2212820. PMID 17571346. doi:10.1038/nature05874.

[113] Created from PDB 1MSW

[114] Pidoux AL, Allshire RC (March 2005). "The role of heterochromatin in centromere function". *Philosophical Transactions of the Royal Society of London. Series B, Biological Sciences*. **360** (1455): 569–79. PMC 1569473. PMID 15905142. doi:10.1098/rstb.2004.1611.

[115] Harrison PM, Hegyi H, Balasubramanian S, Luscombe NM, Bertone P, Echols N, Johnson T, Gerstein M (February 2002). "Molecular fossils in the human genome: identification and analysis of the pseudogenes in chromosomes 21 and 22". *Genome Research*. **12** (2): 272–80. PMC 155275. PMID 11827946. doi:10.1101/gr.207102.

[116] Harrison PM, Gerstein M (May 2002). "Studying genomes through the aeons: protein families, pseudogenes and proteome evolution". *Journal of Molecular Biology*. **318** (5): 1155–74. PMID 12083509. doi:10.1016/S0022-2836(02)00109-2.

[117] Albà M (2001). "Replicative DNA polymerases". *Genome Biology*. **2** (1): REVIEWS3002. PMC 150442. PMID 11178285. doi:10.1186/gb-2001-2-1-reviews3002.

[118] Tani K, Nasu M (2010). "Roles of Extracellular DNA in Bacterial Ecosystems". In Kikuchi Y, Rykova EY. *Extracellular Nucleic Acids*. Springer. pp. 25–38. ISBN 978-3-642-12616-1.

[119] Vlassov VV, Laktionov PP, Rykova EY (July 2007). "Extracellular nucleic acids". *BioEssays*. **29** (7): 654–67. PMID 17563084. doi:10.1002/bies.20604.

[120] Finkel SE, Kolter R (November 2001). "DNA as a nutrient: novel role for bacterial competence gene homologs". *Journal of Bacteriology*. **183** (21): 6288–93. PMC 100116. PMID 11591672. doi:10.1128/JB.183.21.6288-6293.2001.

[121] Mulcahy H, Charron-Mazenod L, Lewenza S (November 2008). "Extracellular DNA chelates cations and induces antibiotic resistance in Pseudomonas aeruginosa biofilms". *PLoS Pathogens*. **4** (11): e1000213. PMC 2581603. PMID 19023416. doi:10.1371/journal.ppat.1000213.

[122] Berne C, Kysela DT, Brun YV (August 2010). "A bacterial extracellular DNA inhibits settling of motile progeny cells within a biofilm". *Molecular Microbiology*. **77** (4): 815–29. PMC 2962764. PMID 20598083. doi:10.1111/j.1365-2958.2010.07267.x.

[123] Whitchurch CB, Tolker-Nielsen T, Ragas PC, Mattick JS (February 2002). "Extracellular DNA required for bacterial biofilm formation". *Science*. **295** (5559): 1487. PMID 11859186. doi:10.1126/science.295.5559.1487.

[124] Hu W, Li L, Sharma S, Wang J, McHardy I, Lux R, Yang Z, He X, Gimzewski JK, Li Y, Shi W (2012). "DNA builds and strengthens the extracellular matrix in Myxococcus xanthus biofilms by interacting with exopolysaccharides". *PloS One*. **7** (12): e51905. PMC 3530553. PMID 23300576. doi:10.1371/journal.pone.0051905.

[125] Hui L, Bianchi DW (February 2013). "Recent advances in the prenatal interrogation of the human fetal genome". *Trends in Genetics*. **29** (2): 84–91. PMC 4378900. PMID 23158400. doi:10.1016/j.tig.2012.10.013.

[126] Sandman K, Pereira SL, Reeve JN (December 1998). "Diversity of prokaryotic chromosomal proteins and the origin of the nucleosome". *Cellular and Molecular Life Sciences*. **54** (12): 1350–64. PMID 9893710. doi:10.1007/s000180050259.

[127] Dame RT (May 2005). "The role of nucleoid-associated proteins in the organization and compaction of bacterial chromatin". *Molecular Microbiology*. **56** (4): 858–70. PMID 15853876. doi:10.1111/j.1365-2958.2005.04598.x.

[128] Luger K, Mäder AW, Richmond RK, Sargent DF, Richmond TJ (September 1997). "Crystal structure of the nucleosome core particle at 2.8 A resolution". *Nature*. **389** (6648): 251–60. Bibcode:1997Natur.389..251L. PMID 9305837. doi:10.1038/38444.

[129] Jenuwein T, Allis CD (August 2001). "Translating the histone code". *Science*. **293** (5532): 1074–80. PMID 11498575. doi:10.1126/science.1063127.

[130] Ito T (2003). "Nucleosome assembly and remodeling". *Current Topics in Microbiology and Immunology*. Current Topics in Microbiology and Immunology. **274**: 1–22. ISBN 978-3-540-44208-0. PMID 12596902. doi:10.1007/978-3-642-55747-7_1.

[131] Thomas JO (August 2001). "HMG1 and 2: architectural DNA-binding proteins". *Biochemical Society Transactions*. **29** (Pt 4): 395–401. PMID 11497996. doi:10.1042/BST0290395.

[132] Grosschedl R, Giese K, Pagel J (March 1994). "HMG domain proteins: architectural elements in the assembly of nucleoprotein structures". *Trends in Genetics*. **10** (3): 94–100. PMID 8178371. doi:10.1016/0168-9525(94)90232-1.

[133] Iftode C, Daniely Y, Borowiec JA (1999). "Replication protein A (RPA): the eukaryotic SSB". *Critical Reviews in Biochemistry and Molecular Biology*. **34** (3): 141–80. PMID 10473346. doi:10.1080/10409239991209255.

[134] Created from PDB 1LMB

[135] Myers LC, Kornberg RD (2000). "Mediator of transcriptional regulation". *Annual Review of Biochemistry.* **69**: 729–49. PMID 10966474. doi:10.1146/annurev.biochem.69.1.729.

[136] Spiegelman BM, Heinrich R (October 2004). "Biological control through regulated transcriptional coactivators". *Cell.* **119** (2): 157–67. PMID 15479634. doi:10.1016/j.cell.2004.09.037.

[137] Li Z, Van Calcar S, Qu C, Cavenee WK, Zhang MQ, Ren B (July 2003). "A global transcriptional regulatory role for c-Myc in Burkitt's lymphoma cells". *Proceedings of the National Academy of Sciences of the United States of America.* **100** (14): 8164–9. Bibcode:2003PNAS..100.8164L. PMC 166200. PMID 12808131. doi:10.1073/pnas.1332764100.

[138] Created from PDB 1RVA

[139] Bickle TA, Krüger DH (June 1993). "Biology of DNA restriction". *Microbiological Reviews.* **57** (2): 434–50. PMC 372918. PMID 8336674.

[140] Doherty AJ, Suh SW (November 2000). "Structural and mechanistic conservation in DNA ligases". *Nucleic Acids Research.* **28** (21): 4051–8. PMC 113121. PMID 11058099. doi:10.1093/nar/28.21.4051.

[141] Schoeffler AJ, Berger JM (December 2005). "Recent advances in understanding structure-function relationships in the type II topoisomerase mechanism". *Biochemical Society Transactions.* **33** (Pt 6): 1465–70. PMID 16246147. doi:10.1042/BST20051465.

[142] Tuteja N, Tuteja R (May 2004). "Unraveling DNA helicases. Motif, structure, mechanism and function". *European Journal of Biochemistry.* **271** (10): 1849–63. PMID 15128295. doi:10.1111/j.1432-1033.2004.04094.x.

[143] Joyce CM, Steitz TA (November 1995). "Polymerase structures and function: variations on a theme?". *Journal of Bacteriology.* **177** (22): 6321–9. PMC 177480. PMID 7592405. doi:10.1128/jb.177.22.6321-6329.1995.

[144] Hubscher U, Maga G, Spadari S (2002). "Eukaryotic DNA polymerases". *Annual Review of Biochemistry.* **71**: 133–63. PMID 12045093. doi:10.1146/annurev.biochem.71.090501.150041.

[145] Johnson A, O'Donnell M (2005). "Cellular DNA replicases: components and dynamics at the replication fork". *Annual Review of Biochemistry.* **74**: 283–315. PMID 15952889. doi:10.1146/annurev.biochem.73.011303.073859.

[146] Tarrago-Litvak L, Andréola ML, Nevinsky GA, Sarih-Cottin L, Litvak S (May 1994). "The reverse transcriptase of HIV-1: from enzymology to therapeutic intervention". *FASEB Journal.* **8** (8): 497–503. PMID 7514143.

[147] Martinez E (December 2002). "Multi-protein complexes in eukaryotic gene transcription". *Plant Molecular Biology.* **50** (6): 925–47. PMID 12516863. doi:10.1023/A:1021258713850.

[148] Created from PDB 1M6G

[149] Cremer T, Cremer C (April 2001). "Chromosome territories, nuclear architecture and gene regulation in mammalian cells". *Nature Reviews. Genetics.* **2** (4): 292–301. PMID 11283701. doi:10.1038/35066075.

[150] Pál C, Papp B, Lercher MJ (May 2006). "An integrated view of protein evolution". *Nature Reviews. Genetics.* **7** (5): 337–48. PMID 16619049. doi:10.1038/nrg1838.

[151] O'Driscoll M, Jeggo PA (January 2006). "The role of double-strand break repair - insights from human genetics". *Nature Reviews. Genetics.* **7** (1): 45–54. PMID 16369571. doi:10.1038/nrg1746.

[152] Vispé S, Defais M (October 1997). "Mammalian Rad51 protein: a RecA homologue with pleiotropic functions". *Biochimie.* **79** (9–10): 587–92. PMID 9466696. doi:10.1016/S0300-9084(97)82007-X.

[153] Neale MJ, Keeney S (July 2006). "Clarifying the mechanics of DNA strand exchange in meiotic recombination". *Nature.* **442** (7099): 153–8. Bibcode:2006Natur.442..153N. PMID 16838012. doi:10.1038/nature04885.

[154] Dickman MJ, Ingleston SM, Sedelnikova SE, Rafferty JB, Lloyd RG, Grasby JA, Hornby DP (November 2002). "The RuvABC resolvasome". *European Journal of Biochemistry.* **269** (22): 5492–501. PMID 12423347. doi:10.1046/j.1432-1033.2002.03250.x.

[155] Joyce GF (July 2002). "The antiquity of RNA-based evolution". *Nature.* **418** (6894): 214–21. Bibcode:2002Natur.418..214J. PMID 12110897. doi:10.1038/418214a.

[156] Orgel LE (2004). "Prebiotic chemistry and the origin of the RNA world". *Critical Reviews in Biochemistry and Molecular Biology.* **39** (2): 99–123. CiteSeerX 10.1.1.537.7679. PMID 15217990. doi:10.1080/10409230490460765.

[157] Davenport RJ (May 2001). "Ribozymes. Making copies in the RNA world". *Science.* **292** (5520): 1278. PMID 11360970. doi:10.1126/science.292.5520.1278a.

[158] Szathmáry E (April 1992). "What is the optimum size for the genetic alphabet?". *Proceedings of the National Academy of Sciences of the United States of America.* **89** (7): 2614–8. Bibcode:1992PNAS...89.2614S. PMC 48712. PMID 1372984. doi:10.1073/pnas.89.7.2614.

[159] Lindahl T (April 1993). "Instability and decay of the primary structure of DNA". *Nature.* **362** (6422): 709–15. Bibcode:1993Natur.362..709L. PMID 8469282. doi:10.1038/362709a0.

5.10. REFERENCES

[160] Vreeland RH, Rosenzweig WD, Powers DW (October 2000). "Isolation of a 250 million-year-old halotolerant bacterium from a primary salt crystal". *Nature.* **407** (6806): 897–900. PMID 11057666. doi:10.1038/35038060.

[161] Hebsgaard MB, Phillips MJ, Willerslev E (May 2005). "Geologically ancient DNA: fact or artefact?". *Trends in Microbiology.* **13** (5): 212–20. PMID 15866038. doi:10.1016/j.tim.2005.03.010.

[162] Nickle DC, Learn GH, Rain MW, Mullins JI, Mittler JE (January 2002). "Curiously modern DNA for a "250 million-year-old" bacterium". *Journal of Molecular Evolution.* **54** (1): 134–7. PMID 11734907. doi:10.1007/s00239-001-0025-x.

[163] Callahan MP, Smith KE, Cleaves HJ, Ruzicka J, Stern JC, Glavin DP, House CH, Dworkin JP (August 2011). "Carbonaceous meteorites contain a wide range of extraterrestrial nucleobases". *Proceedings of the National Academy of Sciences of the United States of America.* **108** (34): 13995–8. Bibcode:2011PNAS..10813995C. PMC 3161613. PMID 21836052. doi:10.1073/pnas.1106493108.

[164] Steigerwald J (8 August 2011). "NASA Researchers: DNA Building Blocks Can Be Made in Space". NASA. Retrieved 10 August 2011.

[165] ScienceDaily Staff (9 August 2011). "DNA Building Blocks Can Be Made in Space, NASA Evidence Suggests". ScienceDaily. Retrieved 9 August 2011.

[166] Marlaire R (3 March 2015). "NASA Ames Reproduces the Building Blocks of Life in Laboratory". *NASA.* Retrieved 5 March 2015.

[167] Goff SP, Berg P (December 1976). "Construction of hybrid viruses containing SV40 and lambda phage DNA segments and their propagation in cultured monkey cells". *Cell.* **9** (4 PT 2): 695–705. PMID 189942. doi:10.1016/0092-8674(76)90133-1.

[168] Houdebine LM (2007). "Transgenic animal models in biomedical research". *Methods in Molecular Biology.* **360**: 163–202. ISBN 1-59745-165-7. PMID 17172731. doi:10.1385/1-59745-165-7:163.

[169] Daniell H, Dhingra A (April 2002). "Multigene engineering: dawn of an exciting new era in biotechnology". *Current Opinion in Biotechnology.* **13** (2): 136–41. PMC 3481857. PMID 11950565. doi:10.1016/S0958-1669(02)00297-5.

[170] Job D (November 2002). "Plant biotechnology in agriculture". *Biochimie.* **84** (11): 1105–10. PMID 12595138. doi:10.1016/S0300-9084(02)00013-5.

[171] Collins A, Morton NE (June 1994). "Likelihood ratios for DNA identification". *Proceedings of the National Academy of Sciences of the United States of America.* **91** (13): 6007–11. Bibcode:1994PNAS...91.6007C. PMC 44126. PMID 8016106. doi:10.1073/pnas.91.13.6007.

[172] Weir BS, Triggs CM, Starling L, Stowell LI, Walsh KA, Buckleton J (March 1997). "Interpreting DNA mixtures". *Journal of Forensic Sciences.* **42** (2): 213–22. PMID 9068179.

[173] Jeffreys AJ, Wilson V, Thein SL (1985). "Individual-specific 'fingerprints' of human DNA". *Nature.* **316** (6023): 76–9. Bibcode:1985Natur.316...76J. PMID 2989708. doi:10.1038/316076a0.

[174] Colin Pitchfork —first murder conviction on DNA evidence also clears the prime suspect Forensic Science Service Accessed 23 December 2006

[175] "DNA Identification in Mass Fatality Incidents". National Institute of Justice. September 2006. Archived from the original on 12 November 2006.

[176] "Paternity Blood Tests That Work Early in a Pregnancy" New York Times June 20, 2012

[177] Breaker RR, Joyce GF (December 1994). "A DNA enzyme that cleaves RNA". *Chemistry & Biology.* **1** (4): 223–9. PMID 9383394. doi:10.1016/1074-5521(94)90014-0.

[178] Chandra M, Sachdeva A, Silverman SK (October 2009). "DNA-catalyzed sequence-specific hydrolysis of DNA". *Nature Chemical Biology.* **5** (10): 718–20. PMC 2746877. PMID 19684594. doi:10.1038/nchembio.201.

[179] Carmi N, Shultz LA, Breaker RR (December 1996). "In vitro selection of self-cleaving DNAs". *Chemistry & Biology.* **3** (12): 1039–46. PMID 9000012. doi:10.1016/S1074-5521(96)90170-2.

[180] Torabi SF, Wu P, McGhee CE, Chen L, Hwang K, Zheng N, Cheng J, Lu Y (May 2015). "In vitro selection of a sodium-specific DNAzyme and its application in intracellular sensing". *Proceedings of the National Academy of Sciences of the United States of America.* **112** (19): 5903–8. PMC 4434688. PMID 25918425. doi:10.1073/pnas.1420361112.

[181] Baldi P, Brunak S (2001). *Bioinformatics: The Machine Learning Approach.* MIT Press. ISBN 978-0-262-02506-5. OCLC 45951728.

[182] Gusfield, Dan. *Algorithms on Strings, Trees, and Sequences: Computer Science and Computational Biology.* Cambridge University Press, 15 January 1997. ISBN 978-0-521-58519-4.

[183] Sjölander K (January 2004). "Phylogenomic inference of protein molecular function: advances and challenges". *Bioinformatics.* **20** (2): 170–9. PMID 14734307. doi:10.1093/bioinformatics/bth021.

[184] Mount DM (2004). *Bioinformatics: Sequence and Genome Analysis* (2 ed.). Cold Spring Harbor, NY: Cold Spring Harbor Laboratory Press. ISBN 0-87969-712-1. OCLC 55106399.

[185] Rothemund PW (March 2006). "Folding DNA to create nanoscale shapes and patterns". *Nature*. **440** (7082): 297–302. Bibcode:2006Natur.440..297R. PMID 16541064. doi:10.1038/nature04586.

[186] Andersen ES, Dong M, Nielsen MM, Jahn K, Subramani R, Mamdouh W, Golas MM, Sander B, Stark H, Oliveira CL, Pedersen JS, Birkedal V, Besenbacher F, Gothelf KV, Kjems J (May 2009). "Self-assembly of a nanoscale DNA box with a controllable lid". *Nature*. **459** (7243): 73–6. Bibcode:2009Natur.459...73A. PMID 19424153. doi:10.1038/nature07971.

[187] Ishitsuka Y, Ha T (May 2009). "DNA nanotechnology: a nanomachine goes live". *Nature Nanotechnology*. **4** (5): 281–2. Bibcode:2009NatNa...4..281I. PMID 19421208. doi:10.1038/nnano.2009.101.

[188] Aldaye FA, Palmer AL, Sleiman HF (September 2008). "Assembling materials with DNA as the guide". *Science*. **321** (5897): 1795–9. Bibcode:2008Sci...321.1795A. PMID 18818351. doi:10.1126/science.1154533.

[189] Wray GA (2002). "Dating branches on the tree of life using DNA". *Genome Biology*. **3** (1): REVIEWS0001. PMC 150454. PMID 11806830. doi:10.1046/j.1525-142X.1999.99010.x.

[190] *Lost Tribes of Israel*, Nova, PBS airdate: 22 February 2000. Transcript available from PBS.org. Retrieved 4 March 2006.

[191] Kleiman, Yaakov. "The Cohanim/DNA Connection: The fascinating story of how DNA studies confirm an ancient biblical tradition". aish.com (13 January 2000). Retrieved 4 March 2006.

[192] Goldman N, Bertone P, Chen S, Dessimoz C, LeProust EM, Sipos B, Birney E (February 2013). "Towards practical, high-capacity, low-maintenance information storage in synthesized DNA". *Nature*. **494** (7435): 77–80. Bibcode:2013Natur.494...77G. PMC 3672958. PMID 23354052. doi:10.1038/nature11875.

[193] Naik, Gautam (24 January 2013). "Storing Digital Data in DNA". *Wall Street Journal*. Retrieved 24 January 2013.

[194] Comment by Dandekar, T., Lopez, D., Schaack, D. (2013) http://www.nature.com/nature/journal/v494/n7435/abs/nature11875.html#comment-57415

[195] Emerging Technology Final, Dandekar T., Lopez, D., Programmable bacterial membranes with active DNA storage; presentation for the University of Würzburg for the Royal Society for Chemistry, London, 2016-06-29

[196] Patent "Molecular highly integrated data storage via active control DNA", DE102013004584 A1, https://www.google.com/patents/DE102013004584A1"

[197] Miescher, Friedrich (1871) "Ueber die chemische Zusammensetzung der Eiterzellen" (On the chemical composition of pus cells), *Medicinisch-chemische Untersuchungen*, **4** : 441–460. From p. 456: "*Ich habe mich daher später mit meinen Versuchen an die ganzen Kerne gehalten, die Trennung der Körper, die ich einstweilen ohne weiteres Präjudiz als lösliches und unlösliches Nuclein bezeichnen will, einem günstigeren Material überlassend.*" (Therefore, in my experiments I subsequently limited myself to the whole nucleus, leaving to a more favorable material the separation of the substances, that for the present, without further prejudice, I will designate as soluble and insoluble nuclear material ("Nuclein").)

[198] Dahm R (January 2008). "Discovering DNA: Friedrich Miescher and the early years of nucleic acid research". *Human Genetics*. **122** (6): 565–81. PMID 17901982. doi:10.1007/s00439-007-0433-0.

[199] See:

- Albrect Kossel (1879) "Ueber Nucleïn der Hefe" (On nuclein in yeast) *Zeitschrift für physiologische Chemie*, **3** : 284–291.
- Albrect Kossel (1880) "Ueber Nucleïn der Hefe II" (On nuclein in yeast, Part 2) *Zeitschrift für physiologische Chemie*, **4** : 290–295.
- Albrect Kossel (1881) "Ueber die Verbreitung des Hypoxanthins im Thier- und Pflanzenreich" (On the distribution of hypoxanthins in the animal and plant kingdoms) *Zeitschrift für physiologische Chemie*, **5** : 267–271.
- Albrect Kossel, *Untersuchungen über die Nucleine und ihre Spaltungsprodukte* [Investigations into nuclein and its cleavage products] (Strassburg, Germany: K.J. Trübne, 1881), 19 pages.
- Albrect Kossel (1882) "Ueber Xanthin und Hypoxanthin" (On xanthin and hypoxanthin), *Zeitschrift für physiologische Chemie*, **6** : 422–431.
- Albrect Kossel (1883) "Zur Chemie des Zellkerns" (On the chemistry of the cell nucleus), *Zeitschrift für physiologische Chemie*, **7** : 7–22.
- Albrect Kossel (1886) "Weitere Beiträge zur Chemie des Zellkerns" (Further contributions to the chemistry of the cell nucleus), *Zeitschrift für Physiologische Chemie*, **10** : 248–264. Available on-line at: Max Planck Institute for the History of Science, Berlin, Germany. On p. 264, Kossel remarked presciently: "*Der Erforschung der quantitativen Verhältnisse der vier stickstoffreichen Basen, der Abhängigkeit ihrer Menge von den physiologischen Zuständen der Zelle, verspricht wichtige Aufschlüsse über die elementaren physiologisch-chemischen Vorgänge.*" (The study of the quantitative relations of the four nitrogenous bases —[and] of the dependence of their quantity on the physiological states of the cell —promises important insights into the fundamental physiological-chemical processes.)

[200] Jones ME (September 1953). "Albrecht Kossel, a biographical sketch". *The Yale Journal of Biology and Medicine*.

National Center for Biotechnology Information. **26** (1): 80–97. PMC 2599350. PMID 13103145.

[201] Levene P (1 December 1919). "The structure of yeast nucleic acid". *J Biol Chem*. **40** (2): 415–24.

[202] See:
- W. T. Astbury and Florence O. Bell (1938) "Some recent developments in the X-ray study of proteins and related structures," *Cold Spring Harbor Symposia on Quantitative Biology*, **6** : 109–121. Available on-line at: University of Leeds.
- Astbury, W. T., (1947) "X-ray studies of nucleic acids," *Symposia of the Society for Experimental Biology*, **1** : 66–76. Available on-line at: Oregon State University.

[203] Koltsov proposed that a cell's genetic information was encoded in a long chain of amino acids. See:
- Н. К. Кольцов, "Физико-химические основы морфологии" (The physical-chemical basis of morphology) – speech given at the 3rd All-Union Meeting of Zoologist, Anatomists, and Histologists at Leningrad, U.S.S.R., December 12, 1927.
- Reprinted in: *Успехи экспериментальной биологии* (Advances in Experimental Biology), series B, 7 (1) : ?-? (1928).
- Reprinted in German as: Nikolaj K. Koltzoff (1928) "Physikalisch-chemische Grundlagen der Morphologie" (The physical-chemical basis of morphology), *Biologisches Zentralblatt*, **48** (6) : 345–369.
- In 1934, Koltsov contended that the proteins that contain a cell's genetic information replicate. See: N. K. Koltzoff (October 5, 1934) "The structure of the chromosomes in the salivary glands of Drosophila," *Science*, **80** (2075) : 312–313. From page 313: "I think that the size of the chromosomes in the salivary glands [of Drosophila] is determined through the multiplication of *genonemes*. By this term I designate the axial thread of the chromosome, in which the geneticists locate the linear combination of genes; … In the normal chromosome there is usually only one genoneme; before cell-division this genoneme has become divided into two strands."

[204] Soyfer VN (September 2001). "The consequences of political dictatorship for Russian science". *Nature Reviews. Genetics*. **2** (9): 723–9. PMID 11533721. doi:10.1038/35088598.

[205] Griffith F (January 1928). "The Significance of Pneumococcal Types". *The Journal of Hygiene*. **27** (2): 113–59. PMC 2167760. PMID 20474956. doi:10.1017/S0022172400031879.

[206] Lorenz MG, Wackernagel W (September 1994). "Bacterial gene transfer by natural genetic transformation in the environment". *Microbiological Reviews*. **58** (3): 563–602. PMC 372978. PMID 7968924.

[207] Avery OT, Macleod CM, McCarty M (February 1944). "Studies on the Chemical Nature of the Substance Inducing Transformation of Pneumococcal Types: Induction of Transformation by a Desoxyribonucleic Acid Fraction Isolated from Pneumococcus Type III". *The Journal of Experimental Medicine*. **79** (2): 137–58. PMC 2135445. PMID 19871359. doi:10.1084/jem.79.2.137.

[208] Hershey AD, Chase M (May 1952). "Independent functions of viral protein and nucleic acid in growth of bacteriophage". *The Journal of General Physiology*. **36** (1): 39–56. PMC 2147348. PMID 12981234. doi:10.1085/jgp.36.1.39.

[209] The B-DNA X-ray pattern on the right of this linked image was obtained by Rosalind Franklin and Raymond Gosling in May 1952 at high hydration levels of DNA and it has been labeled as "Photo 51"

[210] Regis, Ed (2009) *What Is Life?: investigating the nature of life in the age of synthetic biology*. Oxford: Oxford University Press ISBN 0-19-538341-9; p. 52

[211] Nature Archives Double Helix of DNA: 50 Years

[212] "Original X-ray diffraction image". Osulibrary.oregonstate.edu. Retrieved 6 February 2011.

[213] The Nobel Prize in Physiology or Medicine 1962 Nobelprize.org Accessed 22 December 06

[214] Maddox B (January 2003). "The double helix and the 'wronged heroine'" (PDF). *Nature*. **421** (6921): 407–8. Bibcode:2003Natur.421..407M. PMID 12540909. doi:10.1038/nature01399.

[215] Crick, F.H.C. On degenerate templates and the adaptor hypothesis (PDF). Archived 1 October 2008 at the Wayback Machine. genome.wellcome.ac.uk (Lecture, 1955). Retrieved 22 December 2006.

[216] Meselson M, Stahl FW (July 1958). "THE REPLICATION OF DNA IN ESCHERICHIA COLI". *Proceedings of the National Academy of Sciences of the United States of America*. **44** (7): 671–82. Bibcode:1958PNAS...44..671M. PMC 528642. PMID 16590258. doi:10.1073/pnas.44.7.671.

[217] The Nobel Prize in Physiology or Medicine 1968 Nobelprize.org Accessed 22 December 06

[218] Pray, L. (2008) Discovery of DNA structure and function: Watson and Crick. Nature Education 1(1):100

5.11 Further reading

- Berry A, Watson J (2003). *DNA: the secret of life*. New York: Alfred A. Knopf. ISBN 0-375-41546-7.

- Calladine CR, Drew HR, Luisi BF, Travers AA (2003). *Understanding DNA: the molecule & how it works.* Amsterdam: Elsevier Academic Press. ISBN 0-12-155089-3.

- Carina D, Clayton J (2003). *50 years of DNA.* Basingstoke: Palgrave Macmillan. ISBN 1-4039-1479-6.

- Judson HF (1979). *The Eighth Day of Creation: Makers of the Revolution in Biology* (2nd ed.). Cold Spring Harbor Laboratory Press. ISBN 0-671-22540-5.

- Olby RC (1994). *The path to the double helix: the discovery of DNA.* New York: Dover Publications. ISBN 0-486-68117-3., first published in October 1974 by MacMillan, with foreword by Francis Crick; the definitive DNA textbook, revised in 1994 with a 9-page postscript

- Micklas D (2003). *DNA Science: A First Course.* Cold Spring Harbor Press. ISBN 978-0-87969-636-8.

- Ridley M (2006). *Francis Crick: discoverer of the genetic code.* Ashland, OH: Eminent Lives, Atlas Books. ISBN 0-06-082333-X.

- Olby RC (2009). *Francis Crick: A Biography.* Plainview, N.Y: Cold Spring Harbor Laboratory Press. ISBN 0-87969-798-9.

- Rosenfeld I (2010). *DNA: A Graphic Guide to the Molecule that Shook the World.* Columbia University Press. ISBN 978-0-231-14271-7.

- Schultz M, Cannon Z (2009). *The Stuff of Life: A Graphic Guide to Genetics and DNA.* Hill and Wang. ISBN 0-8090-8947-5.

- Stent GS, Watson J (1980). *The Double Helix: A Personal Account of the Discovery of the Structure of DNA.* New York: Norton. ISBN 0-393-95075-1.

- Watson, James (2004). *DNA: The Secret of Life.* Random House. ISBN 978-0-09-945184-6.

- Wilkins M (2003). *The third man of the double helix the autobiography of Maurice Wilkins.* Cambridge, England: University Press. ISBN 0-19-860665-6.

5.12 External links

- DNA at DMOZ

- DNA binding site prediction on protein

- DNA the Double Helix Game From the official Nobel Prize web site

- DNA under electron microscope

- Dolan DNA Learning Center

- Double Helix: 50 years of DNA, *Nature*

- *Proteopedia DNA*

- *Proteopedia Forms_of_DNA*

- ENCODE threads explorer ENCODE home page. Nature (journal)

- Double Helix 1953–2003 National Centre for Biotechnology Education

- Genetic Education Modules for Teachers—*DNA from the Beginning* Study Guide

- PDB Molecule of the Month *DNA*

- Clue to chemistry of heredity found The New York Times June 1953. First American newspaper coverage of the discovery of the DNA structure

- Olby R (January 2003). "Quiet debut for the double helix". *Nature.* **421** (6921): 402–5. Bibcode:2003Natur.421..402O. PMID 12540907. doi:10.1038/nature01397.

- DNA from the Beginning Another DNA Learning Center site on DNA, genes, and heredity from Mendel to the human genome project.

- The Register of Francis Crick Personal Papers 1938 – 2007 at Mandeville Special Collections Library, University of California, San Diego

- Seven-page, handwritten letter that Crick sent to his 12-year-old son Michael in 1953 describing the structure of DNA. See Crick's medal goes under the hammer, Nature, 5 April 2013.

Chapter 6

Palm print

A **palm print** refers to an image acquired of the palm region of the hand. It can be either an online image (i.e. taken by a scanner or CCD) or offline image where the image is taken with ink and paper.*[1]

The palm itself consists of principal lines, wrinkles (secondary lines), and epidermal ridges. It differs to a fingerprint in that it also contains other information such as texture, indents and marks which can be used when comparing one palm to another.

Palm prints can be used for criminal, forensic, or commercial applications. Palm prints, typically from the butt of the palm, are often found at crime scenes as the result of the offender's gloves slipping during the commission of the crime, and thus exposing part of the unprotected hand.*[2]*[3]

6.1 References

Media related to Palm prints at Wikimedia Commons

[1] Zhang, D. (2004). *Palmprint Authentication*, Kluwer Academic Publishers.

[2] Fisher, Barry A.J. *Techniques of Crime Scene Investigation*. Boca Raton, CRC Press. 2004. ISBN 0-8493-1691-X

[3] "FBI —Forensic Spotlight: A New Investigative Biometric Service: The National Palm Print System". *FBI*.

Chapter 7

Hand geometry

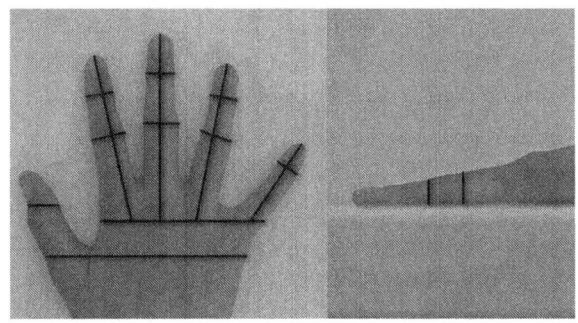

Geometry of a hand and some examples of measurements that can be taken by hand geometry reading devices.[1]

Hand geometry is a biometric that identifies users by the shape of their hands. Hand geometry readers measure a user's hand along many dimensions and compare those measurements to measurements stored in a file.

Viable hand geometry devices have been manufactured since the early 1980s, making hand geometry the first biometric to find widespread computerized use. It remains popular; common applications include access control and time-and-attendance operations.

Since hand geometry is not thought to be as unique as fingerprints, palm veins or irises, fingerprinting, palm veins and iris recognition remain the preferred technology for high-security applications. Hand geometry is very reliable when combined with other forms of identification, such as identification cards or personal identification numbers. In large populations, hand geometry is not suitable for so-called one-to-many applications, in which a user is identified from his biometric without any other identification.

7.1 See also

- Biometrics in schools
- Biometric technology in access control
- INSPASS

A hand geometry reading device with pegs to control the placement of the hand. Angled mirror on the left reflects the side view image of the hand to the camera. A CCD camera is beneath the keypad to take the top view image of the hand and the mirror image.[2]

7.2 References

[1] "Hand Geometry" (PDF). National Science and Technology Council (US). 7 August 2006. Retrieved 28 November 2013.

[2] Miroslav Bača; Petra Grd; Tomislav Fotak (2012). "4: Basic Principles and Trends in Hand Geometry and Hand Shape Biometrics". *New Trends and Developments in Biometrics* (PDF). InTech. Retrieved 1 December 2013.

- Varchol, Peter, and Dusan Levicky. "Using of

hand geometry in biometric security systems." RADIOENGINEERING-PRAGUE- 16.4 (2007): 82. pdf

- Varchol, P., D. Levický, and J. Juhar. "Multimodal biometric authentication using speech and hand geometry fusion." Systems, Signals and Image Processing, 2008. IWSSIP 2008. 15th International Conference on. IEEE, 2008. IEEE

Chapter 8

Iris recognition

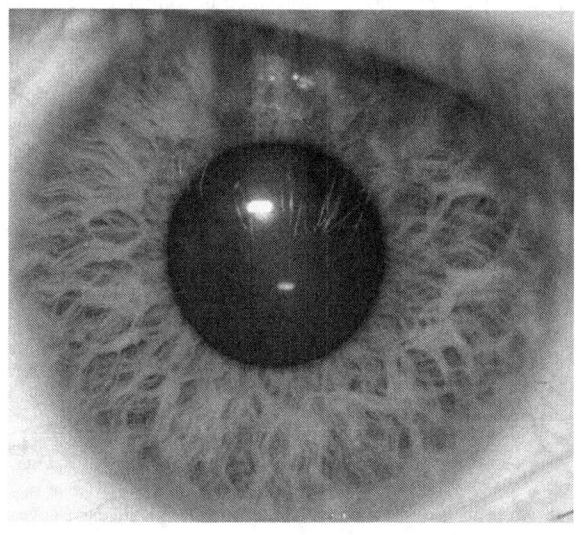

Iris recognition biometric systems apply mathematical pattern-recognition techniques to images of the irises of an individual's eyes.

Iris recognition is an automated method of biometric identification that uses mathematical pattern-recognition techniques on video images of one or both of the irises of an individual's eyes, whose complex patterns are unique, stable, and can be seen from some distance.

Retinal scanning is a different, ocular-based biometric technology that uses the unique patterns on a person's retina blood vessels and is often confused with iris recognition. Iris recognition uses video camera technology with subtle near infrared illumination to acquire images of the detail-rich, intricate structures of the iris which are visible externally. Digital templates encoded from these patterns by mathematical and statistical algorithms allow the identification of an individual or someone pretending to be that individual.*[1] Databases of enrolled templates are searched by matcher engines at speeds measured in the millions of templates per second per (single-core) CPU, and with remarkably low false match rates.

Several hundred million persons in several countries around the world have been enrolled in iris recognition systems for convenience purposes such as passport-free automated border-crossings and some national ID programs. A key advantage of iris recognition, besides its speed of matching and its extreme resistance to false matches, is the stability of the iris as an internal and protected, yet externally visible organ of the eye.

8.1 History

Although John Daugman developed and patented the first actual algorithms to perform iris recognition, published the first papers about it and gave the first live demonstrations, the concept behind this invention has a much longer history and today it benefits from many other active scientific contributors. In a 1953 clinical textbook, F.H. Adler*[2] wrote: "In fact, the markings of the iris are so distinctive that it has been proposed to use photographs as a means of identification, instead of fingerprints." Adler referred to comments by the British ophthalmologist J.H. Doggart,*[3] who in 1949 had written that: "Just as every human being has different fingerprints, so does the minute architecture of the iris exhibit variations in every subject examined. [Its features] represent a series of variable factors whose conceivable permutations and combinations are almost infinite." Later in the 1980s, two American ophthalmologists, L. Flom and A. Safir managed to patent Adler's and Doggart's conjecture that the iris could serve as a human identifier, but they had no actual algorithm or implementation to perform it and so their patent remained conjecture. The roots of this conjecture stretch back even further: in 1892 the Frenchman A. Bertillon had documented nuances in "Tableau de l'iris humain". Divination of all sorts of things based on iris patterns goes back to ancient Egypt, to Chaldea in Babylonia, and to ancient Greece, as documented in stone inscriptions, painted ceramic artefacts, and the writings of Hippocrates. (Iris divination persists today, as "iridology.")

The core theoretical idea in Daugman's algorithms is that the failure of a test of statistical independence can be a very strong basis for pattern recognition, if there is sufficiently high entropy (enough degrees-of-freedom of random variation) among samples from different classes. In 1994 he patented this basis for iris recognition and its underlying computer vision algorithms for image processing, feature extraction, and matching, and published them in a paper.*[4] These algorithms became widely licensed through a series of companies: IriScan (a start-up founded by Flom, Safir, and Daugman), Iridian, Sarnoff, Sensar, LG-Iris, Panasonic, Oki, BI2, IrisGuard, Unisys, Sagem, Enschede, Securimetrics and L-1, now owned by French company Morpho.

With various improvements over the years, these algorithms remain today the basis of all significant public deployments of iris recognition, and they are consistently top performers in NIST tests (implementations submitted by L-1, MorphoTrust and Morpho, for whom Daugman serves as Chief Scientist for Iris Recognition). But research on many aspects of this technology and on alternative methods has exploded, and today there is a rapidly growing academic literature on optics, photonics, sensors, biology, genetics, ergonomics, interfaces, decision theory, coding, compression, protocol, security, mathematical and hardware aspects of this technology.

Most flagship deployments of these algorithms have been at airports, in lieu of passport presentation, and for security screening using watch-lists. In the early years of this century, major deployments began at Amsterdam's Schiphol Airport and at ten UK airport terminals allowing frequent travellers to present their iris instead of their passport, in a programme called IRIS: Iris Recognition Immigration System. Similar systems exist along the US / Canada border, and many others. In the United Arab Emirates, all 32 air, land, and seaports deploy these algorithms to screen all persons entering the UAE requiring a visa. Because a large watch-list compiled among GCC States is exhaustively searched each time, the number of iris cross-comparisons climbed to 62 trillion in 10 years. The Government of India is enrolling the iris codes (as well as fingerprints) of all 1.2 billion citizens within three years for national ID and fraud prevention in entitlements distribution. As of April 2016 the UIDAI (Unique Identification Authority of India) had enrolled more than 1 billion persons in this biometric programme. Iris is one of three biometric identification technologies internationally standardised since 2006 by ICAO for use in e-passports (the other two are fingerprint and face recognition).*[5]

8.2 Visible vs near infrared imaging

All publicly deployed iris recognition systems acquire images of an iris while being illuminated by light in the near infrared wavelength band (NIR: 700–900 nm) of the electromagnetic spectrum. The majority of persons worldwide have "dark brown eyes", the dominant phenotype of the human population, revealing less visible texture in the VW band but appearing richly structured, like the cratered surface of the moon, in the NIR band. (Some examples are shown here.) Using the NIR spectrum also enables the blocking of corneal specular reflections from a bright ambient environment, by allowing only those NIR wavelengths from the narrow-band illuminator back into the iris camera.

Iris melanin, also known as chromophore, mainly consists of two distinct heterogeneous macromolecules, called eumelanin (brown–black) and pheomelanin (yellow–reddish),*[6]*[7] whose absorbance at longer wavelengths in the NIR spectrum is negligible. At shorter wavelengths within the VW spectrum, however, these chromophores are excited and can yield rich patterns. Hosseini, *et al.**[8] provide a comparison between these two imaging modalities. An alternative feature extraction method to encode VW iris images was also introduced, which may offer an alternative approach for multi-modal biometric systems.

8.3 Operating principle

A, now obsolete, IriScan model 2100 iris recognition camera.

First the system has to localize the inner and outer boundaries of the iris (pupil and limbus) in an image of an eye.

Further subroutines detect and exclude eyelids, eyelashes, and specular reflections that often occlude parts of the iris. The set of pixels containing only the iris, normalized by a rubber-sheet model to compensate for pupil dilation or constriction, is then analyzed to extract a bit pattern encoding the information needed to compare two iris images.

In the case of Daugman's algorithms, a Gabor wavelet transform is used. The result is a set of complex numbers that carry local amplitude and phase information about the iris pattern. In Daugman's algorithms, most amplitude information is discarded, and the 2048 bits representing an iris pattern consist of phase information (complex sign bits of the Gabor wavelet projections). Discarding the amplitude information ensures that the template remains largely unaffected by changes in illumination or camera gain, and contributes to the long-term usability of the biometric template.

For identification (one-to-many template matching) or verification (one-to-one template matching),[9] a template created by imaging an iris is compared to stored template(s) in a database. If the Hamming distance is below the decision threshold, a positive identification has effectively been made because of the statistical extreme improbability that two different persons could agree by chance ("collide") in so many bits, given the high entropy of iris templates.

8.4 Advantages

The iris of the eye has been described as the ideal part of the human body for biometric identification for several reasons:

It is an internal organ that is well protected against damage and wear by a highly transparent and sensitive membrane (the cornea). This distinguishes it from fingerprints, which can be difficult to recognize after years of certain types of manual labor. The iris is mostly flat, and its geometric configuration is only controlled by two complementary muscles (the sphincter pupillae and dilator pupillae) that control the diameter of the pupil. This makes the iris shape far more predictable than, for instance, that of the face.

The iris has a fine texture that—like fingerprints—is determined randomly during embryonic gestation. Like the fingerprint, it is very hard (if not impossible) to prove that the iris is unique. However, there are so many factors that go into the formation of these textures (the iris and fingerprint) that the chance of false matches for either is extremely low. Even genetically identical individuals (and the left and right eyes of the same individual) have completely independent iris textures. An iris scan is similar to taking a photograph and can be performed from about 10 cm to a few meters away. There is no need for the person being identified to touch any equipment that has recently been touched by a stranger, thereby eliminating an objection that has been raised in some cultures against fingerprint scanners, where a finger has to touch a surface, or retinal scanning, where the eye must be brought very close to an eyepiece (like looking into a microscope).[10]

The commercially deployed iris-recognition algorithm, John Daugman's IrisCode, has an unprecedented false match rate (better than $10^{*}-11$ if a Hamming distance threshold of 0.26 is used, meaning that up to 26% of the bits in two IrisCodes are allowed to disagree due to imaging noise, reflections, etc., while still declaring them to be a match).[11] While there are some medical and surgical procedures that can affect the colour and overall shape of the iris, the fine texture remains remarkably stable over many decades. Some iris identifications have succeeded over a period of about 30 years.

Iris recognition works with clear contact lenses, eyeglasses, and non-mirrored sunglasses.

8.5 Shortcomings

Many commercial iris scanners can be easily fooled by a high quality image of an iris or face in place of the real thing.[12] The scanners are often tough to adjust and can become bothersome for multiple people of different heights to use in succession. The accuracy of scanners can be affected by changes in lighting. Iris scanners are significantly more expensive than some other forms of biometrics, as well as password and proximity card security systems.

Iris scanning is a relatively new technology and is incompatible with the very substantial investment that the law enforcement and immigration authorities of some countries have already made into fingerprint recognition. Iris recognition is very difficult to perform at a distance larger than a few meters and if the person to be identified is not cooperating by holding the head still and looking into the camera. However, several academic institutions and biometric vendors are developing products that claim to be able to identify subjects at distances of up to 10 meters ("Standoff Iris" or "Iris at a Distance" as well as SRI International's "Iris on the Move" for persons walking at speeds up to 1 meter/sec).[10][13]

As with other photographic biometric technologies, iris recognition is susceptible to poor image quality, with associated failure to enroll rates. As with other identification infrastructure (national residents databases, ID cards, etc.), civil rights activists have voiced concerns that iris-recognition technology might help governments to track individuals beyond their will. Researchers have tricked iris scanners using images generated from digital codes of stored irises. Criminals could exploit this flaw to steal the identities of others.[14]

The first study on surgical patients involved modern cataract surgery and showed that it can change iris texture in such a way that iris pattern recognition is no longer feasible or the probability of falsely rejected subjects is increased.*[15]

8.6 Security considerations

As with most other biometric identification technology, a still not satisfactorily solved problem with iris recognition is the problem of live-tissue verification. The reliability of any biometric identification depends on ensuring that the signal acquired and compared has actually been recorded from a live body part of the person to be identified and is not a manufactured template. Many commercially available iris-recognition systems are easily fooled by presenting a high-quality photograph of a face instead of a real face, which makes such devices unsuitable for unsupervised applications, such as door access-control systems. The problem of live-tissue verification is less of a concern in supervised applications (e.g., immigration control), where a human operator supervises the process of taking the picture.

Methods that have been suggested to provide some defence against the use of fake eyes and irises include changing ambient lighting during the identification (switching on a bright lamp), such that the pupillary reflex can be verified and the iris image be recorded at several different pupil diameters; analysing the 2D spatial frequency spectrum of the iris image for the peaks caused by the printer dither patterns found on commercially available fake-iris contact lenses; analysing the temporal frequency spectrum of the image for the peaks caused by computer displays.

Other methods include using spectral analysis instead of merely monochromatic cameras to distinguish iris tissue from other material; observing the characteristic natural movement of an eyeball (measuring nystagmus, tracking eye while text is read, etc.); testing for retinal retroreflection (red-eye effect) or for reflections from the eye's four optical surfaces (front and back of both cornea and lens) to verify their presence, position and shape. Another proposed method is to use 3D imaging (e.g., stereo cameras) to verify the position and shape of the iris relative to other eye features.

A 2004 report by the German Federal Office for Information Security noted that none of the iris-recognition systems commercially available at the time implemented any live-tissue verification technology. Like any pattern-recognition technology, live-tissue verifiers will have their own false-reject probability and will therefore further reduce the overall probability that a legitimate user is accepted by the sensor.

8.7 Deployed applications

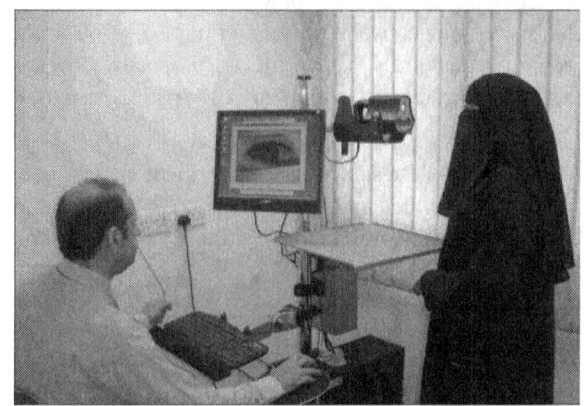

IrisGuard Inc. UAE Enrollment Station

- United Arab Emirates IrisGuard's Homeland Security Border Control has been operating an expellee tracking system in the United Arab Emirates (UAE) since 2003, when the UAE launched a national border-crossing security initiative. Today, all of the UAE's land, air and sea ports of entry are equipped with systems. All foreign nationals who possess a visit visa to enter the UAE are processed through iris cameras installed at all primary and auxiliary immigration inspection points. To date, the system has apprehended over 330,000 persons re-entering the UAE with either another nationality or name, or even fraudulent travel documents.*[16]*[17]

IrisGuard Inc. First Cash Withdrawal on Iris Enabled ATM

- Hashemite Kingdom of Jordan - 2009, IrisGuard deployed the world's first operational iris enabled Automated Teller Machine(ATM) at Cairo Amman Bank, where bank customers can seamlessly withdraw cash from ATM's without a bank card or pin but simply by presenting their eye to the iris recognition camera on the ATM. Since June 2012, IrisGuard is also providing financial inclusion to UNHCR registered Syrian refugees in Jordan on ATM's. The system is designed to facilitate cash-supported interventions that help deliver financial assistance to refugees with speed and dignity while lowering overhead costs and boosting accountability.*[18]

- Aadhaar began operation in 2011 in India, whose government is enrolling the iris patterns (and other biometrics) of all 1.2 billion citizens for the Aadhaar scheme for entitlements distribution, run by the Unique Identification Authority of India (UIDAI).*[19] This programme enrolls about one million persons every day, across 36,000 stations operated by 83 agencies. By October 2015, the number of persons enrolled exceeded 926 million, with each new enrollee being compared to all existing ones for de-duplication checks (hence 926 trillion, i.e. 926 million-million, iris cross-comparisons per day).*[20] Its purpose is to issue each citizen a biometrically provable unique entitlement number (Aadhaar) by which benefits may be claimed, and social inclusion enhanced; thus the slogan of UIDAI is: "To give the poor an identity." Iris technology providers must be granted a STQC (Standardisation Testing and Quality Certification) certificate in order to supply iris scanners for the project. By far, there are providers such as: IriTech Inc. (dual iris scanner IriMagic 100BK), Cogent (CIS-202), Iris ID (icam TD 100), Iris Guard (IG-AD-100), etc.*[21]

- Police forces across America planned to start using BI2 Technologies' mobile MORIS (Mobile Offender Recognition and Information System) in 2012. The New York City Police Department was the first, with a system installed in Manhattan in the fall of 2010.*[22]

- Iris recognition technology has been implemented by BioID Technologies SA in Pakistan for UNHCR repatriation project to control aid distribution for Afghan refugees. Refugees are repatriated by UNHCR in co-operation with Government of Pakistan, and they are paid for their travel. To make sure people do not get paid more than once, their irises are scanned, and the system will detect the refugees on next attempt. The database has more than 1.3 million iris code templates and around 4000 registrations per day. The one-to-many iris comparison takes place within 1.5 seconds against 1.3 million iris codes.

- In early 2013, United Nation High Commissioner for Refugees (UNHCR) also installed a new biometrics identity management system (BIMS) by IriTech Inc. for the refugees in the Malawi Camp. During the pilot program, which lasted four weeks, more than 17,000 people enrolled their iris biometric data and had their identities verified. After the successful pilot in Malawi, Thailand was recently chosen to be the first site of the global roll-out. After 5 months, in June 2015, UNHCR has completed its registration for nearly 110,000 Myanmar refugees in Thailand's border camps with the help of the new system.*[23]

- At Amsterdam Airport Schiphol, Netherlands, iris recognition has permitted expedited, passport-free border security passing since 2001 through the Privium program.*[24]

- Canadian Air Transport Security Authority's Restricted Area Identity Card (RAIC) program is the world's first dual-biometric program deployed around major Canadian airports for staff and aircrews to access the restricted areas using separate channels from passengers.*[25]*[26]

- In a number of Canadian airports, as part of the NEXUS program that facilitates entry into the US and Canada for pre-approved, low-risk travelers.

- In several Canadian airports, as part of the CANPASS Air program that facilitates entry into Canada for pre-approved, low-risk air travelers.*[27]

A U.S. Marine Corps Sergeant uses an iris scanner to positively identify a member of the Baghdadi city council prior to a meeting with local tribal leaders, sheiks, community leaders and U.S. service members.

- UK's Iris Recognition Immigration System, which started operating in 2004 but which was closed to new registrations in 2011 and which has been phased out in 2012 and 2013.*[28]*[29]*[30]

- Used in 2002 to verify the recognition of the "Afghan Girl" (Sharbat Gula) by *National Geographic* photographer Steve McCurry.*[31]

- Since at least 2011, Google uses iris scanners to control access to their datacentres.*[32]

- In 2010, Leon, Mexico, deployed iris scanners in public spaces, that can identify up to fifty people at once.*[33]

- On May 10, 2011, Hoyos Group demonstrated a device called EyeLock using iris-recognition as an alternative to passwords to log people into password-protected Web sites and applications, like Facebook or eBay.*[34]

- SRI International Sarnoff has been developing an "Iris on the Move" system and set of products, primarily for U.S. Government clients, capable of identifying 30 people per minute.*[13] Most recently, they have specialized in a product where drivers can be identified without needing to leave their vehicle.*[35]

- M2SYS Technology has deployed their RightPatient™ biometric patient identification system using iris recognition at 11 Novant Health hospitals in the Charlotte and Winston-Salem markets. The RightPatient™ iris biometric patient identification system is designed to capture both the face and the iris pattern of patients and uniquely link them to their electronic medical record.*[36]

- In March 2015, India's Andhra Pradesh state has launched an iris-based identity management solution developed by IriTech for enhancing pension distribution system. The Chief Minister N. Chandrababu Naidu demonstrated IriShield USB MK2120U device during a launching event of Andhra Pradesh state's iris scanning facility for pension distribution. "The state's decision to use iris technology as a primary method to issue Aadhaar verified DBT (Direct Benefit Transfer) will address concerns of total inclusiveness of its residence as well as providing a more accurate and a hygiene solution," says Binod E. Mathai, Director of Biometronic Technology.*[37]

- On May 28, 2015, Fujitsu released ARROWS NX F-04G the first smartphone with iris scanner.*[38]

- On mid 2015, the Kenya Ministry of Education, Science and Technology in order to provide an accurate attendance tracking for all students in classes (roll-call) or school buses (getting on/off tracking) has implemented iris biometric system. The solution includes IriTech's IriShield camera connecting to a low cost Android phone or tablet via USB cable. Iris matching is done on-board of IriShield whose internal gallery can hold up to 500 identities (expandable to 5,000 identities) which is more than enough for most of the schools. The local matching capability is a particular advantage in the school-bus scenario because it does not require wireless/3G communication between the biometric terminal in the bus and a back-end server.

- At the end of 2015, Microsoft launched two Lumia phones (Lumia 950 and Lumia 950 XL) featuring iris scanning as a way to authenticate the user.

8.8 Iris recognition in television and movies

- *I Origins* (2014), a Hollywood film by writer-director Mike Cahill and winner of the Alfred Sloan Award for best exposition of technology (2014 Sundance Film Festival), uses iris recognition for its core plot. Culminating in India with the UIDAI project to encode and enroll the iris patterns of all 1.2 billion Indian citizens by the end of 2015, the film is described as a *"science fiction love story incorporating spiritualism and reincarnation"*, seeking to reconcile science with religious spirit-world beliefs.

- Steven Spielberg's 2002 science fiction film *Minority Report* depicts a society in which what appears to be a form of iris recognition has become daily practice. The principal character undergoes an eye transplant in order to change his identity but continues to use his original eyes to gain access to restricted locations.*[39]

- In *The Island* (2005), a clone character played by Ewan McGregor uses his eye to gain access through a security door in the home of his DNA donor.

- *The Simpsons Movie* (2007) features a scene that illustrates the difficulty of image acquisition in iris recognition.*[40]

- The TV series *Numb3rs,* features a scene where a robber gets into the CalSci facility by cracking the code assigned to a specific iris.

- *NCIS* uses an iris scanner in the garage, where forensic vehicle investigations are carried out and evidence is stored. There is another scanner at the entrance to MTAC. The sequence of Leroy Jethro Gibbs being verified is shown in the title sequence. The imagery for this sequence has been "enhanced" using special effects. Iris recognition systems do not use the laser like beams shown in the sequence and the light that they do use is near-infrared and nearly invisible.

- The 2010 film RED includes a scene where Bruce Willis' character uses a contact lens to pass an iris scan and gain access to CIA headquarters.

8.9 See also

- Biometric technology in access control

- Iris Recognition Immigration System
- Eye vein verification
- Finger Vein recognition
- Fingerprint recognition
- Samsung Galaxy S8

8.10 References

[1] Zetter, Kim (2012-07-25). "Reverse-Engineered Irises Look So Real, They Fool Eye-Scanners". *Wired Magazine*. Retrieved 25 July 2012.

[2] Adler, F.H., Physiology of the Eye (Chapter VI, page 143), Mosby (1953)

[3] Doggart, J.H., Ocular Signs in Slit-Lamp Microscopy, Kimpton (1949), page 27

[4] Daugman, J., "High confidence visual recognition of persons by a test of statistical independence", IEEE Transactions on Pattern Analysis and Machine Intelligence, 15 (11), pp 1148-1161 (1993)

[5] "ICAO Document 9303: Machine Readable Travel Documents, Part 9: Deployment of Biometric Identification and Electronic Storage of Data in eMRTDs, 7th edition" (PDF). 2015.

[6] Liu Y, Simon JD (February 2005). "Metal-ion interactions and the structural organization of Sepia eumelanin". *Pigment Cell Res*. **18** (1): 42–8. PMID 15649151. doi:10.1111/j.1600-0749.2004.00197.x.

[7] Meredith P, Sarna T (December 2006). "The physical and chemical properties of eumelanin". *Pigment Cell Res*. **19** (6): 572–94. PMID 17083485. doi:10.1111/j.1600-0749.2006.00345.x.

[8] Hosseini, M.S.; Araabi, B.N.; Soltanian-Zadeh, H. (April 2010). "Pigment Melanin: Pattern for Iris Recognition". *IEEE Trans Instrum Meas*. **59** (4): 792–804. doi:10.1109/TIM.2009.2037996.

[9] HRSID Iris Recognition | "rapid and reliable searches in both 1 to 1 (verification) and 1 to many (identification)"

[10] Martin, Zach (2011-03-23). "Biometric Trends: Will emerging modalities and mobile applications bring mass adoption?". *SecureIDNews*. Retrieved 2013-07-14.

[11] "Probing the uniqueness and randomness of IrisCodes: Results from 200 billion iris pair comparisons." Proceedings of the IEEE, vol. 94 (11), 2006, pp. 1927-1935.

[12] "Hacker Finds a Simple Way to Fool IRIS Biometric Security Systems". *thehackernews.com*. 2015-03-06. Retrieved 2017-03-17.

[13] Terdiman, Daniel (2011-05-05). "SRI shows the benefits of shrinking tech". CNET. Retrieved 2013-07-14.

[14] "Iris scanners 'can be tricked'".

[15] R. Roizenblatt, P. Schor et al. Iris recognition as a biometric method after cataract surgery. Biomed Eng Online. 2004; 3: 2

[16] "European Business Review - Identification in the Twinkle of the Eye, Oct 7, 2008"

[17] "Behin IRIS (Automated IRIS-Based Identification System)"

[18] "UNHCR Innovation | Biometric Cash Assistance". *innovation.unhcr.org*. Retrieved 2016-11-03.

[19] https://portal.uidai.gov.in/

[20] "Aadhaar - Unique Identification". *portal.uidai.gov.in*. Retrieved 2015-11-02.

[21] "STQC Certificate granted suppliers" (PDF).

[22] "Police to begin iPhone iris scans amid privacy concerns". *Reuters*. 2011-07-20.

[23] "Biometric Identity Management System". *UNHCR*. Retrieved 2015-11-02.

[24] "Iris scans at Amsterdam Airport Schiphol". Amsterdam Airport Schiphol. Retrieved 2013-07-14.

[25] "Restricted area identity card". Archived from the original on May 14, 2012.

[26] "Backgrounder". Canadian Air Transport Security Authority (CATSA).

[27] CANPASS Air

[28] "IRIS". Archived from the original on May 1, 2008.

[29] "Manchester Airport eye scanners scrapped over delays". *Manchester Evening News*.

[30] Wesley Johnson (16 February 2012). "Airport eye scanners to be reviewed". *The Press Association, printed in The Independent*.

[31] Daugman, John. "How the Afghan Girl was Identified by Her Iris Patterns". University of Cambridge. Retrieved 2013-07-14.

[32] "Google data center security". Google. 2011-04-13. Retrieved 2013-07-14.

[33] Saylor, Michael (2012). *The Mobile Wave: How Mobile Intelligence Will Change Everything*. Perseus Books/Vanguard Press. p. 98. ISBN 978-1593157203.

[34] Whitney, Lance (2011-05-12). "Iris recognition gadget eliminates passwords". CNET. Retrieved 2011-05-12.

[35] "SRI International Sarnoff launches iris biometric vehicle access control system". *Biometric Update*. 2013-04-10. Retrieved 2013-07-15.

[36] "Novant's new iris scan connects patients to records". Archived from the original on April 7, 2014.

[37] "Indian state selects iris based identity management solution by IriTech". *BiometricUpdate*. Retrieved 2015-11-02.

[38] http://www.fujitsu.com/global/about/resources/news/press-releases/2015/0525-01.html

[39] Dolmetsch, Chris (2011-02-01). "'Minority Report' May Come to Real World With Iris Recognition". *Bloomberg L.P.* Retrieved 2013-07-14.

[40] Daugman, John. "Iris Recognition and "The Simpsons Movie"". University of Cambridge. Retrieved 2013-07-15.

8.11 Further reading

- WO 8605018 Leonard Flom, Aran Safir: Iris recognition system. 28 August 1986; also: US 4641349 issued 2/3/1987.

- US 5291560 John Daugman: Biometric personal identification system based on iris analysis. 1 March 1994

- Daugman, John (January 2004). "How iris recognition works" (PDF). *IEEE Transactions on Circuits and Systems for Video Technology*. **14** (1): 21–30. doi:10.1109/TCSVT.2003.818350.

- Daugman, John (2003). "The importance of being random: statistical principles of iris recognition" (PDF). *Pattern Recognition*. **36** (2): 279–291. doi:10.1016/S0031-3203(02)00030-4.

- Daugman, John (June 2005). "Results from 200 billion iris cross-comparisons". *Technical Report UCAM-CL-TR-635*. University of Cambridge Computer Laboratory.

- Zhaofeng He; Tieniu Tan; Zhenan Sun; Xianchao Qiu (15 July 2008). "Towards Accurate and Fast Iris Segmentation for Iris Biometrics". *IEEE Trans Pattern Anal Mach Intell*. **31** (9): 1670–84. PMID 19574626. doi:10.1109/TPAMI.2008.183.

- Zhaofeng He; Tieniu Tan; Zhenan Sun; Xianchao Qiu (June 2008). "Boosting Ordinal Features for Accurate and Fast Iris Recognition". *Proc. of the 26th IEEE Computer Society Conference on Computer Vision and Pattern Recognition (CVPR'08)*. pp. 1–8.

- Kaushik Roy; Prabir Bhattacharya (2008). *Iris Recognition: A Machine Learning Approach*. VDM Verlag Dr. Müller. ISBN 3-639-08259-1.

- K. Roy; P. Bhattacharya (2009). "Variational level set method and game theory applied for nonideal iris recognition". *16th IEEE International Conference on Image Processing (ICIP'09)*. pp. 2721–4. ISBN 978-1-4244-5653-6.

8.12 External links

- Home page of IrisCode inventor John Daugman
- ISO/IEC 19794-6:2011 International standard for iris images
- NIST Iris Challenge Evaluation – a contest for competing iris-recognition algorithms
- NIST IREX – a program for various NIST activities supporting interoperable iris biometrics, including comparison of 19 state-of-the-art iris recognition algorithms from 10 providers
- Iris recognition test results, analysis
- Iris recognition algorithm reidentifies Sharbat Gula – Afghan National Geographic cover girl in 1985 – two decades later
- Bath University Iris Image Database
- John G. Daugman's original patent at USPTO
- First large-scale deployments
- Project Iris an Open Source iris recognition system

Chapter 9

Retinal scan

For diagnostic retinal examination, see Fundus photography and Ophthalmoscopy.

A **retinal scan** is a biometric technique that uses the unique patterns on a person's retina blood vessels. It is not to be confused with another ocular-based technology, iris recognition, commonly called an "iris scanner."

9.1 Introduction

A close of the controls of a Topcon retinal camera

The human retina is a thin tissue composed of neural cells that is located in the posterior portion of the eye. Because of the complex structure of the capillaries that supply the retina with blood, each person's retina is unique. The network of blood vessels in the retina is not entirely genetically determined and thus even identical twins do not share a similar pattern.

Although retinal patterns may be altered in cases of diabetes, glaucoma or retinal degenerative disorders, the retina typically remains unchanged from birth until death. Due to its unique and unchanging nature, the retina appears to be the most precise and reliable biometric, aside from DNA.*[1] The National Center for State Courts estimate that retinal scanning has an error rate of one in ten million.*[2]

A retinal scan is performed by casting an unperceived beam of low-energy infrared light into a person's eye as they look through the scanner's eyepiece. This beam of light traces a standardized path on the retina. Because retinal blood vessels absorb light more readily than the surrounding tissue, the amount of reflection varies during the scan. The pattern of variations is digitized and stored in a database.*[3]

9.2 History

The idea for retinal identification was first conceived by Dr Carleton Simon and Dr Isadore Goldstein and was published in the New York State Journal of Medicine in 1935.*[4] The idea was ahead of its time, but once technology caught up, the concept for a retinal scanning device emerged in 1975. In 1976, Robert "Buzz" Hill formed a corporation named EyeDentify, Inc., and made a full-time effort to research and develop such a device. In 1978, specific means for a retinal scanner was patented, followed by a commercial model in 1981.

9.3 In popular culture

The relative obscurity and "high tech" nature of retinal scans means that they are a frequent device in fiction to suggest that an area has been particularly strongly secured against intrusion. Some notable examples include:

In the movie *Star Trek II: The Wrath of Khan* (1982), Admiral Kirk gains access to top secret computer files by use of a retinal scan.

In the movie *Batman* (1966), Batman describes to Robin how the tiny vessels in the retina are unique to the individual and utilizing the portable retina scan device in the Batmobile they could confirm the identity of the Penguin.

In the *Half-Life* series, the Scientists of Black Mesa are frequently shown operating Retinal Scanners to access locked doors or hidden devices. Characters in the films *GoldenEye* (1995), *Mission: Impossible* (1996), *Barb Wire* (1996), *Entrapment* (1999), *Minority Report* (2002) and *Paycheck* (2003) and *Charlie's Angels (film)* (2000) utilize or try to deceive retinal scanners.*[5]

In the *Splinter Cell* series, retinal scanners are used to identify agents within Third Echelon and guards within military/business complexes.

In the 2012 film *The Avengers*, characters gain access to a quantity of very rare iridium by using two devices: one which apparently hooks onto a victim's eyeball, and another which receives signals from the first to holographically reconstruct the retina to fool the scanner.

9.4 Uses

Retinal scanners are typically used for authentication and identification purposes. Retinal scanning has been utilized by several government agencies including the FBI, CIA, and NASA. However, in recent years, retinal scanning has become more commercially popular. Retinal scanning has been used in prisons, for ATM identity verification and the prevention of welfare fraud.*[6] Retinal scanning also has medical application. Communicable illnesses such as AIDS, syphilis, malaria, chicken pox and Lyme disease as well as hereditary diseases like leukemia, lymphoma, and sickle cell anemia affect the eyes. Pregnancy also affects the eyes. Likewise, indications of chronic health conditions such as congestive heart failure, atherosclerosis, and cholesterol issues first appear in the eyes.*[7]

9.5 Pros and cons

Advantages*[8]

- Low occurrence of false positives
- Extremely low (almost 0%) false negative rates
- Highly reliable because no two people have the same retinal pattern
- Speedy results: Identity of the subject is verified very quickly

Disadvantages*[9]

- Measurement accuracy can be affected by a disease such as cataracts
- Measurement accuracy can also be affected by severe astigmatism
- Scanning procedure is perceived by some as invasive
- Not very user friendly
- Subject being scanned must be very close to the camera optics*[10]
- High equipment cost
- Leads to eye problems that can be identified by CT Scan.
- Cheap equipment used can damage your eyes on regular usage.

9.6 See also

- Vein matching
- Iris recognition

9.7 References

[1] Retina and Iris Scans. Encyclopedia of Espionage, Intelligence, and Security. Copyright © 2004 by The Gale Group, Inc.

[2] Cofta, Piotr; Furnell, Steven (2008). "Use of Biometric Data". *Understanding Public Perceptions: Trust and Engagement in ICT-mediated Services*. International Engineering Consortium. p. 153. ISBN 9781931695954. Retrieved 29 September 2014. The error rate for fingerprint identification can be as high as 1 in 500, whereas a retinal scan boasts an error rate of 1 in 10,000,000.

[3] Retinography: How Retinal Scanning Works. Retrieved on 2007-04-02.

[4] "Eye Prints," TIME Magazine, Dec. 16, 1935. Retrieved on 2008-04-10.

[5] "Why retina scanning works better for James Bond than it ever would for us".

[6] Iris/Retinal Identification. Archived April 26, 2007, at the Wayback Machine. Wcu.Edu. Retrieved on 2007-04-02.

[7] O staff, Courtney. "Retinal Scans Do More Than Let You In The Door." Retrieved on 2007-04-02.

[8] "Iris Recognition vs. Retina Scanning – What are the Differences?", M2SYS Blog on Biometric Technology.

[9] Roberts, Chris. "Biometrics" Retrieved on 2009-06-11.

[10] https://www.bytechplanet.com/blog/iris-recognition-vs-retina-scanning/

Chapter 10

Keystroke dynamics

Keystroke dynamics, **keystroke biometrics** or **typing dynamics**, is the detailed timing information which describes exactly when each key was pressed and when it was released as a person is typing at a computer keyboard.[*][1]

10.1 Science of Keystroke Dynamics

The behavioral biometric of Keystroke Dynamics uses the manner and rhythm in which an individual types characters on a keyboard or keypad.[*][2][*][3][*][4] The keystroke rhythms of a user are measured to develop a unique biometric template of the user's typing pattern for future authentication.[*][5] Raw measurements available from almost every keyboard can be recorded to determine Dwell time (the time a key pressed) and Flight time (the time between "key up" and the next "key down"). The recorded keystroke timing data is then processed through a unique neural algorithm, which determines a primary pattern for future comparison.[*][6] Similarly, vibration information may be used to create a pattern for future use in both identification and authentication tasks.

Data needed to analyze keystroke dynamics is obtained by keystroke logging. Normally, all that is retained when logging a typing session is the sequence of characters corresponding to the order in which keys were pressed and timing information is discarded. When reading email, the receiver cannot tell from reading the phrase "I saw 3 zebras!" whether:

- that was typed rapidly or slowly

- the sender used the left shift key, the right shift key, or the caps-lock key to make the "i" turn into a capitalized letter "I"

- the letters were all typed at the same pace, or if there was a long pause before the letter "z" or the numeral "3" while you were looking for that key

- the sender typed any letters wrong initially and then went back and corrected them, or if they got them right the first time

10.2 Origin of Keystroke Dynamics

On May 24, 1844, the message "What hath God wrought" was sent by telegraph from the U.S. Capitol in Washington, D.C. to the Baltimore and Ohio Railroad "outer depot" in Baltimore, Maryland, a new era in long-distance communications had begun. By the 1860s the telegraph revolution was in full swing and telegraph operators were a valuable resource. With experience, each operator developed their unique "signature" and was able to be identified simply by their tapping rhythm.

As late as World War II the military transmitted messages through Morse Code. Using a methodology called "The Fist of the Sender," Military Intelligence identified that an individual had a unique way of keying in a message's "dots" and "dashes," creating a rhythm that could help distinguish ally from enemy.

10.3 Use as Biometric Data

Researchers are interested in using this keystroke dynamic information, which is normally discarded, to verify or even try to determine the identity of the person who is producing those keystrokes. This is often possible because some characteristics of keystroke production are as individual as handwriting or a signature. The techniques used to do this vary widely in power and sophistication, and range from statistical techniques to AI approaches like neural networks.

Very simple rules can be used to rule out possible users in simple cases. For example, if we know that John types at 20 words per minute, and the person at the keyboard is typing at 70 words per minute, we are able to eliminate John as the typist. This form of test is based simply on raw speed uncorrected for errors. It's only a one-way test, as it's always possible for people to go slower than normal, but it's unusual or impossible for them to go twice their normal speed.

Or, it may be that the mystery user at the keyboard and John both type at 50 words per minute. However, perhaps John never really learned the numbers, and always had to slow down an extra half-second whenever a number had to be entered. If the mystery user did not slow down before numbers were typed, we can eliminate John as the typist.

The time to get to and depress a key (seek-time), and the time the key is held-down (hold-time) may be very characteristic for a person, regardless of how fast they are going overall. Most people have specific letters that take them longer to find or get to than their average seek-time over all letters, but which letters those are may vary dramatically but consistently for different people. Right-handed people may be statistically faster in getting to keys they hit with their right hand fingers than they are with their left hand fingers. Index fingers may be characteristically faster than other fingers to a degree that is consistent for a person day-to-day regardless of their overall speed that day.

In addition, sequences of letters may have characteristic properties for a person. In English, the word "the" is very common, and those three letters may be known as a rapid-fire sequence and not as just three meaningless letters hit in that order. Common endings, such as "ing", may be entered far faster than, say, the same letters in reverse order ("gni") to a degree that varies consistently by person. This consistency may hold and may reveal the person's native language's common sequences even when they are writing entirely in a different language, just as revealing as an accent might in spoken English.

Common "errors" may also be quite characteristic of a person, and there is an entire taxonomy of errors, such as this person's most common "substitutions", "reversals", "drop-outs", "double-strikes", "adjacent letter hits", "homonyms", hold-length-errors (for a shift key held down too short or too long a time). Even without knowing what language a person is working in, by looking at the rest of the text and what letters the person goes back and replaces, these errors might be detected. Again, the patterns of errors might be sufficiently different to distinguish two people.

10.4 Authentication versus identification

Keystroke dynamics is part of a larger class of biometrics known as behavioral biometrics; their patterns are statistical in nature. It is a commonly held belief that behavioral biometrics are not as reliable as physical biometrics used for authentication such as fingerprints or retinal scans or DNA. The reality here is that behavioral biometrics use a *confidence* measurement instead of the traditional *pass/fail* measurements. As such, the traditional benchmarks of False Acceptance Rate (FAR) and False Rejection Rates (FRR) no longer have linear relationships.

The benefit to keystroke dynamics (as well as other behavioral biometrics) is that FRR/FAR can be adjusted by changing the acceptance threshold **at the individual level**. This allows for explicitly defined individual risk mitigation—something physical biometric technologies could never achieve.

Another benefit of keystroke dynamics: they can be captured continuously—not just at the start-up time—and may be adequately accurate to trigger an alarm to another system or person to come double-check the situation.

In some cases, a person at gun-point might be forced to get start-up access by entering a password or having a particular fingerprint, but then that person could be replaced by someone else at the keyboard who was taking over for some bad purpose. In other less dramatic cases, an employee might violate business rules by sharing their password with their secretary, or by logging onto a system but then leaving the computer logged-in while someone else he knows about or doesn't know about uses the system. Keystroke dynamics is one way to detect such problems sufficiently reliably to be worth investigating, because even a 20% true-positive rate would send the word out that this type of behavior is being watched and caught.

Researchers are still a long way from being able to read a keylogger session from a public computer in a library or cafe somewhere and identify the person from the keystroke dynamics, but we may be in a position to confidently **rule out** certain people from being the author, who we are confident is "a left-handed person with small hands who doesn't write in English as their primary language."

10.5 Temporal variation

One of the major problems that keystroke dynamics runs into is that a person's typing varies substantially during a day and between different days. People may get tired, or angry, or have a beer, or switch computers, or move their keyboard tray to a new location, or use a virtual keyboard, or be pasting in information from another source (cut-and-paste), or from a voice-to-text converter. Even while typing, a person, for example, may be on the phone or pausing to talk. And some mornings, perhaps after a long night with little sleep and a lot of drinking, a person's typing may bear little resemblance to the way he or she types when well-rested. Extra doses of medication or missed doses could change the person's rhythm. There are hundreds of confounding circumstances.

Because of these variations, any system will make false-positive and false-negative errors. Some of the successful commercial products have strategies to handle these issues and have proven effective in large-scale use (thousands of users) in real-world settings and applications.

10.6 Legal and regulatory issues

Surreptitious use of key-logging software is on the rise, as of this writing. Use of such software may be in direct and explicit violation of local laws, such as the U.S. Patriot Act, under which such use may constitute wire-tapping. This could have severe penalties including jail time. See spyware for a better description of user-consent issues and various fraud statutes. Spyware and its use for illegal operations such as bank-fraud and identity theft are very much in the news, with even Microsoft issuing new spyware defense products, and tougher laws in the near future being very likely.

Competent legal advice should be obtained before attempting to use or even experiment with such software and keystroke dynamic analysis, if consent is not clearly obtained from the people at the keyboard, even though the actual residual "content" of the message—the resultant text—is never analyzed, read, or retained. The status of the "dynamic context" of the text is probably in legal limbo.

There are some patents in this area. Examples:

- S. Blender and H. Postley. Key sequence rhythm recognition system and method. Patent No. 7 206 938, U.S. Patent and Trademark Office, 2007.

- J. Garcia. Personal identification apparatus. Patent No. 4 621 334, U.S. Patent and Trademark Office, 1986.

- J.R. Young and R.W. Hammon. Method and apparatus for verifying an individual's identity. Patent No. 4 805 222, U.S. Patent and Trademark Office, 1989.

- P. Nordström, J. Johansson. Security system and method for detecting intrusion in a computerized system. Patent No. 2 069 993, European Patent Office, 2009.

- A. Awad and I. Traore. System and method for determining a computer user profile from a motion-based input device. Patent No. 8 230 232, U.S. Patent and Trademark Office, 2012.

10.7 Other uses

Because keystroke timings are generated by human beings, they are not well correlated with external processes, and are frequently used as a source of hardware-generated random numbers for computer systems.

10.8 See also

- Fist (telegraphy)

10.9 References

[1] Robert Moskovitch, Clint Feher, Arik Messerman, Niklas Kirschnick, Tarik Mustafic, Ahmet Camtepe, Bernhard Löhlein, Ulrich Heister, Sebastian Möller, Lior Rokach, Yuval Elovici (2009). *Identity theft, computers and behavioral biometrics* (PDF). Proceedings of the IEEE International Conference on Intelligence and Security Informatics. pp. 155–160.

[2] Deng, Y.; Yu, Y. "Keystroke Dynamics User Authentication Based on Gaussian Mixture Model and Deep Belief Nets". *ISRN Signal Processing*. **2013**: 565183. doi:10.1155/2013/565183].

[3] User authentication through typing biometrics features

[4] Continuous authentication by analysis of keyboard typing characteristics

[5] A modified algorithm for user identification by his typing on the keyboard

[6] User authentication using rhythm click characteristics for nonKeyboard devices

10.9.1 Other references

- Checco, J. (2003). Keystroke Dynamics & Corporate Security. *WSTA Ticker Magazine*, .

- Bergadano, F.; Gunetti, D.; Picardi, C. (2002). "User authentication through Keystroke Dynamics". *ACM Transactions on Information and System Security (TISSEC)*. **5** (4): 367–397. doi:10.1145/581271.581272.

- iMagic Software. (vendor web-site May 2006). Notes: Vendor specializing in keystroke authentication for large enterprises.

- AdmitOne Security - formerly BioPassword. (vendor web-site home [Web Page]. URL . Notes: Vendor specializing in keystroke dynamics

10.9. REFERENCES

- Garcia, J. (Inventor). (1986). Personal identification apparatus. (USA 4621334). Notes: *US Patent Office -*

- Bender, S and Postley, H. (Inventors) (2007). Key sequence rhythm recognition system and method. (USA 7206938), Notes: *US Patent Office -*

- Joyce, R., & Gupta, G. (1990). Identity authorization based on keystroke latencies. *Communications of the ACM*, 33(2), 168-176. Notes: Review up through 1990

- Mahar, D.; Napier, R.; Wagner, M.; Laverty, W.; Henderson, R. D.; Hiron, M. (1995). "Optimizing digraph-latency based biometric typist verification systems: inter and intra typist differences in digraph latency distributions". *International Journal of Human-Computer Studies*. **43** (4): 579–592. doi:10.1006/ijhc.1995.1061.

- Monrose, F., & Rubin Aviel D. (1997). Authentication via Keystroke Dynamics. *ACM Conference on Computer and Communications Security.* Notes: available to subscribers at , much cited

- Monrose, F., & Rubin, A. D. (2000). Keystroke Dynamics as a Biometric for Authentication. *Future Generation Computer Systems*, 16, 351-359. Notes: Review 1990–1999

- Monrose, F. R. M. K., & Wetzel, S. (1999). Password hardening based on keystroke dynamics. *Proceedings of the 6th ACM Conference on Computer and Communications Security*, 73-82. Notes: Kent Ridge Digital Labs, Singapore

- Robinson, J. A., Liang, V. M., Chambers, J. A. M., & MacKenzie, C. L. (1998). Computer user Verification using Login String Keystroke Dynamics. *IEEE Transactions on Systems, Man, and Cybernetics Part A*, 28(2). Notes: Highlights: 10 users were distinguished from 10 "forgers" using 3 classification systems. Hold times were more effective than interkey times for discrimination. Best results used both with a learning classifier. There were a high rate of confounding errors and backspaces in the password samples.

- Young, J. R., & Hammon, R. W. (Inventors). (1989). Method and apparatus for verifying an individual's identity. 4805222). Notes: *US Patent Office -*

- Vertical Company LTD. (vendor web-site October 2006). Notes: Vendor specializing in keystroke authentication solutions for government and commercial agencies.

- Lopatka, M. & Peetz, M.H. (2009). Vibration Sensitive Keystroke Analysis. *Proceedings of the 18th Annual Belgian-Dutch Conference on Machine Learning*, 75-80.

- Coalfire Systems Compliance Validation Assessment (2007) https://web.archive.org/web/20110707084309/http://www.admitonesecurity.com/admitone_library/AOS_Compliance_Functional_Assessment_by_Coalfire.pdf

- Karnan, M.Akila (2011). "Biometric personal authentication using keystroke dynamics: A review". *Applied Soft Computing Journal*. **11** (2).

- Jenkins, J., Nguyen, Q., Reynolds, J., Horner, W., and Szu, H., "The Physiology of Keystroke Dynamics," in SPIE Independent Component Analyses, Wavelets, Neural Networks, Biosystems, and Nanoengineering IX, 2011, vol. 8058, p. 80581N1-10.

Chapter 11

Gait analysis

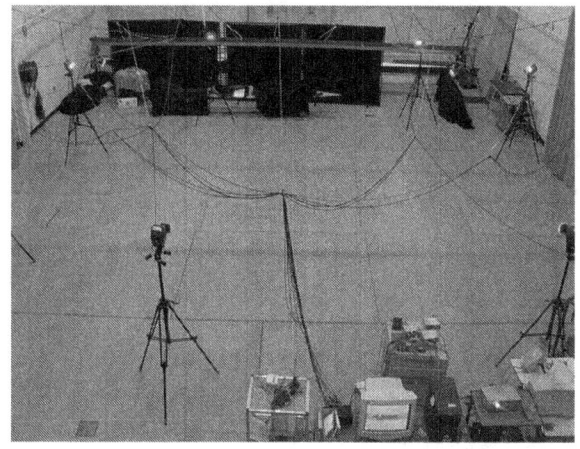

Gait analysis laboratory equipped with infrared cameras and floor mounted force platforms

Gait analysis is the systematic study of animal locomotion, more specifically the study of human motion, using the eye and the brain of observers, augmented by instrumentation for measuring body movements, body mechanics, and the activity of the muscles.[*][1] Gait analysis is used to assess and treat individuals with conditions affecting their ability to walk.[*][2] It is also commonly used in sports biomechanics to help athletes run more efficiently and to identify posture-related or movement-related problems in people with injuries.

The study encompasses quantification (i.e. introduction and analysis of measurable parameters of gaits), as well as interpretation, i.e. drawing various conclusions about the animal (health, age, size, weight, speed etc.) from its gait pattern.

11.1 History

The pioneers of scientific gait analysis were Aristotle in *De Motu Animalium* (On the Gait of Animals)[*][3] and much later in 1680, Giovanni Alfonso Borelli also called *De Motu Animalium (I et II)*. In the 1890s, the German anatomist Christian Wilhelm Braune and Otto Fischer published a series of papers on the biomechanics of human gait under loaded and unloaded conditions.[*][4]

With the development of photography and cinematography, it became possible to capture image sequences that reveal details of human and animal locomotion that were not noticeable by watching the movement with the naked eye. Eadweard Muybridge and Étienne-Jules Marey were pioneers of these developments in the early 1900s. For example, serial photography first revealed the detailed sequence of the horse "gallop", which was usually misrepresented in paintings made prior to this discovery.

Although much early research was done using film cameras, the widespread application of gait analysis to humans with pathological conditions such as cerebral palsy, Parkinson's disease, and neuromuscular disorders, began in the 1970s with the availability of video camera systems that could produce detailed studies of individual patients within realistic cost and time constraints. The development of treatment regimes, often involving orthopaedic surgery, based on gait analysis results, advanced significantly in the 1980s. Many leading orthopaedic hospitals worldwide now have gait labs that are routinely used to design treatment plans and for follow-up monitoring.

Development of modern computer based systems occurred independently during the late 1970s and early 1980s in several hospital based research labs, some through collaborations with the aerospace industry.[*][5] Commercial development soon followed with the emergence of commercial television and later infrared camera systems in the mid-1980s.

11.2 Process and equipment

A typical gait analysis laboratory has several cameras (video and / or infrared) placed around a walkway or a treadmill, which are linked to a computer. The patient has markers located at various points of reference of the body (e.g., iliac spines of the pelvis, ankle malleolus, and the condyles of the

11.3 Factors and parameters

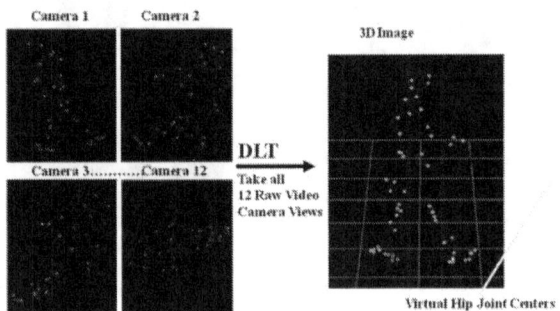

Acquisition of information on the position of the markers in 2D through the chambers of the left and right, this combination of information gives rise to a 3D image on the position of the markers

The gait analysis is modulated or modified by many factors, and changes in the normal gait pattern can be transient or permanent. The factors can be of various types:

- *Extrinsic: such as terrain, footwear, clothing, cargo*
- *Intrinsic: sex (male or female), weight, height, age, etc.*
- *Physical: such as weight, height, physique*
- *Psychological: personality type, emotions*
- *Physiological: anthropometric characteristics, i.e., measurements and proportions of body*
- *Pathological: for example trauma, neurological diseases, musculoskeletal anomalies, psychiatric disorders*

The parameters taken into account for the gait analysis are as follows:

- *Step length*
- *Stride length*
- *Cadence*
- *Speed*
- *Dynamic Base*
- *Progression Line*
- *Foot Angle*
- *Hip Angle*
- *Squat Performance*[7]

knee), or groups of markers applied to half of the body segments. The patient walks down the catwalk or the treadmill and the computer calculates the trajectory of each marker in three dimensions. A model is applied to calculate the movement of the underlying bones. This gives a complete breakdown of the movement of each joint. One common method is to use Helen Hayes Hospital marker set,*[6] in which a total of 15 markers are attached on the lower body. The 15 marker motions are analyzed analytically, and it provides angular motion of each joint.

To calculate the kinetics of gait patterns, most labs have floor-mounted load transducers, also known as force platforms, which measure the ground reaction forces and moments, including the magnitude, direction and location (called the center of pressure). The spatial distribution of forces can be measured with pedobarography equipment. Adding this to the known dynamics of each body segment enables the solution of equations based on the Newton–Euler equations of motion permitting computations of the net forces and the net moments of force about each joint at every stage of the gait cycle. The computational method for this is known as inverse dynamics.

This use of kinetics, however, does not result in information for individual muscles but muscle groups, such as the extensor or flexors of the limb. To detect the activity and contribution of individual muscles to movement, it is necessary to investigate the electrical activity of muscles. Many labs also use surface electrodes attached to the skin to detect the electrical activity or electromyogram (EMG) of muscles. In this way it is possible to investigate the activation times of muscles and, to some degree, the magnitude of their activation —thereby assessing their contribution to gait. Deviations from normal kinematic, kinetic or EMG patterns are used to diagnose specific pathologies, predict the outcome of treatments, or determine the effectiveness of training programs

11.4 Techniques

Gait analysis involves measurement,*[8] where measurable parameters are introduced and analyzed, and interpretation, where conclusions about the subject (health, age, size, weight, speed, etc.) are drawn. The analysis is the measurement of the following:

11.4.1 Temporal / spatial

It consists of the calculation of speed, the length of the rhythm, pitch, and so on. These measurements are carried out through:

- Stopwatch and marks on the ground.

- Walking on a pressure mat.
- Range laser sensors scanning a plane a few centimeters above the floor.[9][10]
- Inertial sensors and software to interpret 3D gyroscopes and 3D accelerometric data.

11.4.2 Kinematics

1. Chronophotography is the most basic method for recording of movement. Strobe lighting at known frequency has been used in the past to aid in the analysis of gait on single photographic images.[11][12]

2. Cine film or video recordings using footage from single or multiple cameras can be used to measure joint angles and velocities. This method has been aided by the development of analysis software that greatly simplifies the analysis process and allows for analysis in three dimensions rather than two dimensions only.

3. Passive marker systems, using reflective markers (typically reflective balls), allows for accurate measurement of movements using multiple cameras (typically five to twelve cameras), simultaneously. The cameras utilize high-powered strobes (typically red, near infrared or infrared) with matching filters to record the reflection from the markers placed on the body. Markers are located at palpable anatomical landmarks. Based on the angle and time delay between the original and reflected signal, triangulation of the marker in space is possible. Software is used to create three dimensional trajectories from these markers that are subsequently given identification labels. A computer model is then used to compute joint angles from the relative marker positions of the labeled trajectories.[13] These are also used for motion capture in the motion picture industry.[14]

4. Active marker systems are similar to the passive marker system but use "active" markers. These markers are triggered by the incoming infra red signal and respond by sending out a corresponding signal of their own. This signal is then used to triangulate the location of the marker. The advantage of this system over the passive one is that individual markers work at predefined frequencies and therefore, have their own "identity". This means that no post-processing of marker locations is required, however, the systems tend to be less forgiving for out-of-view markers than the passive systems.[15]

5. Inertial (cameraless) systems based on MEMS inertial sensors, biomechanical models, and sensor fusion algorithms. These full-body or partial body systems can be used indoors and outdoors regardless of lighting conditions.[16][17][18][19][20]

11.4.3 Markerless Gait Capture

- Markerless gait capture systems utilize one or more color cameras or 2.5D depth sensors (i.e. Kinect) to directly calculate the body joint positions from a sequence of images. The markerless system allows non-invasive human gait analysis in a natural environment without any marker attachment. Eliminating markers can expand the applicability of human gait measurement and analysis techniques, considerably reduce the preparation time, and enable efficient and accurate motion assessment in all kinds of applications. Currently, the main markerless system is the video-based motion capture with monocular camera or multiple camera studio.[21] Nowadays, the depth sensor-based gait analysis for clinical applications becomes more and more popular. Since depth sensors can measure the depth information and provide a 2.5D depth image, they have effectively simplified the task of foreground/background subtraction and significantly reduced pose ambiguities in monocular human pose estimation.[22]

11.4.4 Pressure Measurement

Pressure measurement systems are an additional way to measure gait by providing insights into pressure distribution, contact area, center of force movement and symmetry between sides. These systems typically provide more than just pressure information; additional information available from these systems are force, timing and spatial parameters. Different methods for assessing pressure are available, like a pressure measurement mat or walkway (longer in length to capture more foot strikes), as well as in-shoe pressure measurement systems (where sensors are placed inside the shoe).[23][24] Many pressure measurement systems integrate with additional types of analysis systems, like motion capture, EMG or force plates to provide a comprehensive gait analysis.

11.4.5 Kinetics

Is the study of the forces involved in the production of movements.

11.4.6 Dynamic electromyography

Is the study of patterns of muscle activity during gait.

11.5 Applications

Gait analysis is used to analyze the walking ability of humans and animals, so this technology can be used for the following applications:

11.5.1 Medical diagnostics

Pathological gait may reflect compensations for underlying pathologies, or be responsible for causation of symptoms in itself. Cerebral palsy and stroke patients are commonly seen in gait labs. The study of gait allows diagnoses and intervention strategies to be made, as well as permitting future developments in rehabilitation engineering. Aside from clinical applications, gait analysis is used in professional sports training to optimize and improve athletic performance.

Gait analysis techniques allow for the assessment of gait disorders and the effects of corrective orthopedic surgery. Options for treatment of cerebral palsy include the artificial paralysis of spastic muscles using Botox or the lengthening, re-attachment or detachment of particular tendons. Corrections of distorted bony anatomy are also undertaken (osteotomy).

11.5.2 Chiropractic and Osteopathic Utilizations

Observation of gait is also beneficial for diagnoses in chiropractic and osteopathic professions as hindrances in gait may be indicative of a misaligned pelvis or sacrum. As the sacrum and ilium biomechanically move in opposition to each other, adhesions between the two of them via the sacrospinous or sacrotuberous ligaments (among others) may suggest a rotated pelvis. Both doctors of chiropractic and osteopathic medicine use gait to discern the listing of a pelvis and can employ various techniques in order to restore a full range of motion to areas involved in ambulatory movement. Chiropractic adjustment of the pelvis has shown a trend in helping restore gait patterns*[25]*[26] as has osteopathic manipulative therapy (OMT).*[27]*[28]

11.5.3 Biometric identification and forensics

Minor variations in gait style can be used as a biometric identifier to identify individual people.*[29] The parameters are grouped to spatial-temporal (step length, step width, walking speed, cycle time) and kinematic (joint rotation of the hip, knee and ankle, mean joint angles of the hip/knee/ankle, and thigh/trunk/foot angles) classes. There is a high correlation between step length and height of a person.*[30]*[31]

The approach above belongs to the model-based approach. Another appearance-based approach recognizes individuals through binary gait silhouette sequences. For example, silhouette sequences of full gait cycles can be treated as 3D tensor samples, and multilinear subspace learning, such as the multilinear principal component analysis, can be employed to learning features for classification.

11.5.4 Comparative biomechanics

By studying the gait of non-human animals, more insight can be gained about the mechanics of locomotion, which has diverse implications for understanding the biology of the species in question as well as locomotion more broadly.

11.6 Popular media

- G. K. Chesterton premised one of his Father Brown mysteries, "The Queer Feet", on gait recognition.

- Cory Doctorow makes much of gait recognition as a security technique used in a high school in his book *Little Brother*.

- Arthur Conan Doyle has Sherlock Holmes use gait analysis to identify the height of the *Rache* killer in A Study in Scarlet

- Benji Dunn had to get past a Gait analysis machine in Mission: Impossible – Rogue Nation and had to have Ethan Hunt change the code so he could get past because mask won't work on these types of machines.

11.7 Textbooks

- *Gait Analysis: Normal and Pathological Function, Second Edition.* Authors: Jacquelin Perry and Judith M. Burnfield

- *Gait Analysis.* Authors: David F. Levine, Jim Richards, and Michael Whittle.

- *Observational Gait Analysis.* Author: Los Amigos Research and Education Center

- *Gait Analysis: Theory and Application.* Authors: Rebecca L. Craik and Carol S. Oatis,

- "The Treatment of Gait Problems in Cerebral Palsy" Clinics in Developmental Medicine No. 164-165, edited by James R. Gage, 2004.

11.8 See also

- Biometric points
- *Gait Analysis: Normal and Pathological Function, Second Edition*
- Gait Abnormality Rating Scale
- Multilinear principal component analysis
- Multilinear subspace learning
- Pattern recognition
- Terrestrial locomotion in animals
- Comparison of orthotics

11.9 References

[1] Levine DF, Richards J, Whittle M. (2012). Whittle's Gait Analysis Whittle's Gait Analysis Elsevier Health Sciences. ISBN 978-0702042652

[2] Sejdić, E.; Lowry, K. A.; Bellanca, J.; Redfern, M. S.; Brach, J. S. (May 2014). "A Comprehensive Assessment of Gait Accelerometry Signals in Time, Frequency and Time-Frequency Domains". *IEEE Transactions on Neural Systems and Rehabilitation Engineering*. **22** (3): 603–612. ISSN 1534-4320. doi:10.1109/TNSRE.2013.2265887.

[3] Aristotle (2004). *On the Gait of Animals*. Kessinger Publishing. ISBN 1-4191-3867-7.

[4] Fischer, Otto; Braune, Wilhelm (1895). *Der Gang des Menschen: Versuche am unbelasteten und belasteten Menschen, Band 1*. (in German). Hirzel Verlag.

[5] Sutherland, DH. (2002). The evolution of clinical gait analysis: Part II Kinematics *Gait & Posture*. 16: 159-179.

[6] Kadaba, M. P.; Ramakrishnan, H. K.; Wootten, M. E. (May 1990). "Measurement of lower extremity kinematics during level walking". *Journal of Orthopaedic Research*. **8** (3): 383–392. doi:10.1002/jor.1100080310.

[7] Schweitzer, Eric. "What is a gait analysis?". *IdealRun*.

[8] U. Tasch, P. Moubarak, W. Tang, L. Zhu, R.M. Lovering, J. Roche, R. J. Bloch. (2008). An Instrument that Simultaneously Measures Spatiotemporal Gait Parameters and Ground Reaction Forces in Locomoting Rats, in *Proceeding of 9th Biennial ASME conference on Engineering Systems Design & Analysis, ESDA '08*. Haifa, Israel, pp. 45–49.

[9] Piérard, S.; Azrour, S.; Phan-Ba, R.; Van Droogenbroeck, M. (October 2013). "GAIMS: A reliable non-intrusive gait measuring system". *ERCIM News*. **95**: 26–27.

[10] "The GAIMS project".

[11] Étienne-Jules Marey

[12] Eadweard Muybridge

[13] RB Davis, S Õunpuu, D Tyburski, JR Gage (1991). A gait analysis data collection and reduction technique. *Human Movement Science* 10:575-587.

[14] Robertson DGE, et al. (2004). *Research Methods in Biomechanics*. Champaign IL:Human Kinetics Pubs..

[15] Best, Russell; Begg, Rezaul (2006). "Overview of Movement Analysis and Gait Features". In Begg, Rezaul; Palaniswami, Marimuthu. *Computational Intelligence for Movement Sciences: Neural Networks and Other Emerging Techniques*. Idea Group (published 2006-03-30). pp. 11–18. ISBN 978-1-59140-836-9.

[16] Sejdić, E.; Lowry, K. A.; Bellanca, J.; Perera, S.; Redfern, M. S.; Brach, J. S. (2016). "Extraction of Stride Events From Gait Accelerometry During Treadmill Walking". *IEEE Journal of Translational Engineering in Health and Medicine*. **4**: 1–11. ISSN 2168-2372. doi:10.1109/JTEHM.2015.2504961.

[17] Ambulatory inertial gait analysis

[18] Digital Motion Analysis Systems

[19] Wearable Sensors for Gait Analysis

[20] Sejdić, E.; Millecamps, A.; Teoli, J.; Rothfuss, M. A.; Franconi, N. G.; Perera, S.; Jones, A. K.; Brach, J. S.; Mickle, M. H. (2015-12-01). "Assessing interactions among multiple physiological systems during walking outside a laboratory: An Android based gait monitor". *Computer Methods and Programs in Biomedicine*. **122** (3): 450–461. doi:10.1016/j.cmpb.2015.08.012.

[21] X. Zhang, M. Ding, G. Fan (2016) *Video-based Human Walking Estimation by Using Joint Gait and Pose Manifolds*, IEEE Transactions on Circuits and Systems for Video Technology, 2016

[22] https://sites.google.com/site/mengdingosu/home/research

[23] "Gait Analysis with Pressure Measurement". *Tekscan*. Retrieved 2017-09-29.

[24] Coda, A.; Carline, T.; Santos, D. (2014). "Repeatability and reproducibility of the Tekscan HR-Walkway system in healthy children". *www.ncbi.nlm.nih.gov*. doi:10.1016/j.foot.2014.02.004. Retrieved 2017-09-29.

[25] Herzog, W (1988). "Quantifying the effects of spinal manipulations on gait using patients with low back pain.". *Journal of Manipulative and Physiological Therapeutics*. **11** (3): 151–157.

[26] RO, Robinson; W, Herzog; BM, Nigg (1 August 1987). "Use of force platform variables to quantify the effects of chiropractic manipulation on gait symmetry.". *Journal of manipulative and physiological therapeutics*. **10** (4). ISSN 0161-4754.

[27] MR, Wells; S, Giantinoto; D, D'Agate; RD, Areman; EA, Fazzini; D, Dowling; A, Bosak (1 February 1999). "Standard osteopathic manipulative treatment acutely improves gait performance in patients with Parkinson's disease.". *The Journal of the American Osteopathic Association.* **99** (2). ISSN 0098-6151. doi:10.7556/jaoa.1999.99.2.92.

[28] Vismara, Luca; Cimolin, Veronica; Galli, Manuela; Grugni, Graziano; Ancillao, Andrea; Capodaglio, Paolo (March 2016). "Osteopathic Manipulative Treatment improves gait pattern and posture in adult patients with Prader–Willi syndrome". *International Journal of Osteopathic Medicine.* **19**: 35–43. doi:10.1016/j.ijosm.2015.09.001.

[29] Damaševičius, R.; Maskeliūnas, R.; Venčkauskas, A.; Woźniak, M. Smartphone User Identity Verification Using Gait Characteristics, Symmetry 2016, 8, 100.

[30] journalsip.astm.org/JOURNALS/FORENSIC/PAGES/4706.htm

[31] geradts.com/html/Documents/gait.htm

11.10 External links

- Clinical Gait Analysis

Chapter 12

Speaker recognition

"Voice control" redirects here. For voice controlled devices, see Voice command device.

Speaker recognition is the identification of a person from characteristics of voices (*voice biometrics*).[*][1] It is also called **voice recognition**.[*][2][*][3][*][4][*][5][*][6] There is a difference between *speaker recognition* (recognizing **who** is speaking) and *speech recognition* (recognizing **what** is being said). These two terms are frequently confused, and "voice recognition" can be used for both. In addition, there is a difference between the act of authentication (commonly referred to as **speaker verification** or **speaker authentication**) and identification. Finally, there is a difference between *speaker recognition* (recognizing **who** is speaking) and *speaker diarisation* (recognizing **when** the **same** speaker is speaking). Recognizing the speaker can simplify the task of translating speech in systems that have been trained on specific person's voices or it can be used to authenticate or verify the identity of a speaker as part of a security process.

Speaker recognition has a history dating back some four decades and uses the acoustic features of speech that have been found to differ between individuals. These acoustic patterns reflect both anatomy (e.g., size and shape of the throat and mouth) and learned behavioral patterns (e.g., voice pitch, speaking style). Speaker verification has earned speaker recognition its classification as a "behavioral biometric".

12.1 Verification versus identification

There are two major applications of *speaker recognition* technologies and methodologies. If the speaker claims to be of a certain identity and the voice is used to verify this claim, this is called *verification* or *authentication*. On the other hand, *identification* is the task of determining an unknown speaker's identity. In a sense *speaker verification* is a 1:1 match where one speaker's voice is matched to one template (also called a "voice print" or "voice model") whereas *speaker identification* is a 1:N match where the voice is compared against N templates.

From a security perspective, identification is different from verification. For example, presenting your passport at border control is a verification process: the agent compares your face to the picture in the document. Conversely, a police officer comparing a sketch of an assailant against a database of previously documented criminals to find the closest match(es) is an identification process.

Speaker verification is usually employed as a "gatekeeper" in order to provide access to a secure system (e.g. telephone banking). These systems operate with the users' knowledge and typically require their cooperation. *Speaker identification* systems can also be implemented covertly without the user's knowledge to identify talkers in a discussion, alert automated systems of speaker changes, check if a user is already enrolled in a system, etc.

In forensic applications, it is common to first perform a speaker identification process to create a list of "best matches" and then perform a series of verification processes to determine a conclusive match.

12.2 Variants of speaker recognition

Each *speaker recognition* system has two phases: Enrollment and verification. During enrollment, the speaker's voice is recorded and typically a number of features are extracted to form a *voice print*, *template*, or *model*. In the verification phase, a speech sample or "utterance" is compared against a previously created voice print. For identification systems, the utterance is compared against multiple voice prints in order to determine the best match(es) while verification systems compare an utterance against a single voice print. Because of the process involved, verification is faster than identification.

Speaker recognition systems fall into two categories: text-dependent and text-independent.

Text-Dependent:

If the text must be the same for enrollment and verification this is called text-dependent recognition. In a text-dependent system, prompts can either be common across all speakers (e.g.: a common pass phrase) or unique. In addition, the use of shared-secrets (e.g.: passwords and PINs) or knowledge-based information can be employed in order to create a multi-factor authentication scenario.

Text-Independent:

Text-independent systems are most often used for speaker identification as they require very little if any cooperation by the speaker. In this case the text during enrollment and test is different. In fact, the enrollment may happen without the user's knowledge, as in the case for many forensic applications. As text-independent technologies do not compare what was said at enrollment and verification, verification applications tend to also employ speech recognition to determine what the user is saying at the point of authentication.

In text independent systems both acoustics and speech analysis techniques are used.*[7]

12.3 Technology

Speaker recognition is a pattern recognition problem. The various technologies used to process and store *voice prints* include frequency estimation, hidden Markov models, Gaussian mixture models, pattern matching algorithms, neural networks, matrix representation, Vector Quantization and decision trees. Some systems also use "anti-speaker" techniques, such as cohort models, and world models. Spectral features are predominantly used in representing speaker characteristics.*[8]

Ambient noise levels can impede both collections of the initial and subsequent voice samples. Noise reduction algorithms can be employed to improve accuracy, but incorrect application can have the opposite effect. Performance degradation can result from changes in behavioural attributes of the voice and from enrollment using one telephone and verification on another telephone ("cross channel"). Integration with two-factor authentication products is expected to increase. Voice changes due to ageing may impact system performance over time. Some systems adapt the speaker models after each successful verification to capture such long-term changes in the voice, though there is debate regarding the overall security impact imposed by automated adaptation.

Capture of the biometric is seen as non-invasive. The technology traditionally uses existing microphones and voice transmission technology allowing recognition over long distances via ordinary telephones (wired or wireless).

Digitally recorded audio voice identification and analogue recorded voice identification uses electronic measurements as well as critical listening skills that must be applied by a forensic expert in order for the identification to be accurate.

12.4 Applications

The first international patent was filed in 1983, coming from the telecommunication research in CSELT*[9] (Italy) by Michele Cavazza and Alberto Ciaramella as a basis for both future telco services to final customers and to improve the noise-reduction techniques across the network.

In May 2013 it was announced that Barclays Wealth was to use passive speaker recognition to verify the identity of telephone customers within 30 seconds of normal conversation.*[10] The system used had been developed by voice recognition company Nuance (that in 2011 acquired the company Loquendo, the spin-off from CSELT itself for speech technology), the company behind Apple's Siri technology. A verified voiceprint was to be used to identify callers to the system and the system would in the future be rolled out across the company.

The private banking division of Barclays was the first financial services firm to deploy voice biometrics as the primary means to authenticate customers to their call centers. 93% of customer users had rated the system at "9 out of 10" for speed, ease of use and security.*[11]

Since then, Nuance Voice Biometrics solutions have been deployed across several financial institutions, including Banco Santander, Royal Bank of Canada, Tangerine Bank, and Manulife.*[12]

In August 2014 GoVivace Inc. deployed a speaker identification system that allowed its telecom industry client to positively search for an individual among millions of speakers by using just a single example recording of their voice.*[13]

Speaker recognition may also be used in criminal investigations, such as those of the 2014 executions of, amongst others, James Foley and Steven Sotloff.*[14]

In February 2016 UK high-street bank HSBC and its internet-based retail bank First Direct announced that it would offer 15 million customers its biometric banking software to access online and phone accounts using their fingerprint or voice.*[15]

12.5 See also

- AI effect
- Applications of artificial intelligence
- Speaker diarisation
- Speech recognition
- Voice changer

Lists

- List of emerging technologies
- Outline of artificial intelligence

12.6 Notes

[1] Kinnunen, Tomi; Li, Haizhou (January 1, 2010). "An overview of text-independent speaker recognition: From features to supervectors" (PDF). *Speech Communication.* **52** (1): 12–40. doi:10.1016/j.specom.2009.08.009.

[2] Pollack, Pickett, Sumby (1974). *Experimental phonetics.* MSS Information Corporation. pp. 251–258. ISBN 0-8422-5149-9.

[3] Van Lancker and Kreiman (July 3, 1984). "Familiar voice recognition: Patterns and parameters. Part I: Recognition of backward voices" (PDF). Journal of Phonetics. pp. 19–38. Retrieved February 21, 2012.

[4] "British English definition of voice recognition". Macmillan Publishers Limited. Retrieved February 21, 2012.

[5] "voice recognition, definition of". WebFinance, Inc. Retrieved February 21, 2012.

[6] "Linux Gazette 114". Linux Gazette. Retrieved February 21, 2012.

[7] Lisa Myers (April 19, 2004). "An Exploration of Voice Biometrics".

[8] Sahidullah, Md.; Kinnunen, Tomi (March 2016). "Local spectral variability features for speaker verification". *Digital Signal Processing.* **50**: 1–11. doi:10.1016/j.dsp.2015.10.011.

[9] US4752958 A, Michele Cavazza, Alberto Ciaramella, "Device for speaker's verification" http://www.google.com/patents/US4752958?hl=it&cl=en

[10] International Banking (December 27, 2013). "Voice Biometric Technology in Banking | Barclays". Wealth.barclays.com. Retrieved February 21, 2016.

[11] Matt Warman (May 8, 2013). "Say goodbye to the pin: voice recognition takes over at Barclays Wealth". Retrieved June 5, 2013.

[12] "Voice Biometrics for fast, secure authentication in your IVR and mobile apps.". Nuance. Retrieved February 21, 2016.

[13] "Speaker Identification". Archived from the original on August 15, 2014. Retrieved September 3, 2014.

[14] Ewen MacAskill. "Did 'Jihadi John' kill Steven Sotloff? | Media". *The Guardian.* Retrieved February 21, 2016.

[15] Julia Kollewe (February 19, 2016). "HSBC rolls out voice and touch ID security for bank customers | Business". *The Guardian.* Retrieved February 21, 2016.

12.7 References

- "Biometrics from the movies" –National Institute of Standards and Technology
- Elisabeth Zetterholm (2003), *Voice Imitation. A Phonetic Study of Perceptual Illusions and Acoustic Success*, Phd thesis, Lund University.
- Md Sahidullah (2015), *Enhancement of Speaker Recognition Performance Using Block Level, Relative and Temporal Information of Subband Energies*, PhD thesis, Indian Institute of Technology Kharagpur.
- Homayoon Beigi (2011). Jucheng Yang, ed. *Speaker Recognition Biometrics (Book 1).* INTECH Open Access Publisher. pp. 3–28. ISBN 978-953-307-618-8. Retrieved January 27, 2016.

12.8 Bibliography

- Beigi, Homayoon (2011). *Fundamentals of Speaker Recognition.* New York: Springer. ISBN 978-0-387-77591-3.

12.9 External links

- Circumventing Voice Authentication The PLA Radio podcast recently featured a simple way to fool rudimentary voice authentication systems.
- Speaker recognition – Scholarpedia
- – Voice: Technologies and Algorithms for Biometrics Applications, IEEE eLearning Library (formerly IEEE Expert Now eLearning) Tutuorial, Sep. 2010

12.9. EXTERNAL LINKS

- Voice recognition benefits and challenges in access control

12.9.1 Software

- bob.bio.spear
- ALIZE

Chapter 13

Electroencephalography

Not to be confused with other types of electrography.
"EEG" redirects here. For other uses, see EEG (disambiguation).

Electroencephalography (EEG) is an electrophysiological monitoring method to record electrical activity of the brain. It is typically noninvasive, with the electrodes placed along the scalp, although invasive electrodes are sometimes used such as in electrocorticography. EEG measures voltage fluctuations resulting from ionic current within the neurons of the brain.[1] In clinical contexts, EEG refers to the recording of the brain's spontaneous electrical activity over a period of time,[1] as recorded from multiple electrodes placed on the scalp. Diagnostic applications generally focus either on event-related potentials or on the spectral content of EEG. The former investigates potential fluctuations time locked to an event like stimulus onset or button press. The latter analyses the type of neural oscillations (popularly called "brain waves") that can be observed in EEG signals in the frequency domain.

EEG is most often used to diagnose epilepsy, which causes abnormalities in EEG readings.[2] It is also used to diagnose sleep disorders, depth of anesthesia, coma, encephalopathies, and brain death. EEG used to be a first-line method of diagnosis for tumors, stroke and other focal brain disorders,[3][4] but this use has decreased with the advent of high-resolution anatomical imaging techniques such as magnetic resonance imaging (MRI) and computed tomography (CT). Despite limited spatial resolution, EEG continues to be a valuable tool for research and diagnosis. It is one of the few mobile techniques available (e.g. [5]) and offeres millisecond-range temporal resolution which is not possible with CT, PET or MRI.

Derivatives of the EEG technique include evoked potentials (EP), which involves averaging the EEG activity time-locked to the presentation of a stimulus of some sort (visual, somatosensory, or auditory). Event-related potentials (ERPs) refer to averaged EEG responses that are time-locked to more complex processing of stimuli; this technique is used in cognitive science, cognitive psychology, and psychophysiological research.

13.1 History

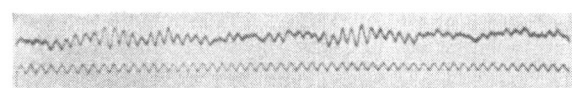

The first human EEG recording obtained by Hans Berger in 1924. The upper tracing is EEG, and the lower is a 10 Hz timing signal.

Hans Berger

The history of EEG is detailed by Barbara E. Swartz in *Electroencephalography and Clinical Neurophysiology*.[6] In 1875, Richard Caton (1842–1926), a physician practicing in Liverpool, presented his findings about electrical phenomena of the exposed cerebral hemispheres of rabbits and monkeys in the *British Medical Journal*. In 1890, Polish physiologist Adolf Beck published an investigation of spontaneous electrical activity of the brain of rabbits and dogs that included rhythmic oscillations altered by light. Beck started experiments on the electrical brain activity of animals. Beck placed electrodes directly on the surface of brain to test for sensory stimulation. His observation of fluctuating brain activity led to the conclusion of brain waves.*[7]

In 1912, Russian physiologist Vladimir Vladimirovich Pravdich-Neminsky published the first animal EEG and the evoked potential of the mammalian (dog).*[8] In 1914, Napoleon Cybulski and Jelenska-Macieszyna photographed EEG recordings of experimentally induced seizures.

German physiologist and psychiatrist Hans Berger (1873–1941) recorded the first human EEG in 1924.*[9] Expanding on work previously conducted on animals by Richard Caton and others, Berger also invented the electroencephalogram (giving the device its name), an invention described "as one of the most surprising, remarkable, and momentous developments in the history of clinical neurology" .*[10] His discoveries were first confirmed by British scientists Edgar Douglas Adrian and B. H. C. Matthews in 1934 and developed by them.

In 1934, Fisher and Lowenback first demonstrated epileptiform spikes. In 1935 Gibbs, Davis and Lennox described interictal spike waves and the three cycles/s pattern of clinical absence seizures, which began the field of clinical electroencephalography. Subsequently, in 1936 Gibbs and Jasper reported the interictal spike as the focal signature of epilepsy. The same year, the first EEG laboratory opened at Massachusetts General Hospital.

Franklin Offner (1911–1999), professor of biophysics at Northwestern University developed a prototype of the EEG that incorporated a piezoelectric inkwriter called a Crystograph (the whole device was typically known as the Offner Dynograph).

In 1947, The American EEG Society was founded and the first International EEG congress was held. In 1953 Aserinsky and Kleitman described REM sleep.

In the 1950s, William Grey Walter developed an adjunct to EEG called EEG topography, which allowed for the mapping of electrical activity across the surface of the brain. This enjoyed a brief period of popularity in the 1980s and seemed especially promising for psychiatry. It was never accepted by neurologists and remains primarily a research tool.

In 1988 report was given on EEG control of a physical object, a robot.*[11]*[12]

13.2 Medical use

A routine clinical EEG recording typically lasts 20–30 minutes (plus preparation time) and usually involves recording from scalp electrodes. Routine EEG is typically used in the

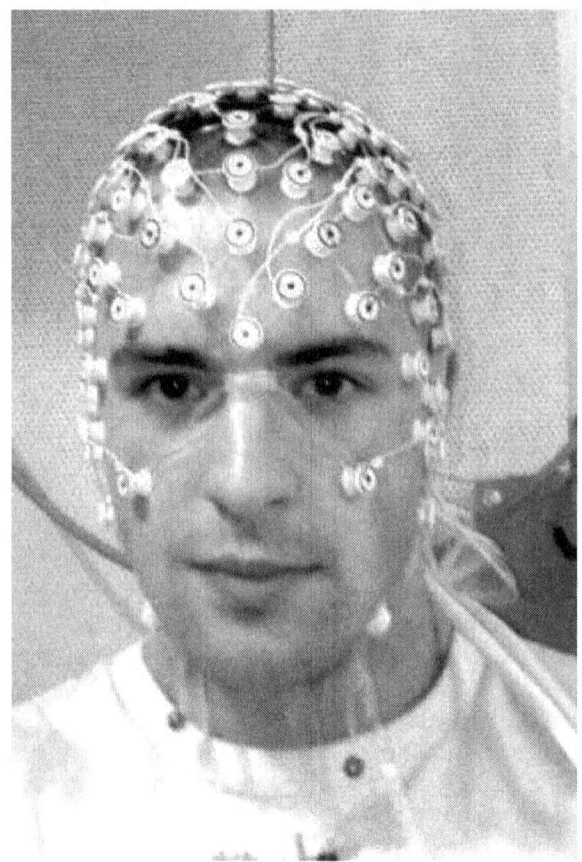

An EEG recording setup

following clinical circumstances:

- to distinguish epileptic seizures from other types of spells, such as psychogenic non-epileptic seizures, syncope (fainting), sub-cortical movement disorders and migraine variants.

- to differentiate "organic" encephalopathy or delirium from primary psychiatric syndromes such as catatonia

- to serve as an adjunct test of brain death

- to prognosticate, in certain instances, in patients with coma

- to determine whether to wean anti-epileptic medications

At times, a routine EEG is not sufficient, particularly when it is necessary to record a patient while he/she is having a seizure. In this case, the patient may be admitted to the hospital for days or even weeks, while EEG is constantly being recorded (along with time-synchronized video and audio recording). A recording of an actual seizure (i.e., an ictal recording, rather than an inter-ictal recording of a possibly epileptic patient at some period between seizures) can give significantly better information about whether or not a spell is an epileptic seizure and the focus in the brain from which the seizure activity emanates.

Epilepsy monitoring is typically done:

- to distinguish epileptic seizures from other types of spells, such as psychogenic non-epileptic seizures, syncope (fainting), sub-cortical movement disorders and migraine variants.

- to characterize seizures for the purposes of treatment

- to localize the region of brain from which a seizure originates for work-up of possible seizure surgery

Additionally, EEG may be used to monitor certain procedures:

- to monitor the depth of anesthesia

- as an indirect indicator of cerebral perfusion in carotid endarterectomy

- to monitor amobarbital effect during the Wada test

EEG can also be used in intensive care units for brain function monitoring:

- to monitor for non-convulsive seizures/non-convulsive status epilepticus

- to monitor the effect of sedative/anesthesia in patients in medically induced coma (for treatment of refractory seizures or increased intracranial pressure)

- to monitor for secondary brain damage in conditions such as subarachnoid hemorrhage (currently a research method)

If a patient with epilepsy is being considered for resective surgery, it is often necessary to localize the focus (source) of the epileptic brain activity with a resolution greater than what is provided by scalp EEG. This is because the cerebrospinal fluid, skull and scalp *smear* the electrical potentials recorded by scalp EEG. In these cases, neurosurgeons typically implant strips and grids of electrodes (or penetrating depth electrodes) under the dura mater, through either a craniotomy or a burr hole. The recording of these signals is referred to as electrocorticography (ECoG), subdural EEG (sdEEG) or intracranial EEG (icEEG)--all terms for the same thing. The signal recorded from ECoG is on a different scale of activity than the brain activity recorded from scalp EEG. Low voltage, high frequency components that cannot be seen easily (or at all) in scalp EEG can be seen clearly in ECoG. Further, smaller electrodes (which cover a smaller parcel of brain surface) allow even lower voltage, faster components of brain activity to be seen. Some clinical sites record from penetrating microelectrodes.*[1] EEG may be done in all pediatric patients presenting with first onset afebrile or complex febrile seizures.*[13] EEG is not indicated for diagnosing headache.*[14] Recurring headache is a common pain problem, and this procedure is sometimes used in a search for a diagnosis, but it has no advantage over routine clinical evaluation.*[14]

13.3 Research use

EEG, and the related study of ERPs are used extensively in neuroscience, cognitive science, cognitive psychology, neurolinguistics and psychophysiological research. Many EEG techniques used in research are not standardised sufficiently for clinical use. But research on mental disabilities, such as Auditory Processing Disorder (APD), ADD, or ADHD, is becoming more widely known and EEGs are used as research and treatment.

13.3.1 Advantages

Several other methods to study brain function exist, including functional magnetic resonance imaging (fMRI), positron emission tomography, magnetoencephalography (MEG), Nuclear magnetic resonance spectroscopy, Electrocorticography, Single-photon emission computed tomography, Near-infrared spectroscopy (NIRS), and Event-related optical signal (EROS). Despite the relatively poor spatial sensitivity of EEG, it possesses multiple advantages over some of these techniques:

- Hardware costs are significantly lower than those of most other techniques *[15]

- EEG prevents limited availability of technologists to provide immediate care in high traffic hospitals.*[16]

- EEG sensors can be used in more places than fMRI, SPECT, PET, MRS, or MEG, as these techniques require bulky and immobile equipment. For example, MEG requires equipment consisting of liquid helium-cooled detectors that can be used only in magnetically shielded rooms, altogether costing upwards of several million dollars;*[17] and fMRI requires the use of a 1-ton magnet in, again, a shielded room.

- EEG has very high temporal resolution, on the order of milliseconds rather than seconds. EEG is commonly recorded at sampling rates between 250 and 2000 Hz

13.3. RESEARCH USE

in clinical and research settings, but modern EEG data collection systems are capable of recording at sampling rates above 20,000 Hz if desired. MEG and EROS are the only other noninvasive cognitive neuroscience techniques that acquire data at this level of temporal resolution.*[17]

- EEG is relatively tolerant of subject movement, unlike most other neuroimaging techniques. There even exist methods for minimizing, and even eliminating movement artifacts in EEG data *[18]

- EEG is silent, which allows for better study of the responses to auditory stimuli.

- EEG does not aggravate claustrophobia, unlike fMRI, PET, MRS, SPECT, and sometimes MEG*[19]

- EEG does not involve exposure to high-intensity (>1 tesla) magnetic fields, as in some of the other techniques, especially MRI and MRS. These can cause a variety of undesirable issues with the data, and also prohibit use of these techniques with participants that have metal implants in their body, such as metal-containing pacemakers*[20]

- EEG does not involve exposure to radioligands, unlike positron emission tomography.*[21]

- ERP studies can be conducted with relatively simple paradigms, compared with IE block-design fMRI studies

- Extremely uninvasive, unlike Electrocorticography, which actually requires electrodes to be placed on the surface of the brain.

EEG also has some characteristics that compare favorably with behavioral testing:

- EEG can detect covert processing (i.e., processing that does not require a response)*[22]

- EEG can be used in subjects who are incapable of making a motor response*[23]

- Some ERP components can be detected even when the subject is not attending to the stimuli

- Unlike other means of studying reaction time, ERPs can elucidate stages of processing (rather than just the final end result)*[24]

- EEG is a powerful tool for tracking brain changes during different phases of life. EEG sleep analysis can indicate significant aspects of the timing of brain development, including evaluating adolescent brain maturation.*[25]

- In EEG there is a better understanding of what signal is measured as compared to other research techniques, i.e. the BOLD response in MRI.

13.3.2 Disadvantages

- Low spatial resolution on the scalp. fMRI, for example, can directly display areas of the brain that are active, while EEG requires intense interpretation just to hypothesize what areas are activated by a particular response.*[26]

- EEG poorly measures neural activity that occurs below the upper layers of the brain (the cortex).

- Unlike PET and MRS, cannot identify specific locations in the brain at which various neurotransmitters, drugs, etc. can be found.*[21]

- Often takes a long time to connect a subject to EEG, as it requires precise placement of dozens of electrodes around the head and the use of various gels, saline solutions, and/or pastes to keep them in place (although a cap can be used). While the length of time differs dependent on the specific EEG device used, as a general rule it takes considerably less time to prepare a subject for MEG, fMRI, MRS, and SPECT.

- Signal-to-noise ratio is poor, so sophisticated data analysis and relatively large numbers of subjects are needed to extract useful information from EEG*[27]

13.3.3 With other neuroimaging techniques

Simultaneous EEG recordings and fMRI scans have been obtained successfully,*[28]*[29] though successful simultaneous recording requires that several technical difficulties be overcome, such as the presence of ballistocardiographic artifact, MRI pulse artifact and the induction of electrical currents in EEG wires that move within the strong magnetic fields of the MRI. While challenging, these have been successfully overcome in a number of studies.*[30]

MRI's produce detailed images created by generating strong magnetic fields that may induce potentially harmful displacement force and torque. These fields produce potentially harmful radio frequency heating and create image artifacts rendering images useless. Due to these potential risks, only certain medical devices can be used in an MR environment.

Similarly, simultaneous recordings with MEG and EEG have also been conducted, which has several advantages over using either technique alone:

- EEG requires accurate information about certain aspects of the skull that can only be estimated, such as skull radius, and conductivities of various skull locations. MEG does not have this issue, and a simultaneous analysis allows this to be corrected for.

- MEG and EEG both detect activity below the surface of the cortex very poorly, and like EEG, the level of error increases with the depth below the surface of the cortex one attempts to examine. However, the errors are very different between the techniques, and combining them thus allows for correction of some of this noise.

- MEG has access to virtually no sources of brain activity below a few centimetres under the cortex. EEG, on the other hand, can receive signals from greater depth, albeit with a high degree of noise. Combining the two makes it easier to determine what in the EEG signal comes from the surface (since MEG is very accurate in examining signals from the surface of the brain), and what comes from deeper in the brain, thus allowing for analysis of deeper brain signals than either EEG or MEG on its own.*[31]

Recently, a combined EEG/MEG (EMEG) approach has been investigated for the purpose of source reconstruction in epilepsy diagnosis.*[32]

EEG has also been combined with positron emission tomography. This provides the advantage of allowing researchers to see what EEG signals are associated with different drug actions in the brain.*[33]

13.4 Mechanisms

The brain's electrical charge is maintained by billions of neurons.*[34] Neurons are electrically charged (or "polarized") by membrane transport proteins that pump ions across their membranes. Neurons are constantly exchanging ions with the extracellular milieu, for example to maintain resting potential and to propagate action potentials. Ions of similar charge repel each other, and when many ions are pushed out of many neurons at the same time, they can push their neighbours, who push their neighbours, and so on, in a wave. This process is known as volume conduction. When the wave of ions reaches the electrodes on the scalp, they can push or pull electrons on the metal in the electrodes. Since metal conducts the push and pull of electrons easily, the difference in push or pull voltages between any two electrodes can be measured by a voltmeter. Recording these voltages over time gives us the EEG.*[35]

The electric potential generated by an individual neuron is far too small to be picked up by EEG or MEG.*[36] EEG activity therefore always reflects the summation of the synchronous activity of thousands or millions of neurons that have similar spatial orientation. If the cells do not have similar spatial orientation, their ions do not line up and create waves to be detected. Pyramidal neurons of the cortex are thought to produce the most EEG signal because they are well-aligned and fire together. Because voltage field gradients fall off with the square of distance, activity from deep sources is more difficult to detect than currents near the skull.*[37]

Scalp EEG activity shows oscillations at a variety of frequencies. Several of these oscillations have characteristic frequency ranges, spatial distributions and are associated with different states of brain functioning (e.g., waking and the various sleep stages). These oscillations represent synchronized activity over a network of neurons. The neuronal networks underlying some of these oscillations are understood (e.g., the thalamocortical resonance underlying sleep spindles), while many others are not (e.g., the system that generates the posterior basic rhythm). Research that measures both EEG and neuron spiking finds the relationship between the two is complex, with a combination of EEG power in the gamma band and phase in the delta band relating most strongly to neuron spike activity.*[38]

13.5 Method

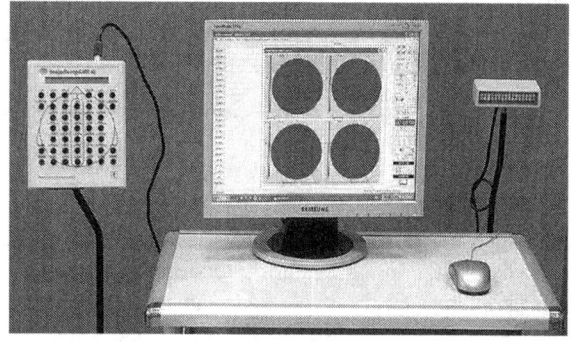

Computer electroencephalograph Neurovisor-BMM 40

In conventional scalp EEG, the recording is obtained by placing electrodes on the scalp with a conductive gel or paste, usually after preparing the scalp area by light abrasion to reduce impedance due to dead skin cells. Many systems typically use electrodes, each of which is attached to an individual wire. Some systems use caps or nets into which electrodes are embedded; this is particularly common when high-density arrays of electrodes are needed.

Electrode locations and names are specified by the International 10–20 system*[39] for most clinical and research applications (except when high-density arrays are used). This system ensures that the naming of electrodes is consistent across laboratories. In most clinical applications, 19 recording electrodes (plus ground and system reference) are used.*[40] A smaller number of electrodes are typically used when recording EEG from neonates. Additional electrodes can be added to the standard set-up when a clinical or research application demands increased spatial resolution for a particular area of the brain. High-density arrays (typically via cap or net) can contain up to 256 electrodes more-or-less evenly spaced around the scalp.

Each electrode is connected to one input of a differential amplifier (one amplifier per pair of electrodes); a common system reference electrode is connected to the other input of each differential amplifier. These amplifiers amplify the voltage between the active electrode and the reference (typically 1,000–100,000 times, or 60–100 dB of voltage gain). In analog EEG, the signal is then filtered (next paragraph), and the EEG signal is output as the deflection of pens as paper passes underneath. Most EEG systems these days, however, are digital, and the amplified signal is digitized via an analog-to-digital converter, after being passed through an anti-aliasing filter. Analog-to-digital sampling typically occurs at 256–512 Hz in clinical scalp EEG; sampling rates of up to 20 kHz are used in some research applications.

During the recording, a series of activation procedures may be used. These procedures may induce normal or abnormal EEG activity that might not otherwise be seen. These procedures include hyperventilation, photic stimulation (with a strobe light), eye closure, mental activity, sleep and sleep deprivation. During (inpatient) epilepsy monitoring, a patient's typical seizure medications may be withdrawn.

The digital EEG signal is stored electronically and can be filtered for display. Typical settings for the high-pass filter and a low-pass filter are 0.5–1 Hz and 35–70 Hz respectively. The high-pass filter typically filters out slow artifact, such as electrogalvanic signals and movement artifact, whereas the low-pass filter filters out high-frequency artifacts, such as electromyographic signals. An additional notch filter is typically used to remove artifact caused by electrical power lines (60 Hz in the United States and 50 Hz in many other countries).*[1]

The EEG signals can be captured with opensource hardware such as OpenBCI and the signal can be processed by freely available EEG software such as EEGLAB or the Neurophysiological Biomarker Toolbox.

As part of an evaluation for epilepsy surgery, it may be necessary to insert electrodes near the surface of the brain, under the surface of the dura mater. This is accomplished via burr hole or craniotomy. This is referred to variously as "electrocorticography (ECoG)", "intracranial EEG (I-EEG)" or "subdural EEG (SD-EEG)". Depth electrodes may also be placed into brain structures, such as the amygdala or hippocampus, structures, which are common epileptic foci and may not be "seen" clearly by scalp EEG. The electrocorticographic signal is processed in the same manner as digital scalp EEG (above), with a couple of caveats. ECoG is typically recorded at higher sampling rates than scalp EEG because of the requirements of Nyquist theorem—the subdural signal is composed of a higher predominance of higher frequency components. Also, many of the artifacts that affect scalp EEG do not impact ECoG, and therefore display filtering is often not needed.

A typical adult human EEG signal is about 10 μV to 100 μV in amplitude when measured from the scalp*[41] and is about 10–20 mV when measured from subdural electrodes.

Since an EEG voltage signal represents a difference between the voltages at two electrodes, the display of the EEG for the reading encephalographer may be set up in one of several ways. The representation of the EEG channels is referred to as a *montage*.

Sequential montage Each channel (i.e., waveform) represents the difference between two adjacent electrodes. The entire montage consists of a series of these channels. For example, the channel "Fp1-F3" represents the difference in voltage between the Fp1 electrode and the F3 electrode. The next channel in the montage, "F3-C3", represents the voltage difference between F3 and C3, and so on through the entire array of electrodes.

Referential montage Each channel represents the difference between a certain electrode and a designated reference electrode. There is no standard position for this reference; it is, however, at a different position than the "recording" electrodes. Midline positions are often used because they do not amplify the signal in one hemisphere vs. the other. Another popular reference is "linked ears", which is a physical or mathematical average of electrodes attached to both earlobes or mastoids.

Average reference montage The outputs of all of the amplifiers are summed and averaged, and this averaged signal is used as the common reference for each channel.

Laplacian montage Each channel represents the difference between an electrode and a weighted average of the surrounding electrodes.*[42]

When analog (paper) EEGs are used, the technologist switches between montages during the recording in order to highlight or better characterize certain features of the EEG. With digital EEG, all signals are typically digitized and stored in a particular (usually referential) montage; since any montage can be constructed mathematically from any other, the EEG can be viewed by the electroencephalographer in any display montage that is desired.

The EEG is read by a clinical neurophysiologist or neurologist (depending on local custom and law regarding medical specialities), optimally one who has specific training in the interpretation of EEGs for clinical purposes. This is done by visual inspection of the waveforms, called graphoelements. The use of computer signal processing of the EEG—so-called quantitative electroencephalography—is somewhat controversial when used for clinical purposes (although there are many research uses).

13.5.1 Limitations

EEG has several limitations. Most important is its poor spatial resolution.*[43] EEG is most sensitive to a particular set of post-synaptic potentials: those generated in superficial layers of the cortex, on the crests of gyri directly abutting the skull and radial to the skull. Dendrites, which are deeper in the cortex, inside sulci, in midline or deep structures (such as the cingulate gyrus or hippocampus), or producing currents that are tangential to the skull, have far less contribution to the EEG signal.

EEG recordings do not directly capture axonal action potentials. An action potential can be accurately represented as a current quadrupole, meaning that the resulting field decreases more rapidly than the ones produced by the current dipole of post-synaptic potentials.*[44] In addition, since EEGs represent averages of thousands of neurons, a large population of cells in synchronous activity is necessary to cause a significant deflection on the recordings. Action potentials are very fast and, as a consequence, the chances of field summation are slim. However, neural backpropagation, as a typically longer dendritic current dipole, can be picked up by EEG electrodes and is a reliable indication of the occurrence of neural output.

Not only do EEGs capture dendritic currents almost exclusively as opposed to axonal currents, they also show a preference for activity on populations of parallel dendrites and transmitting current in the same direction at the same time. Pyramidal neurons of cortical layers II/III and V extend apical dendrites to layer I. Currents moving up or down these processes underlie most of the signals produced by electroencephalography.*[45]

Therefore, EEG provides information with a large bias to select neuron types, and generally should not be used to make claims about global brain activity. The meninges, cerebrospinal fluid and skull "smear" the EEG signal, obscuring its intracranial source.

It is mathematically impossible to reconstruct a unique intracranial current source for a given EEG signal,*[1] as some currents produce potentials that cancel each other out. This is referred to as the inverse problem. However, much work has been done to produce remarkably good estimates of, at least, a localized electric dipole that represents the recorded currents.

13.5.2 EEG vs fMRI, fNIRS and PET

EEG has several strong points as a tool for exploring brain activity. EEGs can detect changes over milliseconds, which is excellent considering an action potential takes approximately 0.5–130 milliseconds to propagate across a single neuron, depending on the type of neuron.*[46] Other methods of looking at brain activity, such as PET and fMRI have time resolution between seconds and minutes. EEG measures the brain's electrical activity directly, while other methods record changes in blood flow (e.g., SPECT, fMRI) or metabolic activity (e.g., PET, NIRS), which are indirect markers of brain electrical activity. EEG can be used simultaneously with fMRI so that high-temporal-resolution data can be recorded at the same time as high-spatial-resolution data, however, since the data derived from each occurs over a different time course, the data sets do not necessarily represent exactly the same brain activity. There are technical difficulties associated with combining these two modalities, including the need to remove the *MRI gradient artifact* present during MRI acquisition and the ballistocardiographic artifact (resulting from the pulsatile motion of blood and tissue) from the EEG. Furthermore, currents can be induced in moving EEG electrode wires due to the magnetic field of the MRI.

EEG can be used simultaneously with NIRS without major technical difficulties. There is no influence of these modalities on each other and a combined measurement can give useful information about electrical activity as well as local hemodynamics.

13.5.3 EEG vs MEG

EEG reflects correlated synaptic activity caused by post-synaptic potentials of cortical neurons. The ionic currents involved in the generation of fast action potentials may not contribute greatly to the averaged field potentials representing the EEG.*[36]*[47] More specifically, the scalp electrical potentials that produce EEG are generally thought to be caused by the extracellular ionic currents caused by dendritic electrical activity, whereas the fields producing magnetoencephalographic signals[17] are associated with intracellular ionic currents.[48]

EEG can be recorded at the same time as MEG so that data from these complementary high-time-resolution techniques can be combined.

Studies on numerical modeling of EEG and MEG have also been done.[49]

13.6 Normal activity

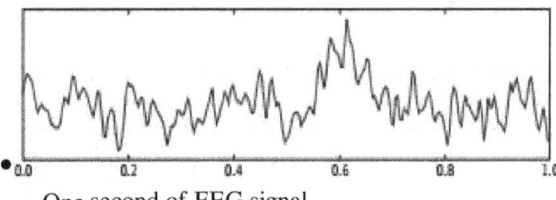

- One second of EEG signal

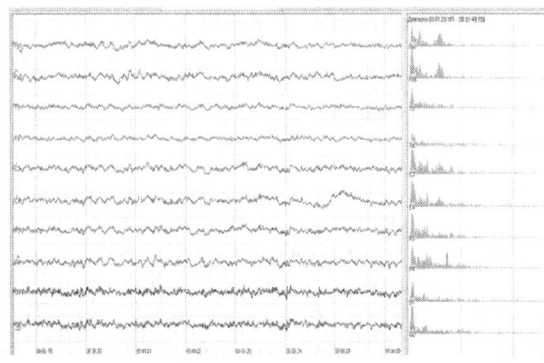

- The sample of human EEG with prominent resting state activity – alpha-rhythm. Left: EEG traces (horizontal – time in seconds; vertical – amplitudes, scale 100 μV). Right: power spectra of shown signals (vertical lines – 10 and 20 Hz, scale is linear). Alpha-rhythm consists of sinusoidal-like waves with frequencies in 8–12 Hz range (11 Hz in this case) more prominent in posterior sites. Alpha range is red at power spectrum graph.

EEG traces (horizontal – time in seconds; vertical – amplitudes, scale 100 μV). Right: power spectra of shown signals (vertical lines – 10 and 20 Hz, scale is linear). 80–90% of people have prominent sinusoidal-like waves with frequencies in 8–12 Hz range – alpha rhythm. Others (like this) lack this type of activity.

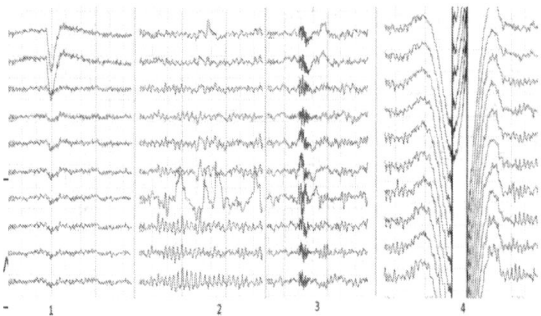

- The sample of human EEG with in resting state. Left:

- samples of main types of artifacts in human EEG. 1: Electrooculographic artifact caused by the excitation of eyeball's muscles (related to blinking, for example). Big-amplitude, slow, positive wave prominent in frontal electrodes. 2: Electrode's artifact caused by bad contact (and thus bigger impedance) between P3 electrode and skin. 3: Swallowing artifact. 4: Common reference electrode's artifact caused by bad contact between reference electrode and skin. Huge wave similar in all channels.

The EEG is typically described in terms of (1) rhythmic activity and (2) transients. The rhythmic activity is divided into bands by frequency. To some degree, these frequency bands are a matter of nomenclature (i.e., any rhythmic activity between 8–12 Hz can be described as "alpha"), but these designations arose because rhythmic activity within a certain frequency range was noted to have a certain distribution over the scalp or a certain biological significance. Frequency bands are usually extracted using spectral methods (for instance Welch) as implemented for instance in freely available EEG software such as EEGLAB or the Neurophysiological Biomarker Toolbox. Computational processing of the EEG is often named quantitative electroencephalography (qEEG).

Most of the cerebral signal observed in the scalp EEG falls in the range of 1–20 Hz (activity below or above this range is likely to be artifactual, under standard clinical recording techniques). Waveforms are subdivided into bandwidths known as alpha, beta, theta, and delta to signify the majority of the EEG used in clinical practice.[50]

The practice of using only whole numbers in the definitions comes from practical considerations in the days when only whole cycles could be counted on paper records. This leads to gaps in the definitions, as seen elsewhere on this page. The theoretical definitions have always been more carefully defined to include all frequencies. Unfortunately there is no agreement in standard reference works on what these ranges should be – values for the upper end of alpha and lower end of beta include 12, 13, 14 and 15. If the threshold is taken as 14 Hz, then the slowest beta wave has about the same duration as the longest spike (70 ms), which makes this the most useful value.

Others sometimes divide the bands into sub-bands for the purposes of data analysis.

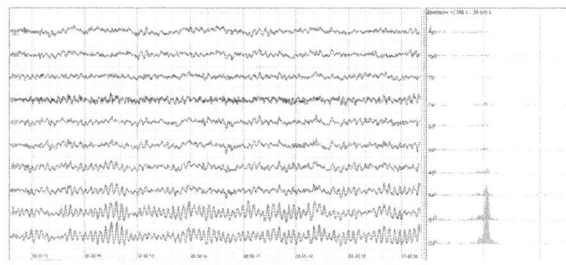

Human EEG with prominent alpha-rhythm

13.6.1 Wave patterns

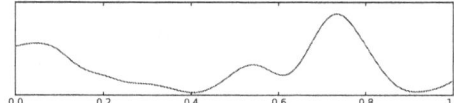

Delta waves

- Delta is the frequency range up to 4 Hz. It tends to be the highest in amplitude and the slowest waves. It is seen normally in adults in slow-wave sleep. It is also seen normally in babies. It may occur focally with subcortical lesions and in general distribution with diffuse lesions, metabolic encephalopathy hydrocephalus or deep midline lesions. It is usually most prominent frontally in adults (e.g. FIRDA – frontal intermittent rhythmic delta) and posteriorly in children (e.g. OIRDA – occipital intermittent rhythmic delta).

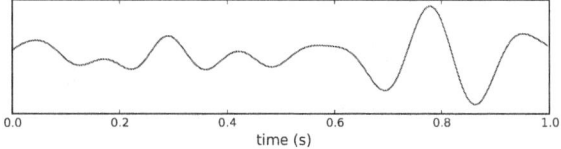

Theta waves

- Theta is the frequency range from 4 Hz to 7 Hz. Theta is seen normally in young children. It may be seen in drowsiness or arousal in older children and adults; it can also be seen in meditation.*[58] Excess theta for age represents abnormal activity. It can be seen as a focal disturbance in focal subcortical lesions; it can be seen in generalized distribution in diffuse disorder or metabolic encephalopathy or deep midline disorders or some instances of hydrocephalus. On the contrary this range has been associated with reports of relaxed, meditative, and creative states.

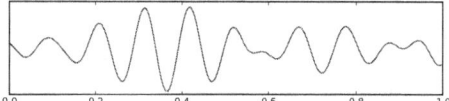

Alpha waves

- Alpha is the frequency range from 7 Hz to 14 Hz. Hans Berger named the first rhythmic EEG activity he saw as the "alpha wave". This was the "posterior basic rhythm" (also called the "posterior dominant rhythm" or the "posterior alpha rhythm"), seen in the posterior regions of the head on both sides, higher in amplitude on the dominant side. It emerges with closing of the eyes and with relaxation, and attenuates with eye opening or mental exertion. The posterior basic rhythm is actually slower than 8 Hz in young children (therefore technically in the theta range).

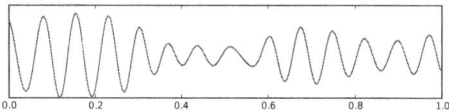

Sensorimotor rhythm aka mu rhythm

In addition to the posterior basic rhythm, there are other normal alpha rhythms such as the mu rhythm (alpha activity in the contralateral sensory and motor cortical areas) that emerges when the hands and arms are idle; and the "third rhythm" (alpha activity in the temporal or frontal lobes).*[59]*[60] Alpha can be abnormal; for example, an EEG that has diffuse alpha occurring in coma and is not responsive to external stimuli is referred to as "alpha coma".

- Beta is the frequency range from 14 Hz to about 30 Hz. It is seen usually on both sides in symmetrical distribution and is most evident frontally. Beta activity is

13.7. ARTIFACTS

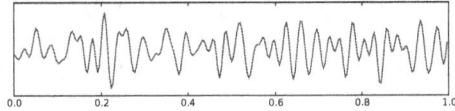

Beta waves

closely linked to motor behavior and is generally attenuated during active movements.*[61] Low-amplitude beta with multiple and varying frequencies is often associated with active, busy or anxious thinking and active concentration. Rhythmic beta with a dominant set of frequencies is associated with various pathologies, such as Dup15q syndrome, and drug effects, especially benzodiazepines. It may be absent or reduced in areas of cortical damage. It is the dominant rhythm in patients who are alert or anxious or who have their eyes open.

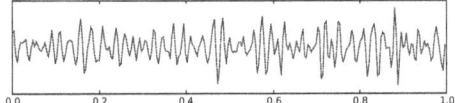

Gamma waves

- Gamma is the frequency range approximately 30–100 Hz. Gamma rhythms are thought to represent binding of different populations of neurons together into a network for the purpose of carrying out a certain cognitive or motor function.*[1]

- Mu range is 8–13 Hz and partly overlaps with other frequencies. It reflects the synchronous firing of motor neurons in rest state. Mu suppression is thought to reflect motor mirror neuron systems, because when an action is observed, the pattern extinguishes, possibly because of the normal neuronal system and the mirror neuron system "go out of sync" and interfere with each other.*[56]

"Ultra-slow" or "near-DC" activity is recorded using DC amplifiers in some research contexts. It is not typically recorded in a clinical context because the signal at these frequencies is susceptible to a number of artifacts.

Some features of the EEG are transient rather than rhythmic. Spikes and sharp waves may represent seizure activity or interictal activity in individuals with epilepsy or a predisposition toward epilepsy. Other transient features are normal: vertex waves and sleep spindles are seen in normal sleep.

Note that there are types of activity that are statistically uncommon, but not associated with dysfunction or disease.

These are often referred to as "normal variants". The mu rhythm is an example of a normal variant.

The normal electroencephalography (EEG) varies by age. The neonatal EEG is quite different from the adult EEG. The EEG in childhood generally has slower frequency oscillations than the adult EEG.

The normal EEG also varies depending on state. The EEG is used along with other measurements (EOG, EMG) to define sleep stages in polysomnography. Stage I sleep (equivalent to drowsiness in some systems) appears on the EEG as drop-out of the posterior basic rhythm. There can be an increase in theta frequencies. Santamaria and Chiappa cataloged a number of the variety of patterns associated with drowsiness. Stage II sleep is characterized by sleep spindles – transient runs of rhythmic activity in the 12–14 Hz range (sometimes referred to as the "sigma" band) that have a frontal-central maximum. Most of the activity in Stage II is in the 3–6 Hz range. Stage III and IV sleep are defined by the presence of delta frequencies and are often referred to collectively as "slow-wave sleep". Stages I–IV comprise non-REM (or "NREM") sleep. The EEG in REM (rapid eye movement) sleep appears somewhat similar to the awake EEG.

EEG under general anesthesia depends on the type of anesthetic employed. With halogenated anesthetics, such as halothane or intravenous agents, such as propofol, a rapid (alpha or low beta), nonreactive EEG pattern is seen over most of the scalp, especially anteriorly; in some older terminology this was known as a WAR (widespread anterior rapid) pattern, contrasted with a WAIS (widespread slow) pattern associated with high doses of opiates. Anesthetic effects on EEG signals are beginning to be understood at the level of drug actions on different kinds of synapses and the circuits that allow synchronized neuronal activity (see: http://www.stanford.edu/group/maciverlab/).

13.7 Artifacts

13.7.1 Biological artifacts

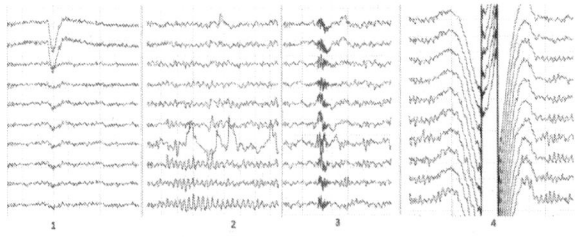

Main types of artifacts in human EEG

Electrical signals detected along the scalp by an EEG, but

that originate from non-cerebral origin are called artifacts. EEG data is almost always contaminated by such artifacts. The amplitude of artifacts can be quite large relative to the size of amplitude of the cortical signals of interest. This is one of the reasons why it takes considerable experience to correctly interpret EEGs clinically. Some of the most common types of biological artifacts include:

- Eye-induced artifacts (includes eye blinks, eye movements and extra-ocular muscle activity)
- ECG (cardiac) artifacts
- EMG (muscle activation)-induced artifacts
- Glossokinetic artifacts

The most prominent eye-induced artifacts are caused by the potential difference between the cornea and retina, which is quite large compared to cerebral potentials. When the eyes and eyelids are completely still, this corneo-retinal dipole does not affect EEG. However, blinks occur several times per minute, the eyes movements occur several times per second. Eyelid movements, occurring mostly during blinking or vertical eye movements, elicit a large potential seen mostly in the difference between the Electrooculography (EOG) channels above and below the eyes. An established explanation of this potential regards the eyelids as sliding electrodes that short-circuit the positively charged cornea to the extra-ocular skin.[62][63] Rotation of the eyeballs, and consequently of the corneo-retinal dipole, increases the potential in electrodes towards which the eyes are rotated, and decrease the potentials in the opposing electrodes.[64] Eye movements called saccades also generate transient electromyographic potentials, known as saccadic spike potentials (SPs).[65] The spectrum of these SPs overlaps the gamma-band (see Gamma wave), and seriously confounds analysis of induced gamma-band responses,[66] requiring tailored artifact correction approaches.[65] Purposeful or reflexive eye blinking also generates electromyographic potentials, but more importantly there is reflexive movement of the eyeball during blinking that gives a characteristic artifactual appearance of the EEG (see Bell's phenomenon).

Eyelid fluttering artifacts of a characteristic type were previously called Kappa rhythm (or Kappa waves). It is usually seen in the prefrontal leads, that is, just over the eyes. Sometimes they are seen with mental activity. They are usually in the Theta (4–7 Hz) or Alpha (7–14 Hz) range. They were named because they were believed to originate from the brain. Later study revealed they were generated by rapid fluttering of the eyelids, sometimes so minute that it was difficult to see. They are in fact noise in the EEG reading, and should not technically be called a rhythm or wave. Therefore, current usage in electroencephalography refers to the phenomenon as an eyelid fluttering artifact, rather than a Kappa rhythm (or wave).[67]

Some of these artifacts can be useful in various applications. The EOG signals, for instance, can be used to detect[65] and track eye-movements, which are very important in polysomnography, and is also in conventional EEG for assessing possible changes in alertness, drowsiness or sleep.

ECG artifacts are quite common and can be mistaken for spike activity. Because of this, modern EEG acquisition commonly includes a one-channel ECG from the extremities. This also allows the EEG to identify cardiac arrhythmias that are an important differential diagnosis to syncope or other episodic/attack disorders.

Glossokinetic artifacts are caused by the potential difference between the base and the tip of the tongue. Minor tongue movements can contaminate the EEG, especially in parkinsonian and tremor disorders.

13.7.2 Environmental artifacts

In addition to artifacts generated by the body, many artifacts originate from outside the body. Movement by the patient, or even just settling of the electrodes, may cause *electrode pops*, spikes originating from a momentary change in the impedance of a given electrode. Poor grounding of the EEG electrodes can cause significant 50 or 60 Hz artifact, depending on the local power system's frequency. A third source of possible interference can be the presence of an IV drip; such devices can cause rhythmic, fast, low-voltage bursts, which may be confused for spikes.

13.7.3 Artifact correction

Recently, independent component analysis (ICA) techniques have been used to correct or remove EEG contaminants.[65][68][69][70][71][72] These techniques attempt to "unmix" the EEG signals into some number of underlying components. There are many source separation algorithms, often assuming various behaviors or natures of EEG. Regardless, the principle behind any particular method usually allow "remixing" only those components that would result in "clean" EEG by nullifying (zeroing) the weight of unwanted components. Fully automated artifact rejection methods, which use ICA, have also been developed.[73]

In the last few years, by comparing data from paralysed and unparalysed subjects, EEG contamination by muscle has been shown to be far more prevalent than had previously been realized, particularly in the gamma range above 20 Hz.[74] However, Surface Laplacian has been shown to be effective in eliminating muscle artefact, particularly for central electrodes, which are further from the strongest contaminants.*[75] The combination of Surface Laplacian with automated techniques for removing muscle components using ICA proved particularly effective in a follow up study.*[76]

13.8 Abnormal activity

Abnormal activity can broadly be separated into epileptiform and non-epileptiform activity. It can also be separated into focal or diffuse.

Focal epileptiform discharges represent fast, synchronous potentials in a large number of neurons in a somewhat discrete area of the brain. These can occur as interictal activity, between seizures, and represent an area of cortical irritability that may be predisposed to producing epileptic seizures. Interictal discharges are not wholly reliable for determining whether a patient has epilepsy nor where his/her seizure might originate. (See focal epilepsy.)

Generalized epileptiform discharges often have an anterior maximum, but these are seen synchronously throughout the entire brain. They are strongly suggestive of a generalized epilepsy.

Focal non-epileptiform abnormal activity may occur over areas of the brain where there is focal damage of the cortex or white matter. It often consists of an increase in slow frequency rhythms and/or a loss of normal higher frequency rhythms. It may also appear as focal or unilateral decrease in amplitude of the EEG signal.

Diffuse non-epileptiform abnormal activity may manifest as diffuse abnormally slow rhythms or bilateral slowing of normal rhythms, such as the PBR.

Intracortical Encephalogram electrodes and sub-dural electrodes can be used in tandem to discriminate and discretize artifact from epileptiform and other severe neurological events.

More advanced measures of abnormal EEG signals have also recently received attention as possible biomarkers for different disorders such as Alzheimer's disease.*[77]

13.8.1 Remote communication

The United States Army Research Office budgeted $4 million in 2009 to researchers at the University of California, Irvine to develop EEG processing techniques to identify correlates of imagined speech and intended direction to enable soldiers on the battlefield to communicate via computer-mediated reconstruction of team members' EEG signals, in the form of understandable signals such as words.*[78]

13.9 Economics

Inexpensive EEG devices exist for the low-cost research and consumer markets. Recently, a few companies have miniaturized medical grade EEG technology to create versions accessible to the general public. Some of these companies have built commercial EEG devices retailing for less than $100 USD.

- In 2004 OpenEEG released its ModularEEG as open source hardware. Compatible open source software includes a game for balancing a ball.

- In 2007 NeuroSky released the first affordable consumer based EEG along with the game NeuroBoy. This was also the first large scale EEG device to use dry sensor technology.*[79]

- In 2008 OCZ Technology developed device for use in video games relying primarily on electromyography.

- In 2008 the Final Fantasy developer Square Enix announced that it was partnering with NeuroSky to create a game, Judecca.*[80]*[81]

- In 2009 Mattel partnered with NeuroSky to release the Mindflex, a game that used an EEG to steer a ball through an obstacle course. By far the best selling consumer based EEG to date.*[80]*[82]

- In 2009 Uncle Milton Industries partnered with NeuroSky to release the Star Wars Force Trainer, a game designed to create the illusion of possessing The Force.*[80]*[83]

- In 2009 Emotiv released the EPOC, a 14 channel EEG device. The EPOC is the first commercial BCI to not use dry sensor technology, requiring users to apply a saline solution to electrode pads (which need remoistening after an hour or two of use).*[84]

- In 2010, NeuroSky added a blink and electromyography function to the MindSet.*[85]

- In 2011, NeuroSky released the MindWave, an EEG device designed for educational purposes and games.*[86] The MindWave won the Guinness Book of World Records award for "Heaviest machine moved using a brain control interface".*[87]

- In 2012, a Japanese gadget project, neurowear, released Necomimi: a headset with motorized cat ears.

The headset is a NeuroSky MindWave unit with two motors on the headband where a cat's ears might be. Slipcovers shaped like cat ears sit over the motors so that as the device registers emotional states the ears move to relate. For example, when relaxed, the ears fall to the sides and perk up when excited again.

- In 2014, OpenBCI released an eponymous open source brain-computer interface after a successful kickstarter campaign in 2013. The basic OpenBCI has 8 channels, expandable to 16, and supports EEG, EKG, and EMG. The OpenBCI is based on the Texas Instruments ADS1299 IC and the Arduino or PIC microcontroller, and costs $399 for the basic version. It uses standard metal cup electrodes and conductive paste.

- In 2015, Mind Solutions Inc released the smallest consumer BCI to date, the NeuroSync. This device functions as a dry sensor at a size no larger than a Bluetooth ear piece.*[88]

- In 2015, A Chinese-based company Macrotellect released BrainLink Pro and BrainLink Lite, a consumer grade EEG wearable product providing 20 brain fitness enhancement Apps on Apple and Android App Stores.*[89]

13.10 Future research

The EEG has been used for many purposes besides the conventional uses of clinical diagnosis and conventional cognitive neuroscience. An early use was during World War II by the U.S. Army Air Corps to screen out pilots in danger of having seizures;*[90] long-term EEG recordings in epilepsy patients are still used today for seizure prediction. Neurofeedback remains an important extension, and in its most advanced form is also attempted as the basis of brain computer interfaces. The EEG is also used quite extensively in the field of neuromarketing. In recent years, researchers have relied on EEG to understand the neural basis of swallowing.*[91]

The EEG is altered by drugs that affect brain functions, the chemicals that are the basis for psychopharmacology. Berger's early experiments recorded the effects of drugs on EEG. The science of pharmaco-electroencephalography has developed methods to identify substances that systematically alter brain functions for therapeutic and recreational use.

Honda is attempting to develop a system to enable an operator to control its Asimo robot using EEG, a technology it eventually hopes to incorporate into its automobiles.*[92]

EEGs have been used as evidence in criminal trials in the Indian state of Maharashtra.*[93]*[94]

A lot of research is currently being carried out in order to make EEG devices smaller, more portable and easier to use. So called "Wearable EEG" is based upon creating low power wireless collection electronics and 'dry' electrodes which do not require a conductive gel to be used.*[95] Wearable EEG aims to provide small EEG devices which are present only on the head and which can record EEG for days, weeks, or months at a time, as ear-EEG. Such prolonged and easy-to-use monitoring could make a step change in the diagnosis of chronic conditions such as epilepsy, and greatly improve the end-user acceptance of BCI systems.*[96] Research is also being carried out on identifying specific solutions to increase the battery lifetime of Werable EEG devices through the use of the data reduction approach. For example, in the context of epilepsy diagnosis, data reduction has been used to extend the battery lifetime of Werable EEG devices by intelligently selecting, and only transmitting, diagnostically relevant EEG data*[97].

13.11 See also

- 10-20 system (EEG)
- Amplitude integrated electroencephalography
- Binaural beats
- Brain-computer interface
- Brainwave synchronization
- CAET-Canadian association of EEG Technology
- Cerebral function monitoring
- Comparison of consumer brain-computer interface devices
- Direct brain interfaces
- EEG measures during anesthesia
- EEG microstates
- Electrocorticography
- Electromagnetic Weapon
- Electroneurogram
- Electropalatograph
- Emotiv Systems
- European data format

- Event-related potential
- Evoked potential
- FieldTrip
- God helmet
- Hemoencephalography
- Hypersynchronization of electrophysiological activity in epilepsy
- Imagined Speech
- Induced activity
- Intracranial EEG
- Local field potentials
- Magnetoencephalography
- Mind machine
- Neural oscillations
- Neurofeedback
- Ongoing brain activity
- Spontaneous potential

13.12 References

[1] Niedermeyer E.; da Silva F.L. (2004). *Electroencephalography: Basic Principles, Clinical Applications, and Related Fields.* Lippincott Williams & Wilkins. ISBN 0-7817-5126-8.

[2] Tatum, William O. (2014). *Handbook of EEG interpretation.* Demos Medical Publishing. pp. 155–190. ISBN 9781617051807. OCLC 874563370.

[3] "EEG".

[4] Chernecky, Cynthia C.; Berger, Barbara J. (2013). *Laboratory tests and diagnostic procedures* (6th ed.). St. Louis, Mo.: Elsevier. ISBN 9781455706945.

[5] Ehinger, BV; Fischer, P; Gert, AL; Kaufhold, L; Weber, F; Pipa, G; König, P ((2014). "Kinesthetic and vestibular information modulate alpha activity during spatial navigation: A mobile EEG study.". *Front Hum Neurosci.* **8**: 71. PMID 24616681. doi:10.3389/fnhum.2014.00071.

[6] Swartz, Barbara E. (1998). "The advantages of digital over analog recording techniques". *Electroencephalography and Clinical Neurophysiology.* **106** (2): 113–7. PMID 9741771. doi:10.1016/S0013-4694(97)00113-2.

[7] Coenen, Anton; Edward Fine; Oksana Zayachkivska (2014). "Adolf Beck: A Forgotten Pioneer In Electroencephalography". *Journal of the History of the Neurosciences.* **23** (3): 276–286. doi:10.1080/0964704x.2013.867600.

[8] Pravdich-Neminsky, VV. (1913). "Ein Versuch der Registrierung der elektrischen Gehirnerscheinungen". *Zentralblatt für Physiologie.* **27**: 951–60.

[9] Haas, L F (2003). "Hans Berger (1873-1941), Richard Caton (1842-1926), and electroencephalography". *Journal of Neurology, Neurosurgery & Psychiatry.* **74** (1): 9. PMC 1738204. PMID 12486257. doi:10.1136/jnnp.74.1.9.

[10] Millet, David (2002). "The Origins of EEG". *International Society for the History of the Neurosciences* (ISHN).

[11] S. Bozinovski, M. Sestakov, L. Bozinovska: Using EEG alpha rhythm to control a mobile robot, In G. Harris, C. Walker (eds.) Proc. IEEE Annual Conference of Medical and Biological Society, p. 1515-1516, New Orleans, 1988

[12] S. Bozinovski: Mobile robot trajectory control: From fixed rails to direct bioelectric control, In O. Kaynak (ed.) Proc. IEEE Workshop on Intelligent Motion Control, p. 63-67, Istanbul, 1990.

[13] Dr Mohammed Ashfaque Tinmaswala, Dr Valinjker S.K, Dr Shilpa Hegde, Dr Parmeshwar Taware Electroencephalographic Abnormalities in First Onset Afebrile and Complex Febrile Seizures and Its Association with Type of Seizures. http://jmscr.igmpublication.org/v3-i8/28%20jmscr.pdf

[14] American Academy of Neurology. "Five Things Physicians and Patients Should Question". *Choosing Wisely: an initiative of the ABIM Foundation.* American Academy of Neurology. Retrieved August 1, 2013, which cites

- Gronseth, G. S.; Greenberg, M. K. (1995). "The utility of the electroencephalogram in the evaluation of patients presenting with headache: A review of the literature". *Neurology.* **45** (7): 1263–1267. PMID 7617180. doi:10.1212/WNL.45.7.1263.

[15] Vespa, Paul M.; Nenov, Val; Nuwer, Marc R. (1999). "Continuous EEG Monitoring in the Intensive Care Unit: Early Findings and Clinical Efficacy". *Journal of Clinical Neurophysiology.* **16** (1): 1–13. PMID 10082088. doi:10.1097/00004691-199901000-00001.

[16] Schultz, Teal L. (2012). "Technical Tips: MRI Compatible EEG Electrodes: Advantages, Disadvantages, And Financial Feasibility In A Clinical Setting.". *Neurodiagnostic Journal* 52.1: 69–81.

[17] Hämäläinen, Matti; Hari, Riitta; Ilmoniemi, Risto J.; Knuutila, Jukka; Lounasmaa, Olli V. (1993). "Magnetoencephalography-theory, instrumentation, and applications to noninvasive studies of the working human brain". *Reviews of Modern Physics.* **65** (2): 413–97. Bibcode:1993RvMP...65..413H. doi:10.1103/RevModPhys.65.413.

[18] O'Regan, S; Faul, S; Marnane, W (2010). "2010 Annual International Conference of the IEEE Engineering in Medicine and Biology": 6353–6. ISBN 978-1-4244-4123-5. doi:10.1109/IEMBS.2010.5627282. |chapter= ignored (help)

[19] Murphy, Kieran J.; Brunberg, James A. (1997). "Adult claustrophobia, anxiety and sedation in MRI". *Magnetic Resonance Imaging*. **15** (1): 51–4. PMID 9084025. doi:10.1016/S0730-725X(96)00351-7.

[20] Schenck, John F. (1996). "The role of magnetic susceptibility in magnetic resonance imaging: MRI magnetic compatibility of the first and second kinds". *Medical Physics*. **23** (6): 815–50. PMID 8798169. doi:10.1118/1.597854.

[21] Yasuno, Fumihiko; Brown, Amira K; Zoghbi, Sami S; Krushinski, Joseph H; Chernet, Eyassu; Tauscher, Johannes; Schaus, John M; Phebus, Lee A; Chesterfield, Amy K; Felder, Christian C; Gladding, Robert L; Hong, Jinsoo; Halldin, Christer; Pike, Victor W; Innis, Robert B (2007). "The PET Radioligand \11C]MePPEP Binds Reversibly and with High Specific Signal to Cannabinoid CB1 Receptors in Nonhuman Primate Brain". *Neuropsychopharmacology*. **33** (2): 259–69. PMID 17392732. doi:10.1038/sj.npp.1301402.

[22] Mulholland, Thomas (2012). "Objective EEG Methods for Studying Covert Shifts of Visual Attention". In McGuigan, F. J.; Schoonover, R. A. *The Psychophysiology of Thinking: Studies of Covert Processes*. pp. 109–51. ISBN 978-0-323-14700-2.

[23] Hinterberger, Thilo; Kübler, Andrea; Kaiser, Jochen; Neumann, Nicola; Birbaumer, Niels (2003). "A brain–computer interface (BCI) for the locked-in: Comparison of different EEG classifications for the thought translation device". *Clinical Neurophysiology*. **114** (3): 416–25. PMID 12705422. doi:10.1016/S1388-2457(02)00411-X.

[24] Sereno, SC; Rayner, K; Posner, MI (1998). "Establishing a time-line of word recognition: Evidence from eye movements and event-related potentials". *NeuroReport*. **9** (10): 2195–200. PMID 9694199. doi:10.1097/00001756-199807130-00009.

[25] Feinberg, I.; Campbell, I. G. (2012). "Longitudinal sleep EEG trajectories indicate complex patterns of adolescent brain maturation". *AJP: Regulatory, Integrative and Comparative Physiology*. **304** (4): R296–303. PMC 3567357. PMID 23193115. doi:10.1152/ajpregu.00422.2012. Lay summary – *ScienceDaily* (March 19, 2013).

[26] Srinivasan, Ramesh (1999). "Methods to Improve the Spatial Resolution of EEG". *International Journal*. **1** (1): 102–11.

[27] Schlögl, Alois; Slater, Mel; Pfurtscheller, Gert (2002). "Presence research and EEG" (PDF).

[28] Horovitz, Silvina G.; Skudlarski, Pawel; Gore, John C. (2002). "Correlations and dissociations between BOLD signal and P300 amplitude in an auditory oddball task: A parametric approach to combining fMRI and ERP". *Magnetic Resonance Imaging*. **20** (4): 319–25. PMID 12165350. doi:10.1016/S0730-725X(02)00496-4.

[29] Laufs, H; Kleinschmidt, A; Beyerle, A; Eger, E; Salek-Haddadi, A; Preibisch, C; Krakow, K (2003). "EEG-correlated fMRI of human alpha activity". *NeuroImage*. **19** (4): 1463–76. PMID 12948703. doi:10.1016/S1053-8119(03)00286-6.

[30] Difrancesco, Mark W.; Holland, Scott K.; Szaflarski, Jerzy P. (2008). "Simultaneous EEG/Functional Magnetic Resonance Imaging at 4 Tesla: Correlates of Brain Activity to Spontaneous Alpha Rhythm During Relaxation". *Journal of Clinical Neurophysiology*. **25** (5): 255–64. PMC 2662486. PMID 18791470. doi:10.1097/WNP.0b013e3181879d56.

[31] Huizenga, HM; Van Zuijen, TL; Heslenfeld, DJ; Molenaar, PC (2001). "Simultaneous MEG and EEG source analysis". *Physics in medicine and biology*. **46** (7): 1737–51. PMID 11474922. doi:10.1088/0031-9155/46/7/301.

[32] Aydin Ü, Vorwerk J, Dümpelmann M, Küpper P, Kugel H, Heers M, Wellmer J, Kellinghaus C, Haueisen J, Rampp S, Stefan H, Wolters CH (2015). "Combined EEG/MEG can outperform single modality EEG or MEG source reconstruction in presurgical epilepsy diagnosis". *PLOS ONE* (Review). **10** (3): e0118753. PMC 4356563. PMID 25761059. doi:10.1371/journal.pone.0118753.

[33] Schreckenberger, Mathias; Lange-Asschenfeldt, Christian; Lochmann, Matthias; Mann, Klaus; Siessmeier, Thomas; Buchholz, Hans-Georg; Bartenstein, Peter; Gründer, Gerhard (2004). "The thalamus as the generator and modulator of EEG alpha rhythm: A combined PET/EEG study with lorazepam challenge in humans". *NeuroImage*. **22** (2): 637–44. PMID 15193592. doi:10.1016/j.neuroimage.2004.01.047.

[34] The Human Brain in NumbersHerculano-Houzel S (2009). "The Human Brain in Numbers". *NIH*. **3**: 31. PMC 2776484. PMID 19915731. doi:10.3389/neuro.09.031.2009.

[35] Tatum, W. O., Husain, A. M., Benbadis, S. R. (2008) "Handbook of EEG Interpretation" Demos Medical Publishing.

[36] Nunez PL, Srinivasan R (1981). *Electric fields of the brain: The neurophysics of EEG*. Oxford University Press.

[37] Klein, S.; Thorne, B. M. (3 October 2006). *Biological psychology*. New York, N.Y.: Worth. ISBN 0-7167-9922-7.

[38] Whittingstall, Kevin; Logothetis, Nikos K. (2009). "Frequency-Band Coupling in Surface EEG Reflects Spiking Activity in Monkey Visual Cortex"

. *Neuron*. **64** (2): 281–9. PMID 19874794. doi:10.1016/j.neuron.2009.08.016.

[39] Towle, Vernon L.; Bolaños, José; Suarez, Diane; Tan, Kim; Grzeszczuk, Robert; Levin, David N.; Cakmur, Raif; Frank, Samuel A.; Spire, Jean-Paul (1993). "The spatial location of EEG electrodes: Locating the best-fitting sphere relative to cortical anatomy". *Electroencephalography and Clinical Neurophysiology*. **86** (1): 1–6. PMID 7678386. doi:10.1016/0013-4694(93)90061-Y.

[40] &Na; (1994). "Guideline Seven A Proposal for Standard Montages to Be Used in Clinical EEG". *Journal of Clinical Neurophysiology*. **11** (1): 30–6. PMID 8195424. doi:10.1097/00004691-199401000-00008.

[41] Aurlien, H; Gjerde, I.O; Aarseth, J.H; Eldøen, G; Karlsen, B; Skeidsvoll, H; Gilhus, N.E (2004). "EEG background activity described by a large computerized database". *Clinical Neurophysiology*. **115** (3): 665–73. PMID 15036063. doi:10.1016/j.clinph.2003.10.019.

[42] Nunez, Paul L.; Pilgreen, Kenneth L. (1991). "The Spline-Laplacian in Clinical Neurophysiology". *Journal of Clinical Neurophysiology*. **8** (4): 397–413. PMID 1761706. doi:10.1097/00004691-199110000-00005.

[43] Kondylis, Efstathios D. (2014). "Detection Of High-Frequency Oscillations By Hybrid Depth Electrodes In Standard Clinical Intracranial EEG Recordings.". *Frontiers in Neurology*. **5**: 1–10. doi:10.3389/fneur.2014.00149.

[44] Hämäläinen, Matti; Hari, Riitta; Ilmoniemi, Risto J.; Knuutila, Jukka; Lounasmaa, Olli V. (1993). "Magnetoencephalography —theory, instrumentation, and applications to noninvasive studies of the working human brain". *Reviews of Modern Physics*. **65** (2): 413–497. Bibcode:1993RvMP...65..413H. doi:10.1103/RevModPhys.65.413.

[45] Murakami, S.; Okada, Y. (13 April 2006). "Contributions of principal neocortical neurons to magnetoencephalography and electroencephalography signals". *The Journal of Physiology*. **575** (3): 925–936. PMC 1995687. PMID 16613883. doi:10.1113/jphysiol.2006.105379.

[46] Anderson, J. (22 October 2004). *Cognitive Psychology and Its Implications* (Hardcover) (6th ed.). New York, NY: Worth. p. 17. ISBN 0-7167-0110-3.

[47] Creutzfeldt, Otto D.; Watanabe, Satoru; Lux, Hans D. (1966). "Relations between EEG phenomena and potentials of single cortical cells. I. Evoked responses after thalamic and epicortical stimulation". *Electroencephalography and Clinical Neurophysiology*. **20** (1): 1–18. PMID 4161317. doi:10.1016/0013-4694(66)90136-2.

[48] Buzsaki G (2006). *Rhythms of the brain*. Oxford University Press. ISBN 0-19-530106-4.

[49] Tanzer Oguz I. (2006). *Numerical Modeling in Electro- and Magnetoencephalography, Ph.D. Thesis*. Helsinki University of Technology. ISBN 9512280914.

[50] Tatum, William O. (2014). "Ellen R. Grass Lecture: Extraordinary EEG.". *Neurodiagnostic Journal 54.1*: 3–21.

[51] Kirmizi-Alsan, Elif; Bayraktaroglu, Zubeyir; Gurvit, Hakan; Keskin, Yasemin H.; Emre, Murat; Demiralp, Tamer (2006). "Comparative analysis of event-related potentials during Go/NoGo and CPT: Decomposition of electrophysiological markers of response inhibition and sustained attention". *Brain Research*. **1104** (1): 114–28. PMID 16824492. doi:10.1016/j.brainres.2006.03.010.

[52] Frohlich, Joel; Senturk, Damla; Saravanapandian, Vidya; Golshani, Peyman; Reiter, Lawrence; Sankar, Raman; Thibert, Ronald; DiStefano, Charlotte; Cook, Edwin; Jeste, Shafali (December 2016). "A Quantitative Electrophysiological Biomarker of Duplication 15q11.2-q13.1 Syndrome". *PLOS One*. **11**: e0167179. doi:10.1371/journal.pone.0167179. Retrieved 6 January 2017.

[53] Kisley, Michael A.; Cornwell, Zoe M. (2006). "Gamma and beta neural activity evoked during a sensory gating paradigm: Effects of auditory, somatosensory and cross-modal stimulation". *Clinical Neurophysiology*. **117** (11): 2549–63. PMC 1773003. PMID 17008125. doi:10.1016/j.clinph.2006.08.003.

[54] Kanayama, Noriaki; Sato, Atsushi; Ohira, Hideki (2007). "Crossmodal effect with rubber hand illusion and gamma-band activity". *Psychophysiology*. **44** (3): 392–402. PMID 17371495. doi:10.1111/j.1469-8986.2007.00511.x.

[55] Gastaut, H (1952). "Electrocorticographic study of the reactivity of rolandic rhythm". *Revue neurologique*. **87** (2): 176–82. PMID 13014777.

[56] Oberman, Lindsay M.; Hubbard, Edward M.; McCleery, Joseph P.; Altschuler, Eric L.; Ramachandran, Vilayanur S.; Pineda, Jaime A. (2005). "EEG evidence for mirror neuron dysfunction in autism spectrum disorders". *Cognitive Brain Research*. **24** (2): 190–8. PMID 15993757. doi:10.1016/j.cogbrainres.2005.01.014.

[57] Recommendations for the Practice of Clinical Neurophysiology: Guidelines of the International Federation of Clinical Physiology (EEG Suppl. 52) Editors: G. Deuschl and A. Eisen q 1999 International Federation of Clinical Neurophysiology. All rights reserved. Published by Elsevier Science B.V.

[58] Cahn, B. Rael; Polich, John (2006). "Meditation states and traits: EEG, ERP, and neuroimaging studies". *Psychological Bulletin*. **132** (2): 180–211. PMID 16536641. doi:10.1037/0033-2909.132.2.180.

[59] Niedermeyer, E. (1997). "Alpha rhythms as physiological and abnormal phenomena". *International Journal of Psychophysiology*. **26** (1–3): 31–49. PMID 9202993. doi:10.1016/S0167-8760(97)00754-X.

[60] Feshchenko, Vladimir A.; Reinsel, Ruth A.; Veselis, Robert A. (2001). "Multiplicity of the α Rhythm in Normal Humans". *Journal of Clinical Neurophysiology*. **18** (4): 331–44. PMID 11673699. doi:10.1097/00004691-200107000-00005.

[61] Pfurtscheller, G.; Lopes da Silva, F. H. (1999). "Event-related EEG/MEG synchronization and desynchronization: Basic principles". *Clinical Neurophysiology*. **110** (11): 1842–57. PMID 10576479. doi:10.1016/S1388-2457(99)00141-8.

[62] Barry, W; Jones, GM (1965). "Influence of Eye Lid Movement Upon Electro-Oculographic Recording of Vertical Eye Movements". *Aerospace medicine*. **36**: 855–8. PMID 14332336.

[63] Iwasaki, Masaki; Kellinghaus, Christoph; Alexopoulos, Andreas V.; Burgess, Richard C.; Kumar, Arun N.; Han, Yanning H.; Lüders, Hans O.; Leigh, R. John (2005). "Effects of eyelid closure, blinks, and eye movements on the electroencephalogram". *Clinical Neurophysiology*. **116** (4): 878–85. PMID 15792897. doi:10.1016/j.clinph.2004.11.001.

[64] Lins, Otavio G.; Picton, Terence W.; Berg, Patrick; Scherg, Michael (1993). "Ocular artifacts in EEG and event-related potentials I: Scalp topography". *Brain Topography*. **6** (1): 51–63. PMID 8260327. doi:10.1007/BF01234127.

[65] Keren, Alon S.; Yuval-Greenberg, Shlomit; Deouell, Leon Y. (2010). "Saccadic spike potentials in gamma-band EEG: Characterization, detection and suppression". *NeuroImage*. **49** (3): 2248–63. PMID 19874901. doi:10.1016/j.neuroimage.2009.10.057.

[66] Yuval-Greenberg, Shlomit; Tomer, Orr; Keren, Alon S.; Nelken, Israel; Deouell, Leon Y. (2008). "Transient Induced Gamma-Band Response in EEG as a Manifestation of Miniature Saccades". *Neuron*. **58** (3): 429–41. PMID 18466752. doi:10.1016/j.neuron.2008.03.027.

[67] Epstein, Charles M. (1983). *Introduction to EEG and evoked potentials*. J. B. Lippincott Co. ISBN 0-397-50598-1.

[68] Jung, Tzyy-Ping; Makeig, Scott; Humphries, Colin; Lee, Te-Won; McKeown, Martin J.; Iragui, Vicente; Sejnowski, Terrence J. (2000). "Removing electroencephalographic artifacts by blind source separation". *Psychophysiology*. **37** (2): 163–78. PMID 10731767. doi:10.1017/S0048577200980259.

[69] Jung, Tzyy-Ping; Makeig, Scott; Westerfield, Marissa; Townsechesne, Eric; Sejnowski, Terrence J. (2000). "Removal of eye activity artifacts from visual event-related potentials in normal and clinical subjects". *Clinical Neurophysiology*. **111** (10): 1745–58. PMID 11018488. doi:10.1016/S1388-2457(00)00386-2.

[70] Joyce, Carrie A.; Gorodnitsky, Irina F.; Kutas, Marta (2004). "Automatic removal of eye movement and blink artifacts from EEG data using blind component separation". *Psychophysiology*. **41** (2): 313–25. PMID 15032997. doi:10.1111/j.1469-8986.2003.00141.x.

[71] Fitzgibbon, Sean P; Powers, David M W; Pope, Kenneth J; Clark, C Richard (2007). "Removal of EEG noise and artifact using blind source separation". *Journal of Clinical Neurophysiology*. **24** (3): 232–243. PMID 17545826. doi:10.1097/WNP.0b013e3180556926.

[72] Shackman, Alexander J.; McMenamin, Brenton W.; Maxwell, Jeffrey S.; Greischar, Lawrence L.; Davidson, Richard J. (2010). "Identifying robust and sensitive frequency bands for interrogating neural oscillations". *NeuroImage*. **51** (4): 1319–33. PMC 2871966. PMID 20304076. doi:10.1016/j.neuroimage.2010.03.037.

[73] Nolan, H.; Whelan, R.; Reilly, R.B. (2010). "FASTER: Fully Automated Statistical Thresholding for EEG artifact Rejection". *Journal of Neuroscience Methods*. **192** (1): 152–62. PMID 20654646. doi:10.1016/j.jneumeth.2010.07.015.

[74] Whitham, Emma M; Pope, Kenneth J; Fitzgibbon, Sean P; Lewis, Trent W; Clark, C Richard; Loveless, Stephen; Broberg, Marita; Wallace, Angus; DeLosAngeles, Dylan; Lillie, Peter; et al. (2007). "Scalp electrical recording during paralysis: Quantitative evidence that EEG frequencies above 20Hz are contaminated by EMG". *Clinical Neurophysiology*. Elsevier. **118** (8): 1877–1888. PMID 17574912. doi:10.1016/j.clinph.2007.04.027.

[75] Fitzgibbon, Sean P; Lewis, Trent W; Powers, David M W; Whitham, Emma M; Willoughby, John O; Pope, Kenneth J (2013). "Surface Laplacian of Central Scalp Electrical Signals is Insensitive to Muscle Contamination". *IEEE Transactions on Biomedical Engineering*. IEEE. **60** (1): 4–9. PMID 22542648. doi:10.1109/TBME.2012.2195662.

[76] Fitzgibbon, Sean P; DeLosAngeles, Dylan; Lewis, Trent W; Powers, David MW; Whitham, Emma M; Willoughby, John O; Pope, Kenneth J (2014). "Surface Laplacian of scalp electrical signals and independent component analysis resolve EMG contamination of electroencephalogram". *Journal International Journal of Psychophysiology*. Elsevier.

[77] Montez, Teresa; Poil, S.-S.; Jones, B. F.; Manshanden, I.; Verbunt, J. P. A.; Van Dijk, B. W.; Brussaard, A. B.; Van Ooyen, A.; Stam, C. J.; Scheltens, P.; Linkenkaer-Hansen, K. (2009). "Altered temporal correlations in parietal alpha and prefrontal theta oscillations in early-stage Alzheimer disease". *Proceedings of the National Academy of Sciences*. **106** (5): 165–70. Bibcode:2009PNAS..106.1614M. PMC 2635782. PMID 19164579. doi:10.1073/pnas.0811699106.

[78] MURI: Synthetic Telepathy. *Cnslab.ss.uci.ed m mm m m Retrieved 2011-07-19.*

[79] "Mind Games". The Economist. 2007-03-23.

[80] Li, Shan (2010-08-08). "Mind reading is on the market". Los Angeles Times.

[81] "Brains-on with NeuroSky and Square Enix's Judecca mind-control game". Engadget. Retrieved 2010-12-02.

[82] "New games powered by brain waves". Physorg.com. Archived from the original on 2011-06-06. Retrieved 2010-12-02.

[83] Snider, Mike (2009-01-07). "Toy trains 'Star Wars' fans to use The Force". USA Today. Retrieved 2010-05-01.

[84] "Emotiv Systems Homepage". Emotiv.com. Retrieved 2009-12-29.

[85] "News - NeuroSky Upgrades SDK, Allows For Eye Blink, Brainwave-Powered Games". Gamasutra. 2010-06-30. Retrieved 2010-12-02.

[86] Fiolet, Eliane. "NeuroSky MindWave Brings Brain-Computer Interface to Education". www.ubergizmo.com. Ubergizmo.

[87] "NeuroSky MindWave Sets Guinness World Record for "Largest Object Moved Using a Brain-Computer Interface"". NeuroGadget.com. NeuroGadget.

[88] "Product Launch! Neurosync - The World's Smallest Brain-Computer-Interface". www.prnewswire.com. July 15, 2015. Retrieved July 21, 2017.

[89] "APP - Macrotellect". o.macrotellect.com. Retrieved 2016-12-08.

[90] Keiper, Adam. "The Age of Neuroelectronics". The New Atlantis. Archived from the original on 2016-02-12.

[91] Jestrović I.; Coyle J. L.; Sejdić E. (2015). "Decoding human swallowing via electroencephalography: a state-of-the-art review". Journal of Neural Engineering. 12 (5): 051001. doi:10.1088/1741-2560/12/5/051001.

[92] Mind over matter: Brain waves control Asimo Archived 2009-04-03 at the Wayback Machine. 1 Apr 2009, Japan Times

[93] This brain test maps the truth 21 Jul 2008, 0348 hrs IST, Nitasha Natu, TNN

[94] "Puranik, D.A., Joseph, S.K., Daundkar, B.B., Garad, M.V. (2009). Brain Signature profiling in India. Its status as an aid in investigation and as corroborative evidence – as seen from judgments. Proceedings of XX All India Forensic Science Conference, 815 – 822, November 15 – 17, Jaipur." (PDF).

[95] "Casson, A.J., Yates, D.C., Smith, S., Duncan, J.S., Rodriguez-Villegas, E. (2010). Wearable electroencephalography. IEEE Engineering in Medicine and Biology Magazine, 44 – 56, May 2010.".

[96] Looney, D.; Kidmose, P.; Park, C.; Ungstrup, M.; Rank, M. L.; Rosenkranz, K.; Mandic, D. P. (2012-11-01). "The In-the-Ear Recording Concept: User-Centered and Wearable Brain Monitoring". IEEE Pulse. 3 (6): 32–42. ISSN 2154-2287. doi:10.1109/MPUL.2012.2216717.

[97] Iranmanesh, Saam; Rodriguez-Villegas, Esther. "A 950 nW Analog-Based Data Reduction Chip for Wearable EEG Systems in Epilepsy". IEEE Journal of Solid-State Circuits.

65. Keiper, A. (2006). The age of neuroelectronics. *The New Atlantis*, 11, 4-41.

13.13 Further reading

- Nunez Paul L., Srinivasan Ramesh. "PDF". *Scholarpedia*. **2** (2): 1348. doi:10.4249/scholarpedia.1348.

13.14 External links

- Tanzer Oguz I., (2006) Numerical Modeling in Electro- and Magnetoencephalography, Ph.D. Thesis, Helsinki University of Technology, Finland.

- A tutorial on simulating and estimating EEG sources in Matlab

- A tutorial on analysis of ongoing, evoked, and induced neuronal activity: Power spectra, wavelet analysis, and coherence

Chapter 14

Electrocardiography

"ECG" redirects here. For other uses, see ECG (disambiguation).
"EKG" redirects here. For the album by Edyta Górniak, see EKG (album).
Not to be confused with other types of electrography or with echocardiography.

Electrocardiography (**ECG** or **EKG**[*][lower-alpha 1]) is the process of recording the electrical activity of the heart over a period of time using electrodes placed on the skin. These electrodes detect the tiny electrical changes on the skin that arise from the heart muscle's electrophysiologic pattern of depolarizing and repolarizing during each heartbeat. It is a very commonly performed cardiology test.

In a conventional 12-lead ECG, ten electrodes are placed on the patient's limbs and on the surface of the chest. The overall magnitude of the heart's electrical potential is then measured from twelve different angles ("leads") and is recorded over a period of time (usually ten seconds). In this way, the overall magnitude and direction of the heart's electrical depolarization is captured at each moment throughout the cardiac cycle.[*][4] The graph of voltage versus time produced by this noninvasive medical procedure is referred to as an **electrocardiogram**.

During each heartbeat, a healthy heart has an orderly progression of depolarization that starts with pacemaker cells in the sinoatrial node, spreads out through the atrium, passes through the atrioventricular node down into the bundle of His and into the Purkinje fibers, spreading down and to the left throughout the ventricles. This orderly pattern of depolarization gives rise to the characteristic ECG tracing. To the trained clinician, an ECG conveys a large amount of information about the structure of the heart and the function of its electrical conduction system.[*][5] Among other things, an ECG can be used to measure the rate and rhythm of heartbeats, the size and position of the heart chambers, the presence of any damage to the heart's muscle cells or conduction system, the effects of cardiac drugs, and the function of implanted pacemakers.[*][6]

14.1 History

An early commercial ECG device (1911)

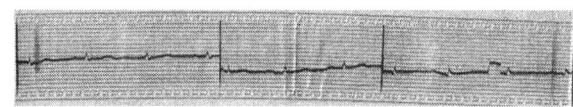

Electrocardiography (1957)

The etymology of the word is derived from the Greek *electro*, because it is related to electrical activity, *kardio*, Greek for heart, and *graph*, a Greek root meaning "to write".

Alexander Muirhead is reported to have attached wires to a feverish patient's wrist to obtain a record of the patient's heartbeat in 1872 at St Bartholomew's Hospital.[*][7] Another early pioneer was Augustus Waller, of St Mary's Hospital in London.[*][8] His electrocardiograph machine consisted of a Lippmann capillary electrometer fixed to a projector. The trace from the heartbeat was projected onto a photographic plate that was itself fixed to a toy train. This allowed a heartbeat to be recorded in real time.

An initial breakthrough came when Willem Einthoven, working in Leiden, the Netherlands, used the string galvanometer (the first practical electrocardiograph) he invented in 1901.*[9] This device was much more sensitive than both the capillary electrometer Waller used and the string galvanometer that had been invented separately in 1897 by the French engineer Clément Ader.*[10] Einthoven had previously, in 1895, assigned the letters P, Q, R, S, and T to the deflections in the theoretical waveform he created using equations which corrected the actual waveform obtained by the capillary electrometer to compensate for the imprecision of that instrument. Using letters different from A, B, C, and D (the letters used for the capillary electrometer's waveform) facilitated comparison when the uncorrected and corrected lines were drawn on the same graph.*[11] Einthoven probably chose the initial letter P to follow the example set by Descartes in geometry.*[11] When a more precise waveform was obtained using the string galvanometer, which matched the corrected capillary electrometer waveform, he continued to use the letters P, Q, R, S, and T,*[11] and these letters are still in use today. Einthoven also described the electrocardiographic features of a number of cardiovascular disorders. In 1924, he was awarded the Nobel Prize in Medicine for his discovery.*[12]

In 1937, Taro Takemi invented the first portable electrocardiograph machine.*[13]

Though the basic principles of that era are still in use today, many advances in electrocardiography have been made over the years. Instrumentation has evolved from a cumbersome laboratory apparatus to compact electronic systems that often include computerized interpretation of the electrocardiogram.*[14]

14.2 Medical uses

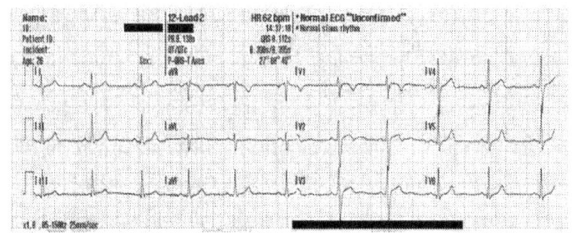

A 12-lead ECG of a 26-year-old male with an incomplete RBBB

The overall goal of performing electrocardiography is to obtain information about the structure and function of the heart. Medical uses for this information are varied and generally relate to having a need for knowledge of the structure and/or function. Some indications for performing electrocardiography include:

- Suspected myocardial infarction (heart attack) or new chest pain

- Suspected pulmonary embolism or new shortness of breath

- A third heart sound, fourth heart sound, a cardiac murmur*[15] or other findings to suggest structural heart disease

- Perceived cardiac dysrhythmias*[15] either by pulse or palpitations

- Monitoring of known cardiac dysrhythmias

- Fainting or collapse*[15]

- Seizures*[15]

- Monitoring the effects of a heart medication (e.g. drug-induced QT prolongation)

- Assessing severity of electrolyte abnormalities, such as hyperkalemia

- Hypertrophic cardiomyopathy screening in adolescents as part of a sports physical out of concern for sudden cardiac death (varies by country)

- Perioperative monitoring in which any form of anesthesia is involved (e.g. monitored anesthesia care, general anesthesia); typically both intraoperative and postoperative

- As a part of a pre-operative assessment some time before a surgical procedure (especially for those with known cardiovascular disease or who are undergoing invasive or cardiac, vascular or pulmonary procedures, or who will receive general anesthesia)

- Cardiac stress testing

- Computed tomography angiography (CTA) and Magnetic resonance angiography (MRA) of the heart (ECG is used to "gate" the scanning so that the anatomical position of the heart is steady)

- Biotelemetry of patients for any of the above reasons and such monitoring can include internal and external defibrillators and pacemakers

The United States Preventive Services Task Force does not recommend electrocardiography for routine screening procedure in patients without symptoms and those at low risk for coronary heart disease.*[16]*[17] This is because an

ECG may falsely indicate the existence of a problem, leading to misdiagnosis, the recommendation of invasive procedures, or overtreatment. However, persons employed in certain critical occupations, such as aircraft pilots,*[18] may be required to have an ECG as part of their routine health evaluations.

Continuous ECG monitoring is used to monitor critically ill patients, patients undergoing general anesthesia,*[15] and patients who have an infrequently occurring cardiac dysrhythmia that would be unlikely to be seen on a conventional ten second ECG.

Performing a 12-lead ECG in the United States is commonly performed by specialized technicians that may be certified electrocardiogram technicians. ECG interpretation is a component of many healthcare fields (nurses and physicians and cardiac surgeons being the most obvious) but anyone trained to interpret an ECG is free to do so. However, "official" interpretation is performed by a cardiologist. Certain fields such as anesthesia utilize continuous ECG monitoring and knowledge of interpreting ECGs is crucial to their jobs.

One additional form of electrocardiography is used in clinical cardiac electrophysiology in which a catheter is used to measure the electrical activity. The catheter is inserted through the femoral vein and can have several electrodes along its length to record the direction of electrical activity from within the heart.

14.3 Electrocardiographs

An electrocardiograph is a machine that is used to perform electrocardiography, and produces the electrocardiogram. The first electrocardiographs are discussed above and are electrically primitive compared to today's machines.

The fundamental component to electrocardiograph is the Instrumentation amplifier, which is responsible for taking the voltage difference between leads (see below) and amplifying the signal. ECG voltages measured across the body are on the order of hundreds of microvolts up to 1 millivolt (the small square on a standard ECG is 100 microvolts). This low voltage necessitates a low noise circuit and instrumentation amplifiers are key.

Early electrocardiographs were constructed with analog electronics and the signal could drive a motor to print the signal on paper. Today, electrocardiographs use analog-to-digital converters to convert to a digital signal that can then be manipulated with digital electronics. This permits digital recording of ECGs and use on computers.

There are other components to the electrocardiograph:*[19]

An electrocardiograph with integrated display and keyboard on a wheeled cart

- Safety features that include voltage protection for the patient and operator. Since the machines are powered by mains power, it is conceivable that either person could be subjected to voltage capable of causing death. Additionally, the heart is sensitive to the AC frequencies typically used for mains power (50 or 60 Hz).

- Defibrillation protection. Any ECG used in healthcare may be attached to a person who requires defibrillation and the electrocardiograph needs to protect itself from this source of energy.

- Electrostatic discharge is similar to defibrillation discharge and requires voltage protection up to 18,000 volts.

- Additionally circuitry called the right leg driver can be used to reduce common-mode interference (typically the 50/60 Hz mains power).

Typical design for a portable electrocardiograph is a combined unit that includes a screen, keyboard, and printer on a small wheeled cart. The unit connects to a long cable that branches to each lead which attaches to a conductive pad on the patient.

Lastly, the electrocardiograph may include a rhythm analysis algorithm that produces a computerized interpretation of the electrocardiogram. The results from these algorithms are considered "preliminary" until verified and/or modified by someone trained in interpreting electrocardiograms. Included in this analysis is computation of common parameters that include PR interval, QT duration, corrected QT (QTc) duration, PR axis, QRS axis, and more. Earlier designs recorded each lead sequentially but current designs employ circuits that can record all leads simultaneously. The former introduces problems in interpretation since there may be beat-to-beat changes in the rhythm that makes it unwise to compare across beats.

14.4 Electrodes and leads

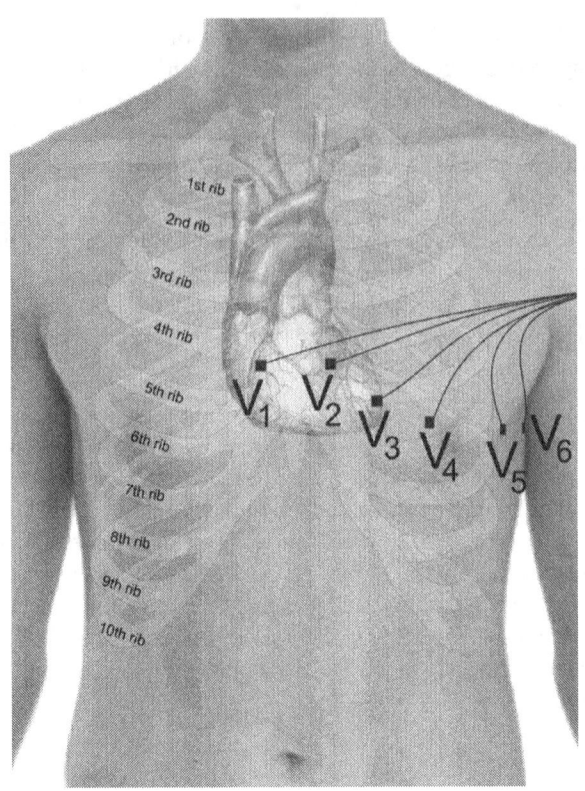

Placement of the precordial electrodes.

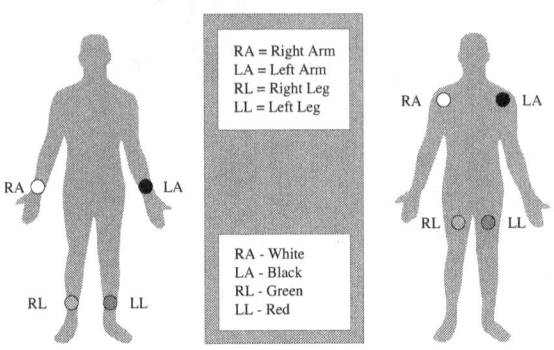

Proper placement of the limb electrodes. The limb electrodes can be far down on the limbs or close to the hips/shoulders as long as they are placed symmetrically.[20]

A "lead" is not the same as an "electrode". Whereas an electrode is a conductive pad in contact with the body that makes an electrical circuit with the electrocardiograph, a lead is a connector to an electrode. Since leads can share the same electrode, a standard 12-lead EKG happens to need only 10 electrodes (as listed in the table below).

A lead is slightly more abstract and is the source of measurement of a vector. For the limb leads, they are "bipolar" and are the comparison between two electrodes. For the precordial leads, they are "unipolar" and compared to a common lead (commonly the *Wilson's central terminal*), as described below.*[21]

Leads are broken down into three sets: limb; augmented limb; and precordial. The 12-lead EKG has a total of three *limb leads* and three *augmented limb leads* arranged like spokes of a wheel in the coronal plane (vertical) and six *precordial leads* that lie on the perpendicular transverse plane (horizontal).

In medical settings, the term *leads* is also sometimes used to refer to the electrodes themselves, although this is not technically a correct usage of the term, which complicates the understanding of difference between the two.

The 10 electrodes in a 12-lead EKG are listed below.*[22]

Two common electrodes used are a flat paper-thin sticker and a self-adhesive circular pad. The former are typically used in a single ECG recording while the latter are for continuous recordings as they stick longer. Each electrode consists of an electrically conductive electrolyte gel and a silver/silver chloride conductor.*[23] The gel typically contains potassium chloride —sometimes silver chloride as well —to permit electron conduction from the skin to the wire and to the electrocardiogram.

The common lead, Wilson's central terminal V_W, is produced by averaging the measurements from the electrodes RA, LA, and LL to give an average potential across the body:

$$V_W = \frac{1}{3}(RA + LA + LL)$$

In a 12-lead ECG, all leads except the limb leads are unipolar (aVR, aVL, aVF, V_1, V_2, V_3, V_4, V_5, and V_6). The measurement of a voltage requires two contacts and so, electrically, the unipolar leads are measured from the common lead (negative) and the unipolar lead (positive). This averaging for the common lead and the abstract unipolar lead concept makes for a more challenging understanding and is complicated by sloppy usage of "lead" and "electrode".

14.4.1 Limb leads

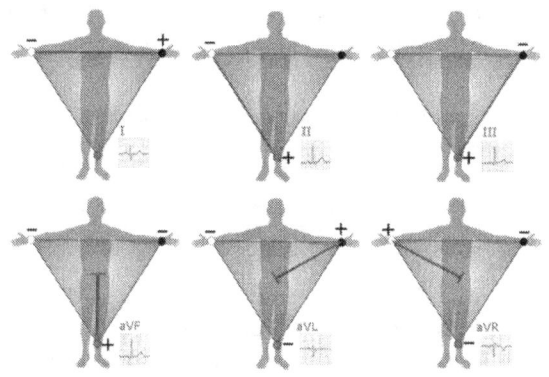

The limb leads and augmented limb leads

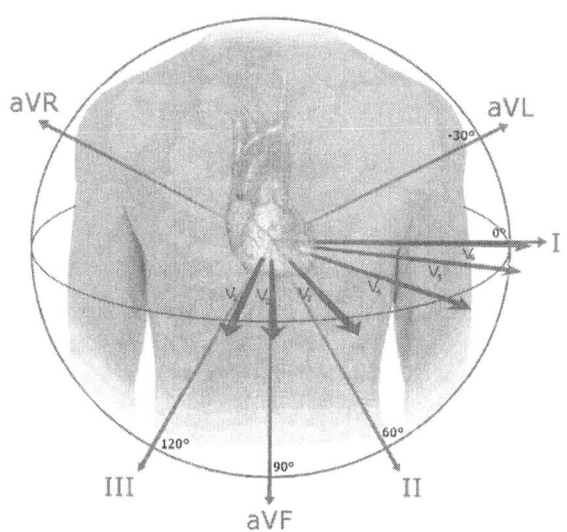

Leads I, II and III are called the *limb leads*. The electrodes that form these signals are located on the limbs—one on each arm and one on the left leg.[24][25][26] The limb leads form the points of what is known as Einthoven's triangle.[27]

- Lead I is the voltage between the (positive) left arm (LA) electrode and right arm (RA) electrode:

$I = LA - RA$

- Lead II is the voltage between the (positive) left leg (LL) electrode and the right arm (RA) electrode:

$II = LL - RA$

- Lead III is the voltage between the (positive) left leg (LL) electrode and the left arm (LA) electrode:

$III = LL - LA$

14.4.2 Augmented limb leads

Leads aVR, aVL, and aVF are the *augmented limb leads*. They are derived from the same three electrodes as leads I, II, and III, but they use Goldberger's central terminal as their negative pole. Goldberger's central terminal is a combination of inputs from two limb electrodes, with a different combination for each augmented lead. It is referred to immediately below as "the negative pole".

- Lead *augmented vector right* (aVR)' has the positive electrode on the right arm. The negative pole is a combination of the left arm electrode and the left leg electrode:

$aVR = RA - \frac{1}{2}(LA + LL) = \frac{3}{2}(RA - V_W)$

- Lead *augmented vector left* (aVL) has the positive electrode on the left arm. The negative pole is a combination of the right arm electrode and the left leg electrode:

$aVL = LA - \frac{1}{2}(RA + LL) = \frac{3}{2}(LA - V_W)$

- Lead *augmented vector foot* (aVF) has the positive electrode on the left leg. The negative pole is a combination of the right arm electrode and the left arm electrode:

$aVF = LL - \frac{1}{2}(RA + LA) = \frac{3}{2}(LL - V_W)$

Together with leads I, II, and III, augmented limb leads aVR, aVL, and aVF form the basis of the hexaxial reference system, which is used to calculate the heart's electrical axis in the frontal plane.

14.4.3 Precordial leads

The precordial leads lie in the transverse (horizontal) plane, perpendicular to the other six leads. The six precordial electrodes act as the positive poles for the six corresponding precordial leads: (V_1, V_2, V_3, V_4, V_5 and V_6). Wilson's central terminal is used as the negative pole.

14.4.4 Specialized leads

Additional electrodes may rarely be placed to generate other leads for specific diagnostic purposes. *Right-sided* precordial leads may be used to better study pathology of the right ventricle or for dextrocardia (and are denoted with an R (e.g., V5R)). *Posterior leads* (V7 to V9) may be used to demonstrate the presence of a posterior myocardial infarction. A *Lewis lead* (requiring an electrode at the right sternal border in the second intercostal space) can be used to study pathological rhythms arising in the right atrium.

An *esophogeal lead* can be inserted to a part of the esophagus where the distance to the posterior wall of the left atrium is only approximately 5–6 mm (remaining constant in people of different age and weight).[28] An esophageal lead avails for a more accurate differentiation between certain cardiac arrhythmias, particularly atrial flutter, AV nodal reentrant tachycardia and orthodromic atrioventricular reentrant tachycardia.[29] It can also evaluate the risk in people with Wolff-Parkinson-White syndrome, as well as terminate supraventricular tachycardia caused by re-entry.[29]

An intracardiac electrogram (ICEG) is essentially an ECG with some added *intracardiac leads* (that is, inside the heart). The standard ECG leads (external leads) are I, II, III, aVL, V1, and V6. Two to four intracardiac leads are added via cardiac catheterization. The word "electrogram" (EGM) without further specification usually means an intracardiac electrogram.

14.4.5 Lead locations on an ECG report

A standard 12-lead ECG report (an electrocardiograph) shows a 2.5 second tracing of each of the twelve leads. The tracings are most commonly arranged in a grid of four columns and three rows. the first column is the limb leads (I, II, and III), the second column is the augmented limb leads (aVR, aVL, and aVF), and the last two columns are the precordial leads (V1-V6). Additionally, a rhythm strip may be included as a fourth or fifth row.

The timing across the page is continuous and not tracings of the 12 leads for the same time period. In other words, if the output were traced by needles on paper, each row would switch which leads as the paper is pulled under the needle. For example, the top row would first trace lead I, then switch to lead aVR, then switch to V1, and then switch to V4 and so none of these four tracings of the leads are from the same time period as they are traced in sequence through time.

I Lateral	aVR	V1 Septal	V4 Anterior
II Inferior	aVL Lateral	V2 Septal	V5 Lateral
III Inferior	aVF Inferior	V3 Anterior	V6 Lateral

Diagram showing the contiguous leads in the same color in the standard 12-lead layout

14.4.6 Contiguity of leads

Each of the 12 ECG leads records the electrical activity of the heart from a different angle, and therefore align with different anatomical areas of the heart. Two leads that look at neighboring anatomical areas are said to be *contiguous*.

In addition, any two precordial leads next to one another are considered to be contiguous. For example, though V4 is an anterior lead and V5 is a lateral lead, they are contiguous because they are next to one another.

14.5 Electrophysiology

Main article: Cardiac electrophysiology

The formal study of the electrical conduction system of the heart is called cardiac electrophysiology (EP). An electrophysiology study involves a formal study of the conduction system and can be done for various reasons. During such a study, catheters are used to access the heart and some of these catheters include electrodes that can be placed anywhere in the heart to record the electrical activity from within the heart. Some catheters contain several electrodes and can record the propagation of electrical activity.

14.6 Interpretation

Interpretation of the ECG is fundamentally about understanding the electrical conduction system of the heart. Normal conduction starts and propagates in a predictable pattern, and deviation from this pattern can be a normal variation or be pathological. An ECG does not equate with mechanical pumping activity of the heart, for example, pulseless electrical activity produces an ECG that should pump blood but no pulses are felt (and constitutes a medical emergency and CPR should be performed). Ventricular fibrillation produces an ECG but is too dysfunctional to produce a life-sustaining cardiac output. Certain rhythms are known to have good cardiac output and some are known to have bad cardiac output. Ultimately, an echocardiogram or other anatomical imaging modality is useful in assessing the mechanical function of the heart.

Like all medical tests, what constitutes "normal" is based on population studies. The heart rate range of between 60 and 100 is considered normal since data shows this to be the usual resting heart rate.

14.6.1 Theory

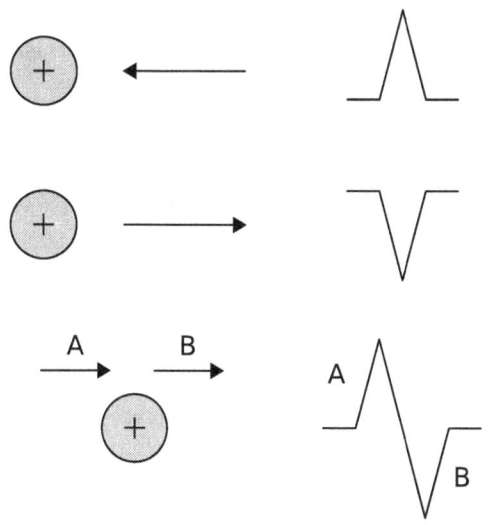

QRS is upright in a lead when its axis is aligned with that lead's vector

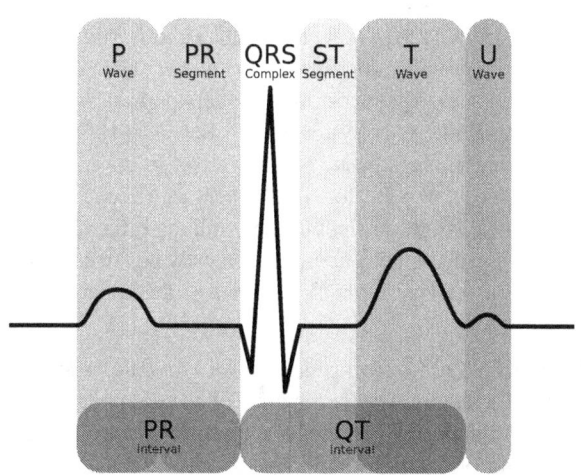

Schematic representation of normal ECG

Interpretation of the ECG is ultimately that of pattern recognition. In order to understand the patterns found, it is helpful to understand the theory of what ECGs represent. The theory is rooted in electromagnetics and boils down to the four following points:

- depolarization of the heart *toward* the positive electrode produces a positive deflection
- depolarization of the heart *away* from the positive electrode produces a negative deflection
- repolarization of the heart *toward* the positive electrode produces a negative deflection
- repolarization of the heart *away* from the positive electrode produces a positive deflection

Thus, the overall direction of depolarization and repolarization produces a vector that produces positive or negative deflection on the ECG depending on which lead it points to. For example, depolarizing from right to left would produce a positive deflection in lead I because the two vectors point in the same direction. In contrast, that same depolarization would produce minimal deflection in V1 and V2 because the vectors are perpendicular and this phenomenon is called isoelectric.

Normal rhythm produces four entities —a P wave, a QRS complex, a T wave, and a U wave —that each have a fairly unique pattern.

- The P wave represents atrial depolarization.
- The QRS complex represents ventricular depolarization.
- The T wave represents ventricular repolarization.
- The U wave represents papillary muscle repolarization.

However, the U wave is not typically seen and its absence is generally ignored. Changes in the structure of the heart and its surroundings (including blood composition) change the patterns of these four entities.

14.6.2 Electrocardiogram grid

ECGs are normally printed on a grid. The horizontal axis represents time and the vertical axis represents voltage. The standard values on this grid are shown in the adjacent image:

- A small box is 1 mm x 1 mm big and represents 0.1 mV x 0.04 seconds.

14.6. INTERPRETATION

- A large box is 5 mm x 5mm big and represents 0.5 mV x 0.2 seconds wide.

The "large" box is represented by a heavier line weight than the small boxes.

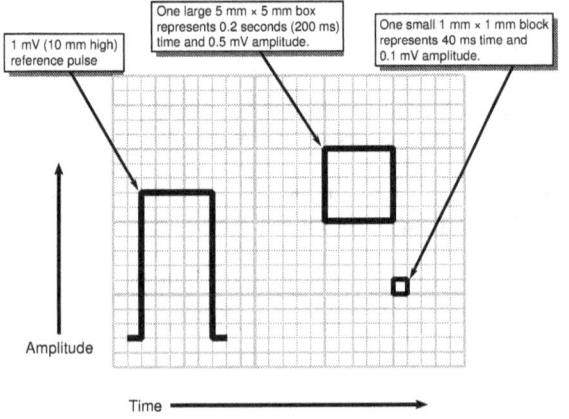

Measuring time and voltage with ECG graph paper

Not all aspects of an ECG rely on precise recordings or having a known scaling of amplitude or time. For example, determining if the tracing is a sinus rhythm only requires feature recognition and matching, and not measurement of amplitudes or times (i.e., the scale of the grids are irrelevant). An example to the contrary, the voltage requirements of left ventricular hypertrophy require knowing the grid scale.

14.6.3 Rate and rhythm

In a normal heart, the heart rate is the rate in which the sinoatrial node depolarizes as it is the source of depolarization of the heart. Heart rate, like other vital signs like blood pressure and respiratory rate, change with age. In adults, a normal heart rate is between 60 and 100 beats per minute (normocardic) where in children it is higher. A heart rate less than normal is called bradycardia (<60 in adults) and higher than normal is tachycardia (>100 in adults). A complication of this is when the atria and ventricles are not in synchrony and the "heart rate" must be specified as atrial or ventricular (e.g., atrial rate in atrial fibrillation is 300–600 bpm, whereas ventricular rate can be normal (60-100) or faster (100–150)).

In normal resting hearts, the physiologic rhythm of the heart is normal sinus rhythm (NSR). Normal sinus rhythm produces the prototypical pattern of P wave, QRS complex, and T wave. Generally, deviation from normal sinus rhythm is considered a cardiac arrhythmia. Thus, the first question in interpreting an ECG is whether or not there is a sinus rhythm. A criterion for sinus rhythm is that P waves and QRS complexes appear 1-to-1, thus implying that the P wave causes the QRS complex.

Once sinus rhythm is established, or not, the second question is the rate. For a sinus rhythm this is either the rate of P waves or QRS complexes since they are 1-to-1. If the rate is too fast then it is sinus tachycardia and if it is too slow then it is sinus bradycardia.

If it is not a sinus rhythm, then determining the rhythm is necessary before proceeding with further interpretation. Some arrhythmias with characteristic findings:

- Absent P waves with "irregularly irregular" QRS complexes is the hallmark of atrial fibrillation
- A "saw tooth" pattern with QRS complexes is the hallmark of atrial flutter
- Sine wave pattern is the hallmark of ventricular flutter
- Absent P waves with wide QRS complexes with fast rate is ventricular tachycardia

Determination of rate and rhythm is necessary in order to make sense of further interpretation.

14.6.4 Axis

The heart has several axes, but the most common by far is the axis of the QRS complex (references to "the axis" implicitly means the QRS axis). Each axis can be computationally determined to result in a number representing degrees of deviation from zero, or it can be categorized into a few types.

The QRS axis is the general direction of the ventricular depolarization wavefront (or mean electrical vector) in the frontal plane. It is often sufficient to classify the axis as one of three types: normal, left deviated, or right deviated. Population data shows that normal QRS axis is from −30° to 105° with 0° being along lead I and positive being inferior and negative being superior (best understood graphically as the hexaxial reference system).*[30] Beyond +105° is right axis deviation and beyond −30° is left axis deviation (the third quadrant of −90° to −180° is very rare and is an indeterminate axis). A shortcut for determining if the QRS axis is normal is if the QRS complex is mostly positive in lead I and lead II (or lead I and aVF if +90° is the upper limit of normal).

The normal QRS axis is generally *down and to the left*, following the anatomical orientation of the heart within the chest. An abnormal axis suggests a change in the physical shape and orientation of the heart, or a defect in its conduction system that causes the ventricles to depolarize in an abnormal way.

The extent of normal axis can be +90° or 105° depending on the source.

14.6.5 Amplitudes and intervals

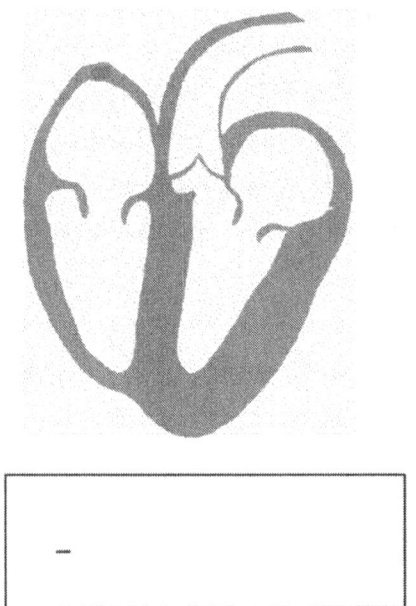

Animation of a normal ECG wave

All of the waves on an EKG tracing and the intervals between them have a predictable time duration, a range of acceptable amplitudes (voltages), and a typical morphology. Any deviation from the normal tracing is potentially pathological and therefore of clinical significance.

For ease of measuring the amplitudes and intervals, an EKG is printed on graph paper at a standard scale: each 1 mm (one small box on the standard EKG paper) represents 40 milliseconds of time on the x-axis, and 0.1 millivolts on the y-axis.

14.6.6 Ischemia and infarction

Main article: Electrocardiography in myocardial infarction

Ischemia or non-ST elevation myocardial infarctions may manifest as ST depression or inversion of T waves. It may also affect the high frequency band of the QRS.

ST elevation myocardial infarctions have different characteristic ECG findings based on the amount of time elapsed since the MI first occurred. The earliest sign is *hyperacute T waves*, peaked T-waves due to local hyperkalemia in ischemic myocardium. This then progresses over a period of minutes to elevations of the ST segment by at least 1 mm. Over a period of hours, a pathologic Q wave may appear and the T wave will invert. Over a period of days the ST elevation will resolve. Pathologic q waves generally will remain permanently.*[33]

The coronary artery that has been occluded can be identified in an ST-elevation myocardial infarction based on the location of ST elevation. The LAD supplies the anterior wall of the heart, and therefore causes ST elevations in anterior leads (V1 and V2). The LCx supplies the lateral aspect of the heart and therefore causes ST elevations in lateral leads (I, aVL and V6). The RCA usually supplies the inferior aspect of the heart, and therefore causes ST elevations in inferior leads (II, III and aVF).

14.6.7 Artifacts

An EKG tracing is affected by patient motion. Some rhythmic motions (such as shivering or tremors) can create the illusion of cardiac dysrhythmia.*[34] Artifacts are distorted signals caused by a secondary internal or external sources, such as muscle movement or interference from an electrical device.*[35]*[36]

Distortion poses significant challenges to healthcare providers,*[35] who employ various techniques*[37] and strategies to safely recognize*[38] these false signals. Accurately separating the ECG artifact from the true ECG signal can have a significant impact on patient outcomes and legal liabilities.*[39]

Improper lead placement (for example, reversing two of the limb leads) has been estimated to occur in 0.4% to 4% of all EKG recordings,*[40] and has resulted in improper diagnosis and treatment including unnecessary use of thrombolytic therapy.*[41]*[42]

14.7 Diagnosis

Numerous diagnosis and findings can be made based upon electrocardiography and many are discussed above. The following is an organized list of these and more.

Rhythm disturbances/ Arrhythmias:

- Atrial fibrillation & atrial flutter without rapid ventricular response

- Premature atrial contraction (PACs) & Premature ventricular contraction (PVCs)

- Sinus arrhythmia

- Sinus bradycardia & sinus tachycardia
- Sinus pause & sinoatrial arrest
- Sick sinus syndrome: bradycardia-tachycardia syndrome
- Supraventricular tachycardia
    - Atrial fibrillation (afib) with rapid ventricular response
    - Atrial flutter with rapid ventricular response
    - AV nodal reentrant tachycardia
    - Atrioventricular reentrant tachycardia
    - Junctional ectopic tachycardia
    - Atrial tachycardia
        - Ectopic atrial tachycardia (unicentric)
        - Multifocal atrial tachycardia
        - Paroxysmal atrial tachycardia
    - Sinoatrial nodal reentrant tachycardia
- Torsades de pointes (polymorphic ventricular tachycardia)
- Wide complex tachycardia
    - Ventricular flutter
    - Ventricular fibrillation
    - Ventricular tachycardia (monomorphic ventricular tachycardia)
- Pre-excitation syndrome
    - Wolff–Parkinson–White syndrome

Heart block and conduction problems:

- Aberration
- Brugada syndrome
- First-degree AV block, Second-degree AV block (Mobitz I & II), Third-degree AV block
- Left anterior & left posterior fascicular block; bifasciular block and trifasciular blocks
- Incomplete and complete right bundle branch block (RBBB)
- Incomplete and complete Left bundle branch block (LBBB)
- Long QT syndrome
- Right and left atrial abnormality

Electrolytes disturbances & intoxication:

- Digitalis intoxication
- Calcium: hypocalcemia and hypercalcemia
- Potassium: hypokalemia and hyperkalemia

Ischemia and infarction:

- ST elevation and ST depression
- High Frequency QRS changes
- Myocardial infarction (heart attack)
    - Non-Q wave myocardial infarction
    - NSTEMI
    - STEMI

Structural:

- Acute pericarditis
- Right and left ventricular hypertrophy
- Right ventricular strain / S1Q3T3

14.8 See also

- Electrical conduction system of the heart
- Electrogastrogram
- Electropalatography
- Electroretinography
- Heart rate monitor
- Emergency medicine

14.9 Notes

[1] The version with -K-, which is rarer in British English than in American English, is an early-20th-century loanword from the German acronym *EKG* for *Elektrokardiogramm* (*electrocardiogram*),[1] which reflects that German physicians were pioneers in the field at the time. Today AMA style and, under its stylistic influence, most American medical publications use *ECG* instead of *EKG*.[2] The German term *Elektrokardiogramm* as well as the English equivalent *electrocardiogram* consist of the New Latin/international scientific vocabulary elements *elektro-* (cognate *electro-*) and *kardi-* (cognate *cardi-*), the latter from Greek *kardia* (heart).[3] The -K- version is more often retained under circumstances where there may be verbal confusion between *ECG* and *EEG* (electroencephalography) due to similar pronunciation.

14.10 References

[1] EKG. Oxford Online Dictionaries

[2] American Medical Association, "15.3.1 Electrocardiographic Terms", *AMA Manual of Style*

[3] Merriam-Webster. "Merriam-Webster's Collegiate Dictionary". Merriam-Webster.

[4] "ECG- simplified. Aswini Kumar M.D.". LifeHugger. Retrieved 11 February 2010.

[5] Walraven, G. (2011). *Basic arrhythmias* (7th ed.), pp. 1–11

[6] Braunwald E. (ed) (1997), *Heart Disease: A Textbook of Cardiovascular Medicine, Fifth Edition*, p. 108, Philadelphia, W.B. Saunders Co.. ISBN 0-7216-5666-8.

[7] Ronald M. Birse,rev. Patricia E. Knowlden Oxford Dictionary of National Biography 2004 (Subscription required) – (original source is his biography written by his wife – Elizabeth Muirhead. Alexandernn Muirhead 1848–1920. Oxford, Blackwell: privately printed 1926.)

[8] Waller AD (1887). "A demonstration on man of electromotive changes accompanying the heart's beat". *J Physiol.* **8** (5): 229–34. PMC 1485094. PMID 16991463.

[9] Rivera-Ruiz M, Cajavilca C, Varon J (29 September 1927). "Einthoven's String Galvanometer: The First Electrocardiograph". *Texas Heart Institute journal / from the Texas Heart Institute of St. Luke's Episcopal Hospital, Texas Children's Hospital*. **35** (2): 174–8. PMC 2435435. PMID 18612490.

[10] Interwoven W (1901). "Un nouveau galvanometre". *Arch Neerl Sc Ex Nat.* **6**: 625.

[11] Hurst JW (3 November 1998). "Naming of the Waves in the ECG, With a Brief Account of Their Genesis". *Circulation.* **98** (18): 1937–42. PMID 9799216. doi:10.1161/01.CIR.98.18.1937.

[12] Cooper JK (1986). "Electrocardiography 100 years ago. Origins, pioneers, and contributors". *N Engl J Med.* **315** (7): 461–4. PMID 3526152. doi:10.1056/NEJM198608143150721.

[13] "Dr. Taro Takemi". 27 August 2012.

[14] Mark, Jonathan B. (1998). *Atlas of cardiovascular monitoring.* New York: Churchill Livingstone. ISBN 0-443-08891-8.

[15] Masters, Jo; Bowden, Carole; Martin, Carole (2003). *Textbook of veterinary medical nursing.* Oxford: Butterworth-Heinemann. p. 244. ISBN 0-7506-5171-7.

[16] Moyer VA (2 October 2012). "Screening for coronary heart disease with electrocardiography: U.S. Preventive Services Task Force recommendation statement." *Annals of Internal Medicine.* **157** (7): 512–8. PMID 22847227. doi:10.7326/0003-4819-157-7-201210020-00514.

[17] Consumer Reports; American Academy of Family Physicians; ABIM Foundation (April 2012), "EKGs and exercise stress tests: When you need them for heart disease — and when you don't" (PDF), *Choosing Wisely*, Consumer Reports, retrieved 14 August 2012

[18] "Summary of Medical Standards" (PDF). U.S. Federal Aviation Administration. 2006. Retrieved 27 December 2013.

[19] "Mitigation Strategies for ECG Design Challenges" (PDF). *Analog Devices*. Retrieved 24 April 2016.

[20] RESTING 12-LEAD ECG ELECTRODE PLACEMENT AND ASSOCIATED PROBLEMS Archived 2 April 2015 at the Wayback Machine..DrTanzil

[21] "Electrocardiogram Leads". CV Physiology. 26 March 2007. Retrieved 2009-08-15.

[22] "12-Lead ECG Placement Guide with Illustrations". *Cables and Sensors*. Retrieved 11 July 2017.

[23] Kavuru, Madhav S.; Vesselle, Hubert; Thomas, Cecil W. (1987). "Advances in Body Surface Potential Mapping (BSPM) Instrumentation". *Pediatric and Fundamental Electrocardiography*. Developments in Cardiovascular Medicine. **56**: 315–327. ISBN 978-1-4612-9428-3. ISSN 0166-9842. doi:10.1007/978-1-4613-2323-5_15.

[24] "Lead Placement". *Univ. of Maryland School of Medicine Emergency Medicine Interest Group*. Archived from the original on 20 July 2011. Retrieved 15 August 2009.

[25] "Limb Leads – ECG Lead Placement – Normal Function of the Heart – Cardiology Teaching Package – Practice Learning – Division of Nursing – The University of Nottingham". Nottingham.ac.uk. Retrieved 15 August 2009.

[26] "Lesson 1: The Standard 12 Lead ECG". Library.med.utah.edu. Archived from the original on 22 March 2009. Retrieved 15 August 2009.

[27] "Electrocardiogram explanation image". Retrieved 28 February 2014.

[28] Meigas, K; Kaik, J; Anier, A (2008). "Device and methods for performing transesophageal stimulation at reduced pacing current threshold". *Estonian Journal of Engineering.* **57** (2): 154. doi:10.3176/eng.2008.2.05.

[29] Pehrson, Steen M.; Blomströ-LUNDQVIST, Carina; Ljungströ, Erik; Blomströ, Per (1994). "Clinical value of transesophageal atrial stimulation and recording in patients with arrhythmia-related symptoms or documented supraventricular tachycardia-correlation to clinical history and invasive studies". *Clinical Cardiology.* **17** (10): 528–534. PMID 8001299. doi:10.1002/clc.4960171004.

[30] Surawicz, Borys; Knillans, Timothy (2008). *Chou's electrocardiography in clinical practice : adult and pediatric* (6th ed.). Philadelphia, PA: Saunders/Elsevier. p. 12. ISBN 1416037748.

[31] Otero J, Lenihan DJ. "The "normothermic" Osborn wave induced by severe hypercalcemia". *Tex Heart Inst J*. **27**: 316–7. PMC 101092. PMID 11093425.

[32] Houghton, Andrew R; Gray, David (2012). *Making Sense of the ECG, Third Edition*. Hodder Education. p. 214. ISBN 978-1-4441-6654-5.

[33] Alpert JS, Thygesen K, Antman E, Bassand JP (2000). "Myocardial infarction redefined—a consensus document of The Joint European Society of Cardiology/American College of Cardiology Committee for the redefinition of myocardial infarction". *J Am Coll Cardiol*. **36** (3): 959–69. PMID 10987628. doi:10.1016/S0735-1097(00)00804-4.

[34] Segura-Sampedro, Juan José; Parra-López, Loreto; Sampedro-Abascal, Consuelo; Muñoz-Rodríguez, Juan Carlos (2015). "Atrial flutter EKG can be useless without the proper electrophysiological basis". *International Journal of Cardiology*. **179**: 68–9. PMID 25464416. doi:10.1016/j.ijcard.2014.10.076.

[35] Takla, George; Petre, John H.; Doyle, D John; Horibe, Mayumi; Gopakumaran, Bala (2006). "The Problem of Artifacts in Patient Monitor Data During Surgery: A Clinical and Methodological Review". *Anesthesia & Analgesia*. **103** (5): 1196–1204. doi:10.1213/01.ane.0000247964.47706.5d.

[36] Kligfield, Paul; Gettes, Leonard S.; Bailey, James J.; Childers, Rory; Deal, Barbara J.; Hancock, E. William; van Herpen, Gerard; Kors, Jan A.; Macfarlane, Peter (2007-03-13). "Recommendations for the standardization and interpretation of the electrocardiogram: part I: The electrocardiogram and its technology: a scientific statement from the American Heart Association Electrocardiography and Arrhythmias Committee, Council on Clinical Cardiology; the American College of Cardiology Foundation; and the Heart Rhythm Society: endorsed by the International Society for Computerized Electrocardiology". *Circulation*. **115** (10): 1306–1324. PMID 17322457. doi:10.1161/CIRCULATIONAHA.106.180200.

[37] "northstarcpr.com".

[38] Jafary, Fahim H (2007). "The "incidental" episode of ventricular fibrillation: A case report". *Journal of Medical Case Reports*. **1**: 72. PMC 2000884. PMID 17760955. doi:10.1186/1752-1947-1-72.

[39] Mangalmurti, Sandeep; Seabury, Seth A.; Chandra, Amitabh; Lakdawalla, Darius; Oetgen, William J.; Jena, Anupam B. (2014). "Medical professional liability risk among US cardiologists". *American Heart Journal*. **167** (5): 690–6. PMC 4153384. PMID 24766979. doi:10.1016/j.ahj.2014.02.007.

[40] Incorrect electrode cable connection during electrocardiographic recording (2007) Velislav N. Batchvarov, Marek Malik, A. John Camm, Europace, Oct 2007

[41] Chanarin, N., Caplin, J., & Peacock, A. (1990). "Pseudo reinfarction": a consequence of electrocardiogram lead transposition following myocardial infarction. Clinical cardiology, 13(9), 668–669.

[42] Guijarro-Morales A., Gil-Extremera B., Maldonado-Martín A. (1991). "ECG diagnostic errors due to improper connection of the right arm and leg cables". *International Journal of Cardiology*. **30** (2): 233–235. doi:10.1016/0167-5273(91)90103-v.

14.11 External links

- The whole ECG course on 1 A4 paper from ECGpedia, a wiki encyclopedia for a course on interpretation of ECG

- Wave Maven – a large database of practice ECG questions provided by Beth Israel Deaconess Medical Center

- PysioBank – a free scientific database with physiologic signals (here ecg)

- EKG Academy – free EKG lectures, drills and quizzes

Chapter 15

BioAPI

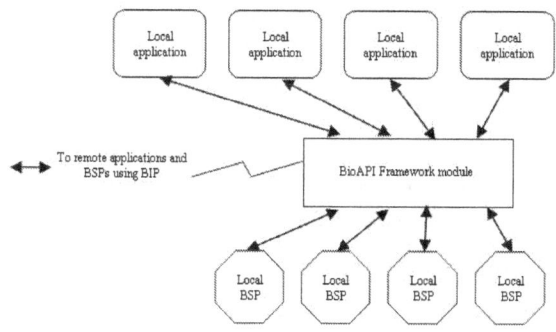

BioAPI architecture

BioAPI (Biometric Application Programming Interface) is a key part of the International Standards that support systems that perform biometric enrollment and verification (or identification). It defines interfaces between modules that enable software from multiple vendors to be integrated together to provide a biometrics application within a system, or between one or more systems using a defined Biometric Interworking Protocol (BIP) - see below.

Biometrics (measurements of physical characteristics of a person) are increasingly being used to provide verification of the identity of an individual, once they have been enrolled (one or more of their physical characteristics has been measured).

Computer systems that perform biometric enrollment, verification, or identification are becoming increasingly used. The BioAPI specification enables such systems to be produced by the integration of modules from multiple independent vendors.

15.1 Origins

The BioAPI specification is one of a set of International Standards produced jointly by the International Organisation for Standardisation (ISO) and the International Electrotechnical Commission (IEC) under their Joint Technical Committee 1 (JTC1), Subcommittee 37 on Biometrics.

The Standard was based on some early work done in the United States of America and by the BioAPI Consortium which was called BioAPI 1.0 and BioAPI 1.1, but these specifications were heavily revised to correct bugs and to provide enhancements when the work was introduced to ISO/IEC. The first international version was therefore called BioAPI 2.0. A subsequent international version of BioAPI containing extensions of the user interface-related features and other enhancements produced a BioApi 2.1. Further enhancements to BioAPI are expected.

BioAPI 2.0 is specified in ISO/IEC 19784-1 and was first published on 1 May 2006.

15.2 What and why?

The purpose of the BioAPI specification is to define an architecture and all necessary interfaces (using C programming language specifications) to allow biometric applications (perhaps distributed across a network) to be integrated from modules provided by different vendors.

The ability for system integrators to produce complete systems using components from multiple vendors is essential in the rapidly changing technology of biometrics. It gives flexibility in the provision of modules, avoids vendor lock-in, provides a degree of future-proofing as the best available biometrics technologies change.

The modules being integrated may be software components containing capture devices, such as fingerprint readers, cameras for face recognition, iris scanners, signature recognition devices, vascular imaging systems, etc.

They can also be modules that provide support for image processing of biometric data, feature extraction (a form of compression that is specific to a given biometric technology and allows direct matching of the compressed formats - for example, the relative distances on the face of eyes, nose, mouth, or the number of ridges between identifiable ridge endings or ridge bifurcations).

In addition, modules that provide archiving and retrieval of biometric records to support matching or searching for a match are also a recognised part of the BioAPI architecture.

Applications can be concerned with personal identification (for example for credit cards), or with more specific areas such as identity card verification, checks for duplicate enrollment, passports, or physical access control in a commercial environment or for airport employees or merchant seamen wishing to go on-shore at their arrival port.

Whilst today a system is commonly built using a single device for a single application, it is likely that in the long term many such applications will interact (securely, and via a network) with a common set of trusted devices (with various security policies and certificates).

It is also expected that future biometrics applications will use multiple biometric modalities (for example, fingerprint, iris, and face), both to improve the accuracy of identification and to cope with people that are missing a finger, or have disability problems that prevent use of iris or face recognition.

BioAPI supports all these use cases.

15.3 The basic architecture

The basic architecture of BioAPI 2.0 is illustrated in the figure at the top of this page. There are multiple possible (independent) *biometric applications* that interact with a *BioAPI Framework*, which in turn routes their messages to *Biometric Service Providers (BSPs)* that support the various biometric capture devices, image enhancement modules, feature extraction, matching, searching, etc.

A later extension of the architecture introduces the concept of a *Biometric Function Provider (BFP)* and defines further lower-level interfaces between a BFP and a controlling BSP. This minimises the amount of software that a biometric device vendor needs to develop, allowing (other) software vendors to do most of the work of producing the BSP with an interface to the framework.

15.4 Procurement issues

The *BioAPI Framework* is the heart of BioAPI. Procurements of biometric systems need to consider the merits of basing their invitations-to-tender on systems conforming to the BioAPI Standard, which contain a BioAPI Framework module.

The importance of this Framework module is recognised by the BioAPI Consortium, which identifies an implementation of this Framework from BioFoundry.

15.5 Distributed systems

It might be uncommon to find multiple biometric applications and multiple biometric devices on a single computer system, but the long-term aim of telebiometrics is to allow multiple biometrics applications on multiple systems on the Internet to interwork with multiple other systems that support biometrics devices.

BioAPI has already laid the foundations for this, with its architecture.

Another ISO/IEC JTC1/SC37 Standard - BioAPI Interworking Protocol (BIP) - specifies an enhancement of the BioAPI Framework that essentially maps all API calls into network messages (defined using ASN.1) to provide a distributed BioAPI system.

BIP is also being progressed as a Recommendation in ITU-T as Joint text with ISO/IEC.

15.6 References

ISO/IEC 19784-1, *Information Technology – BioAPI – Biometric Application Programming Interface – Part 1: BioAPI Specification*

ISO/IEC 24708, *Information Technology —BioAPI Interworking Protocol (BIP)*

Chapter 16

Biometric passport

This biometric symbol is usually printed on the cover of biometric passports.

A **biometric passport** (also known as an **e-passport**, **ePassport** or a **digital passport**) is a traditional passport that has an embedded electronic microprocessor chip which contains biometric information that can be used to authenticate the identity of passport holder. It uses contactless smart card technology, including a microprocessor chip (computer chip) and antenna (for both power to the chip and communication) embedded in the front or back cover, or center page, of the passport. The passport's critical information is both printed on the data page of the passport and stored in the chip. Public Key Infrastructure (PKI) is used to authenticate the data stored electronically in the passport chip making it expensive and difficult to forge when all security mechanisms are fully and correctly implemented. Many countries are moving towards the issue of biometric passports. As of December 2008, 60 countries were issuing such passports,*[1] and this number was 96 as of 5 April 2017.

The currently standardized biometrics used for this type of identification system are facial recognition, fingerprint recognition, and iris recognition. These were adopted after assessment of several different kinds of biometrics including retinal scan. Document and chip characteristics are documented in the International Civil Aviation Organization's (ICAO) Doc 9303.*[2] The ICAO defines the biometric file formats and communication protocols to be used in passports. Only the digital image (usually in JPEG or JPEG2000 format) of each biometric feature is actually stored in the chip. The comparison of biometric features is performed outside the passport chip by electronic border control systems (e-borders). To store biometric data on the contactless chip, it includes a minimum of 32 kilobytes of EEPROM storage memory, and runs on an interface in accordance with the ISO/IEC 14443 international standard, amongst others. These standards intend interoperability between different countries and different manufacturers of passport books.

Some national identity cards (for example in the Netherlands, Albania and Brazil) are fully ICAO9303 compliant biometric travel documents; however others, such as the United States Passport Card, are not.

16.1 Availability

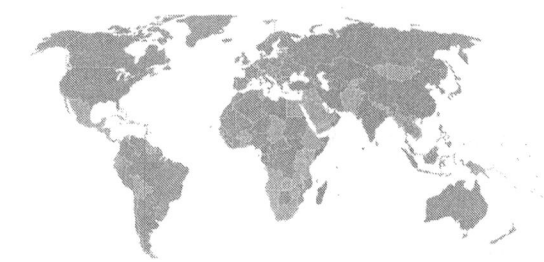

Availability of biometric passports

16.2 Data protection

Biometric passports have protection mechanisms to avoid and/or detect attacks:

- Non-traceable chip characteristics. Random chip identifiers reply to each request with a different chip

16.3. INSPECTION PROCESS

number. This prevents tracing of passport chips. Using random identification numbers is optional.

- Basic Access Control (BAC). BAC protects the communication channel between the chip and the reader by encrypting transmitted information. Before data can be read from a chip, the reader needs to provide a key which is derived from the Machine Readable Zone: the date of birth, the date of expiry and the document number. If BAC is used, an attacker cannot (easily) eavesdrop transferred information without knowing the correct key. Using BAC is optional.

- Passive Authentication (PA). PA is aimed at identifying modification of passport chip data. The chip contains a file (SOD) that stores hash values of all files stored in the chip (picture, fingerprint, etc.) and a digital signature of these hashes. The digital signature is made using a document signing key which itself is signed by a country signing key. If a file in the chip (e.g. the picture) is changed, this can be detected since the hash value is incorrect. Readers need access to all used public country keys to check whether the digital signature is generated by a trusted country. Using PA is mandatory. According to a September 2011 United States Central Intelligence Agency document released by Wikileaks in December 2014, "Although falsified e-passports will not have the correct digital signature, inspectors may not detect the fraud if the passports are from countries that do not participate in the International Civil Aviation Organization's Public Key Directory (ICAO PKD)."[3] As of January 2017, 55 of over 60 e-passport-issuing countries belong to the PKD program.*[4]

- Active Authentication (AA). AA prevents cloning of passport chips. The chip contains a private key that cannot be read or copied, but its existence can easily be proven. Using AA is optional.

- Extended Access Control (EAC). EAC adds functionality to check the authenticity of both the chip (chip authentication) and the reader (terminal authentication). Furthermore, it uses stronger encryption than BAC. EAC is typically used to protect fingerprints and iris scans. Using EAC is optional. In the European Union, using EAC is mandatory for all documents issued starting 28 June 2009.

- Supplemental Access Control (SAC) was introduced by ICAO in 2009 for addressing BAC weaknesses. It was introduced as a supplement to BAC (for keeping compatibility), but will replace it in the future.

- Shielding the chip. This prevents unauthorized reading. Some countries – including at least the US – have integrated a very thin metal mesh into the passport's cover to act as a shield when the passport cover is closed.*[5] The use of shielding is optional.

- To assure interoperability and functionality of the security mechanisms listed above, ICAO and German Federal Office for Information Security (BSI) have specified several test cases. These test specifications are updated with every new protocol and are covering details starting from the paper used and ending in the chip that is included.*[6]

16.3 Inspection process

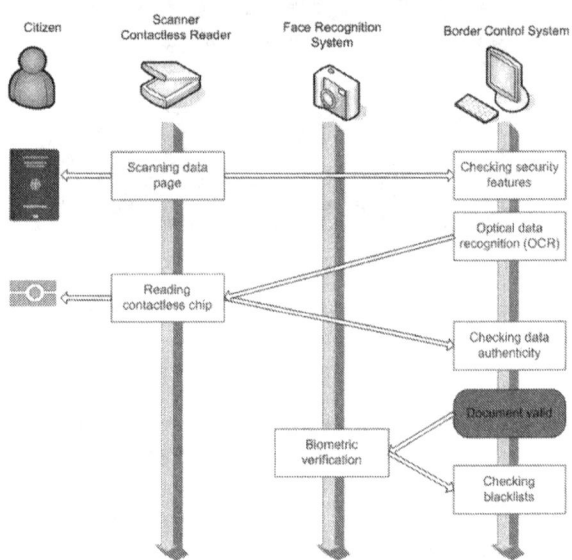

The typical work-flow of an automatic border control system (eGate)[7]*

16.4 Attacks

Since the introduction of biometric passports several attacks have been presented and demonstrated:

- Non-traceable chip characteristics. In 2008 a Radboud/Lausitz University team demonstrated that it's possible to determine which country a passport chip is from without knowing the key required for reading it.*[8] The team fingerprinted error messages of passport chips from different countries. The resulting lookup table allows an attacker to determine from where a chip originated. In 2010 Tom Chothia and Vitaliy Smirnov documented an attack that allows an individual passport to be traced,*[9]*[10] by sending specific BAC authentication requests.

- Basic Access Control (BAC). In 2005 Marc Witteman showed that the document numbers of Dutch passports were predictable,*[11] allowing an attacker to guess/crack the key required for reading the chip. In 2006 Adam Laurie wrote software that tries all known passport keys within a given range, thus implementing one of Witteman's attacks. Using online flight booking sites, flight coupons and other public information it's possible to significantly reduce the number of possible keys. Laurie demonstrated the attack by reading the passport chip of a Daily Mail's reporter in its envelope without opening it.*[12] Note that in some early biometric passports BAC wasn't used at all, allowing attacker to read the chip's content without providing a key.*[13]

- Passive Authentication (PA). In 2006 Lukas Grunwald demonstrated that it is trivial to copy passport data from a passport chip into a standard ISO/IEC 14443 smartcard using a standard contactless card interface and a simple file transfer tool.*[14] Grunwald used a passport that did not use Active Authentication (anti-cloning) and did not change the data held on the copied chip, thus keeping its cryptographic signature valid. In 2008 Jeroen van Beek demonstrated that not all passport inspection systems check the cryptographic signature of a passport chip. For his demonstration Van Beek altered chip information and signed it using his own document signing key of a non-existing country. This can only be detected by checking the country signing keys that are used to sign the document signing keys. To check country signing keys the ICAO PKD*[15] can be used. Only 5 out of 60+ countries are using this central database.*[16] Van Beek did not update the original passport chip: instead an ePassport emulator was used.*[17] Also in 2008, The Hacker's Choice implemented all attacks and published code to verify the results.*[18] The release included a video clip that demonstrated problems by using a forged Elvis Presley passport that is recognized as a valid US passport.*[19]*[20]

- Active Authentication (AA). In 2005 Marc Witteman showed that the secret Active Authentication key can be retrieved using power analysis.*[11] This may allow an attacker to clone passport chips that use the optional Active Authentication anti-cloning mechanism on chips – if the chip design is susceptible to this attack. In 2008 Jeroen van Beek demonstrated that optional security mechanisms can be disabled by removing their presence from the passport index file.*[21] This allows an attacker to remove – amongst others – anti-cloning mechanisms (Active Authentication). The attack is documented in supplement 7 of Doc 9303 (R1-p1_v2_sIV_0006)*[22] and can be solved by patching inspection system software. Note that supplement 7 features vulnerable examples in the same document that – when implemented – result in a vulnerable inspection process.

- Extended Access Control (EAC). In 2007 Luks Grunwald presented an attack that can make EAC-enabled passport chips unusable.*[23] Grunwald states that if an EAC-key – required for reading fingerprints and updating certificates – is stolen or compromised, an attacker can upload a false certificate with an issue date far in the future. The affected chips block read access until the future date is reached.

16.5 Opposition

Privacy proponents in many countries question and protest the lack of information about exactly what the passports' chip will contain, and whether they impact civil liberties. The main problem they point out is that data on the passports can be transferred with wireless RFID technology, which can become a major vulnerability. Although this could allow ID-check computers to obtain a person's information without a physical connection, it may also allow anyone with the necessary equipment to perform the same task. If the personal information and passport numbers on the chip are not encrypted, the information might wind up in the wrong hands.

On 15 December 2006, the BBC published an article*[24] on the British ePassport, citing the above stories and adding that:

> "Nearly every country issuing this passport has a few security experts who are yelling at the top of their lungs and trying to shout out: 'This is not secure. This is not a good idea to use this technology'", citing a specialist who states "It is much too complicated. It is in places done the wrong way round – reading data first, parsing data, interpreting data, then verifying whether it is right. There are lots of technical flaws in it and there are things that have just been forgotten, so it is basically not doing what it is supposed to do. It is supposed to get a higher security level. It is not."

and adding that the Future of Identity in the Information Society (FIDIS) network's research team (a body of IT security experts funded by the European Union) has "also come out against the ePassport scheme... [stating that] European governments have forced a document on its people that dramatically decreases security and increases the risk of identity theft." *[25]

Most security measures are designed against untrusted citizens (the "provers"), but the scientific security community recently also addressed the threats from untrustworthy verifiers, such as corrupt governmental organizations, or nations using poorly implemented, unsecure electronic systems.*[26] New cryptographic solutions such as private biometrics are being proposed to mitigate threats of mass theft of identity. These are under scientific study, but not yet implemented in biometric passports.

16.6 Countries using biometric passports

16.6.1 European Union

Main article: Passports of the European Union

It was planned that, except for Denmark, Ireland and the UK, EU passports would have digital imaging and fingerprint scan biometrics placed on their RFID chips.*[27] This combination of biometrics aims to create an unrivaled level of security and protection against fraudulent identification papers. Technical specifications for the new passports have been established by the European Commission.*[28] The specifications are binding for the Schengen agreement parties, i.e. the EU countries, except Ireland and the UK, and three of the four European Free Trade Association countries – Iceland, Norway and Switzerland.*[29] These countries are obliged to implement machine readable facial images in the passports by 28 August 2006, and fingerprints by 29 June 2009. The European Data Protection Supervisor has stated that the current legal framework fails to "address all the possible and relevant issues triggered by the inherent imperfections of biometric systems".*[30] Currently, the British and Irish biometric passports only use a digital image and not fingerprinting. German passports printed after 1 November 2007 contain two fingerprints, one from each hand, in addition to a digital photograph. Romanian passports will also contain two fingerprints, one from each hand. The Netherlands also takes fingerprints and is the only EU member that plans to store these fingerprints centrally.*[31] According to EU requirements, only nations that are signatories to the Schengen acquis are required to add fingerprint biometrics.*[32]

In the EU nations, passport prices will be

- Austria (available since 16 June 2006): an adult passport costs €75.90,*[33] while a chip-free child's version costs €30.*[34] As of March 2009 all newly issued adult passports contain fingerprints.*[35]
- Belgium (introduced in October 2004): €71 or €41 for children + local taxes. As of May 2014, passports for adults are valid for 7 years.
- Bulgaria (introduced in July 2009; available since 29 March 2010): 40 BGN (€20) for adults. Passports are valid for 5 years.*[36]
- Croatia (available since 1 July 2009): 390 HRK (€53). The chip contains two fingerprints and a digital photo of the holder. Since 18 January 2010 only biometric passports can be obtained at issuing offices inside Croatia. Diplomatic missions and consular offices must implement new issuing system until 28 June 2010.
- Cyprus (available since 13 December 2010): €70, valid for 10 years
- Czech Republic (available since 1 September 2006): 600 CZK for adults (valid 10 years), CZK100 for children (valid 5 years). Passports contain fingerprints.
- Denmark (available since 1 August 2006): DKK600 for adults (valid for 10 years), DKK115 for children (valid for 5 years) and DKK350 for over 65 (valid for 10 years). As of January 2012 all newly issued passports contain fingerprints.*[37]
- Estonia (available since 22 May 2007): EEK450 (€28.76) (valid for 5 years). As of 29 June 2009, all newly issued passports contain fingerprints.*[38]
- Finland (available since 21 August 2006) €53 (valid for up to 5 years). As of 29 June 2009, all newly issued passports contain fingerprints.*[39]
- France (available since April 2006): €86 or €89 (depending whether applicant provides photographs), valid for 10 years. As of 16 June 2009, all newly issued passports contain fingerprints.*[40]
- Germany (available since November 2005): ≤23-year-old applicants (valid for 6 years) €37.50, >24 years (valid 10 years) €59 Passports issued from 1 November 2007 on include fingerprints.*[41]*[42]
- Greece (available since 26 August 2006) €84.40 (valid for 5 years).*[43] Since June 2009, passports contain fingerprints.*[44]
- Hungary (available since 29 August 2006): HUF7,500 (€26), valid for 5 years, HUF14,000 (€48.50) valid for 10 years. As of 29 June 2009, all newly issued passports contain fingerprints.*[45]*[46]*[47]
- Ireland Biometric passport booklets have been available since 16 October 2006, and Biometric passport cards since October 2015.

32-page passport booklets are priced at €80, 66-page booklets at €110, both valid for 10 years. For children aged between 3 and 18 years the price is €26.50 and the passport booklets are valid for 5 years. Infants' passport booklets for those under 3 years cost €16 and expire 3 years after issue.

Irish biometric passport cards are only available to adults of 18 years and over who already have an Irish passport booklet and cost €35. They expire on the same date as the holder's Irish passport booklet or 5 years after issue, whichever is the shorter period.

(Ireland is not a signatory to the Schengen Acquis and has no obligation or plans to implement fingerprint biometrics)

- Italy (available since 26 October 2006): €116,*[48] valid for 10 years. As of January 2010 newly issued passports contain fingerprints.*[49]

- Latvia (available since 20 November 2007): an adult passport costs Ls15 (€21.36 [prior to 16, July 2012]), valid for 10 or 5 years.

- Lithuania (available since 28 August 2006): LTL150 (€43). For children up to 16 years old, valid max 5 years. For persons over 16 years old, valid for 10 years.*[50]

- Luxembourg (available since 28 August 2006): €30. Valid for 5 years. As of 29 June 2009, all newly issued passports contain fingerprints.*[51]

- Malta (available since 8 October 2008): €70 for persons over 16 years old, valid for 10 years, €35 for children between 10–16 years (valid for 5 years) and €14 for children under 10 years (valid for 2 years).

- Netherlands (available since 28 August 2006): Approximately €11 on top of regular passport (€38.33) cost €49.33. Passports issued from 21 September 2009 include fingerprints. Dutch identity cards are lookalike versions of the holder's page of the passport but don't contain fingerprints.*[52]

- Poland (available since 28 August 2006): 140PLN (€35) for adults, PLN70 for children aged under 13, free for seniors 70+ years, valid 10 years (5 years for children aged below 13). Passports issued from 29 June 2009 include fingerprints of both index fingers.*[53]

- Portugal (available since 31 July 2006 – special passport; 28 August 2006 – ordinary passport): €65 for all citizens valid for 5 years. All passports have 32 pages.*[54]

- Romania (available since 31 December 2008): 302 RON (€67),*[55] valid for 5 years for those over the age of 6, and for 3 years for those under 6. As of 19 January 2010, new passport includes both facial images and fingerprints.*[56]

- Slovakia (available since 15 January 2008): an adult passport (>13 years) costs €33.19 valid for 10 years, while a chip-free child's (5–13 years) version costs €13.27 valid for 5 years and for children under 5 years €8.29, but valid only for 2 years.

- Slovenia (available since 28 August 2006): €42.05 for adults, valid for 10 years. €35.25 for children from 3 to 18 years of age, valid for 5 years. €31.17 for children up to 3 years of age, valid for 3 years. All passports have 32 pages, a 48-page version is available at a €2.50 surcharge. As of 29 June 2009, all newly issued passports contain fingerprints.*[57]

- Spain (available since 28 August 2006) at a price of €25 (price at the 22 April 2012). They include fingerprints of both index fingers as of October 2009. (Aged 30 or less a Spanish passport is valid for 5 years, otherwise they remain valid for 10 years).

- Sweden (available since October 2005): SEK 350 (valid for 5 years). As of 1 January 2012, new passport includes both facial images and fingerprints.*[58]

- UK (introduced March 2006): £72.50 for adults (valid for 10 years) and £46*[59] for children under the age of 16 (valid for 5 years).*[60] (Not Signatory to Schengen Acquis, no obligation to fingerprint biometrics.)

Unless otherwise noted, none of the issued biometric passports mentioned above include fingerprints as of 5 May 2010.

16.6.2 Albania

See also: Albanian passport

The Albanian biometric passport has been available since May 2009, costs 6000 Lekë (€50) and is valid for 10 years. The microchip contains ten fingerprints, the bearer's photo and all the data written on the passport.

16.6.3 Algeria

See also: Algerian passport

Algerian biometric passports were introduced on 5 January 2012 with a validity of 10 years for adults.

16.6. COUNTRIES USING BIOMETRIC PASSPORTS

16.6.4 Argentina

See also: Argentine passport

On 15 June 2012, the government announced the availability of a new biometric passport at a cost of 400 pesos, valid for 10 years*[61]

16.6.5 Armenia

See also: Armenian passport

In July 2012 Armenia introduced two new identity documents to replace ordinary passports of Armenian citizens. One of the documents – ID card with electronic signature and other personal data, is used locally within the country, and the biometric passport with an electronic chip is used for traveling abroad. Electronic chip of biometric passport contains digital images of fingerprints, photo and electronic signature of the passport holder. The passport will be valid for 10 years.*[62]*[63]*[64]

16.6.6 Australia

See also: Australian passport

Biometric Australian passports were introduced in October 2005. A microchip contains the same personal information that is on the colour photo page of the ePassport, including a digitized photograph. SmartGates have been installed in Australian airports to allow Australian ePassport holders and ePassport holders of several other countries to clear immigration controls more rapidly, and facial recognition technology has been installed at immigration gates.*[65]

16.6.7 Azerbaijan

See also: Azerbaijani passport

Biometric Azerbaijan passports were introduced in September 2013. Biometric passports include information about the passport holder's facial features, as well as finger and palm prints. Each passport will also include a personal identification number. The program covers the development of the appropriate legislative framework and information systems to ensure information security.

16.6.8 Bangladesh

See also: Bangladeshi passport

Bangladesh introduced biometric passports in April 2010 and costs BDT3450. All traditional non-MR passports must be withdrawn from circulation by November 2015.

16.6.9 Bosnia and Herzegovina

See also: Bosnia and Herzegovina passport

Available since 15 October 2009 and costing 50 KM (€25.65). Valid 10 years for adults and 5 years for younger than 18. Produced by Bundesdruckerei. On 1 June 2010 Bosnia and Herzegovina issued its first EAC passport.

16.6.10 Botswana

See also: Botswana passport

Botswana began to issue biometric e-passports to its citizens on 8 March 2010. *[66] *[67]

16.6.11 Brazil

See also: Brazilian passport

Brazil started issuing ICAO compliant passports in December 2006. However just in December 2010 it began to issue passports with microchips, first in the capital Brasília and Goiás state. Since the end of January 2011 this last is available to be issued all over Brazil. It is valid for 5 years for adults and costs R$156.07 (approximately €80).*[68] In December 2014, the Federal Police Department extended the validity of the document, from five to ten years.*[69]

16.6.12 Brunei

See also: Bruneian passport

The Bruneian biometric passport was introduced on 17 February 2007. It was produced by German printer Giesecke & Devrient (G&D) following the Visa Waiver Program's requirements. The Bruneian ePassport has the same functions as the other biometric passports.*[70]

16.6.13 Cambodia

See also: Cambodian passport

Cambodia began to issue biometric passports to its citizens on 17 July 2014. The cost for a 5-year passport, issued only to children aged five and under, is 80 USD; while the 10-year passport, issued to all people older than five, costs 100 USD.*[71]

16.6.14 Canada

See also: Canadian passport

All Canadian passports issued on or after 1 July 2013 have been ePassports containing an electronic chip encoded with the bearer's name, gender, and date and place of birth and a digital portrait of their face.*[72]

16.6.15 Cape Verde

See also: Cape Verdean passport

Cape Verde started to issue biometric passports on 26 January 2016. The cost of a biometric passport is 50 euros with a processing time of 30 days. It is noted that the scheme will gradually expand to Cape Verdean diplomatic missions in Boston and Lisbon in the future.*[73]

16.6.16 Chile

See also: Chilean passport

Chile introduced new biometric passports and national ID cards on 2 September 2013. The newly designed passport booklet has a validity of 5 years.*[74]

16.6.17 China

See also: Chinese passport

On 30 January 2011, the Ministry of Foreign Affairs of the People's Republic of China launched a trial issuance of e-passports for public affairs. The face, fingerprint and other biometric features of the passport holder will be digitalized and stored in pre-installed contactless smart chip in the passport.*[75]*[76] On 1 July 2011, the Ministry began issuing biometric passports to all individuals conducting public affairs work overseas on behalf of the Chinese government.*[77]

Ordinary biometric passports have been introduced by the Ministry of Public Security starting from 15 May 2012.*[78] The cost of a passport is 200 CNY (approximately US$31) for first-time applicants in China and 220 CNY (or US$35) for renewals and passports issued abroad.

16.6.18 Colombia

See also: Colombian passport

The Colombian foreign ministry announced that, starting 1 September 2015, new biometric passports will be issued. The only visible change will be that ordinary Colombian passports will now carry the standard biometric symbol (▬◦▬) at the bottom of the front cover of the booklet.*[79] the cost of the passport is COP 163.000 (approx. USD 56).*[80]

16.6.19 Dominican Republic

See also: Dominican Republic passport

In the Dominican Republic, biometric passports began to be issued in May 2004. However the Dominican biometric passports do not carry the "chip inside" symbol ▬◦▬. In January 2010, the cost of the passport was 1,250 DOP, about 35–40 USD at that date.

16.6.20 Egypt

See also: Egyptian passport

The Egyptian Government has, from 5 February 2007, introduced the electronic Passport (e-Passport) and electronic Document of Identity for Visa Purposes (e-Doc/I) which are compliant with the standard of the International Civil Aviation Organization (ICAO). Digital data including holder's personal data and facial image will be contained in the contactless chip embedded in the back cover of e-Passport and e-Doc/I.

16.6.21 Gabon

See also: Gabonese passport

Available since 23 January 2014. The Gabonese biometric passports carry the "chip inside" symbol (▬◦▬).*[81]

16.6.22 Ghana

See also: Ghanaian passport

Available since 1 March 2010 and costing GH₵ 50.00–100.00 for adults and children. The passports contain several other technological characteristics other than biometric technology. However the Ghanaian biometric passports do not carry the "chip inside" symbol (▬●▬), which is mandatory for ICAO-standard electronic passports.

16.6.23 Hong Kong

See also: Hong Kong Special Administrative Region passport

In 2006, the Immigration Department announced that Unihub Limited (a PCCW subsidiary company heading a consortium of suppliers, including Keycorp) had won the tender to provide the technology to produce biometric passports. In February 2007, the first biometric passport was introduced. The cover of the new biometric passport remains essentially the same as that of previous versions, with the addition of the "electronic passport" logo at the bottom. However, the design of the inner pages has changed substantially. The design conforms with the document design recommendations of the International Civil Aviation Organization. The new ePassport featured in the 2008 Stockholm Challenge Event and was a finalist for the Stockholm Challenge Award in the Public Administration cateogeory. The Hong Kong SAR ePassport design was praised on account of the "multiple state-of-the-art technologies [which] are seamlessly integrated in the sophisticated Electronic Passport System (e-Passport System)".[*][82] The cost for a HKSAR passport is HK$370 (or US$48) for a 32-page passport and HK$460 (or US$59) for a 48-page passport.[*][83]

16.6.24 Iceland

See also: Icelandic passport

Available since 23 May 2006 and costing ISK5100 (ISK1900 for under 18 and over 67).

16.6.25 India

See also: Indian passport

India has recently initiated first phase deployment of Biometric e-Passport for Diplomatic passport holders in India and abroad. The new passports have been designed indigenously by the Central Passport Organization, the India Security Press, Nashik and IIT Kanpur. The passport contains a security chip with personal data and digital images. Initially, the new passports will have a 64KB chip with a photograph of passport holder and subsequently include the holder's fingerprint(s). The biometric passport has been tested with passport readers abroad and is noted to have a 4-second response time – less than that of a US Passport (10 seconds). The passport need not be carried in a metal jacket for security reasons as it first needs to be passed through a reader, after which generates access keys to unlock the chip data for reader access.[*][84]

India has also given out a contract to Tata Consultancy Services for issuing e-passports through passport seva kendra. India plans to open 77 such centers across the country to issue these passports.

On 25 June 2008 Indian Passport Authority issued first e-passport to the then President of India, Pratibha Patil. The e-passport is under the first phase of deployment and will be initially restricted to diplomatic passport holders. It is expected to be made available to ordinary citizens from 2017 onwards[*][85]

16.6.26 Indonesia

See also: Indonesian passport

Indonesia started issuing e-Passports on 26 January 2011. The passport costs Rp655,000 (US$66) for the 48-page valid for 5 years, and Rp405,000 (USD41) for the 24-page passport valid for 5 years.[*][86]

16.6.27 Iran

See also: Iranian passport

Iran started issuing biometric diplomatic and service passports in July 2007. Ordinary biometric passports began to be issued on 20 February 2011. The cost of a new passport was approximately US$37 (IRR1,125,000) .[*][87]

16.6.28 Iraq

See also: Iraqi passport

Starting February 1, 2010 the Iraqi Ministry of Interior introduced new electronic system to issue the new A series biometric passports in contract with the German SAFE ID Solutions, the new series is a machine-readable biometric passport available to the public which cost 25,000 dinars or about USD20.*[88]

16.6.29 Ireland

On October 16, 2006, the Minister of Foreign Affairs presented the first biometric passports.

16.6.30 Israel

See also: Israeli passport

Since July 2013, the Israeli Ministry of the Interior has been issuing biometric passports for those citizens who wish to receive them. For a 2-year pilot period under the Biometric Database Law, this was optional. As of August 2013, any passport expiring in more than 2 years can be replaced with a biometric one upon request, free of charge. Passports expiring within 2 years will be charged the full fee. The program review that was supposed to be concluded in 2015 was postponed by order of the Minister of the Interior to a later date, due to the controversy regarding the creation of the Biometric Database rather than storing the biometric data only within the passport's chip, as is the practice in many other countries. Since 2015, the pilot period has been extended until 2017. In May 2017, the pilot period ended. Newly issued passports are required to be biometric.*[89]

To obtain a biometric passport, an applicant must appear in an Interior Ministry office "to be photographed by the special camera which records information such as facial bone structure, distance between one's eyes, ears to eyes and ratio of facial features one from another. One will also be fingerprinted and all this information will be contained in the new high-tech electronic passport." *[90]

16.6.31 Japan

See also: Japanese passport

The Japanese government started issuing biometric passports in March 2006. With this, Japan has met requirements under the US Visa Waiver Program which calls for countries to roll out their biometric passports before 26 October 2006.

16.6.32 Kazakhstan

See also: Kazakhstani passport

Kazakhstan introduced biometric passport in 2009.

16.6.33 Kosovo

See also: Kosovan passport

In May 2011, the Ministry of Interior of the Republic of Kosovo announced that biometric passports would be issued in the summer of 2011 after the winning firm is chosen and awarded the production of the passports.*[91] The first biometric passports were issued in October 2011.

16.6.34 Kuwait

See also: Kuwaiti passport

In March 2017, the Ministry of Interior of the State of Kuwait began issuing biometric passports. The Interior Minisry announced that by late 2018 older non-biometric passports will no longer be valid for use.

16.6.35 Laos

See also: Laotian passport

In September 2016, the Ministry of Foreign Affair of Laos has announced that the biometric passports will roll out after 30 June 2016

16.6.36 Lebanon

See also: Lebanese passport

The Lebanese Directorate General of General Security (*La Sûreté Générale*) started issuing biometric passports as of August 1st, 2016. All new Lebanese passports (*Passeport de la république Libanaise*) issued are biometric passports and machine-readable contain a contactless smart RFID chip embedded inside the bottom of the front cover under the word "PASSEPORT."

The French state-run printing firm, Imprimerie Nationale, carries on the official printing works of both the Lebanese and French governments. *[92]

16.6.37 Lesotho

Date of introduction is uncertain. However, the enabling statute was tabled in November 2016. *[93]

16.6.38 Macau

See also: MSAR passport

Applications for electronic passports and electronic travel permits have been started and processed since 1 September 2009.

16.6.39 Macedonia

See also: Macedonian passport

Available since 2 April 2007 and costs 1500 MKD or c. €22.

16.6.40 Madagascar

See also: Malagasy passport

The passport is available since 2014 and costs 110,000 Ariary. Since September 2014, it is mandatory for Malagasy citizens to depart the country with a biometric passport.*[94]

16.6.41 Malaysia

See also: Malaysian passport

Malaysia was the first country in the world to issue biometric passports in 1998, after a local company, IRIS Corporation, developed the technology. Malaysia is however not a member of the Visa Waiver Program (VWP) and its first biometric passport did not conform to the same standards as the VWP biometric document because the Malaysian biometric passport was issued several years ahead of the VWP requirement. The difference lies in the storage of fingerprint template instead of fingerprint image in the chip, the rest of the technologies are the same. Also the biometric passport was designed to be read only if the receiving country has the authorization from the Malaysian Immigration Department. Malaysia started issuing ICAO compliant passports from February 2010.

Malaysia used to issue passports with validity for 2 years and 5 years, but the passport with 2 years validity was withdrawn since Jan, 2015.

The pricing for a Malaysian passport are RM 200 with 5 years validity, RM100 for senior citizens, children below 12, Haj pilgrims, and students below 21 and studying abroad is RM100 and is free for disabled citizens.

16.6.42 Maldives

See also: Maldivian passport

Maldives started rolling out its new ePassport to its citizens on 26 July 2006. The new passport follows a completely new design, and features the passport holder's facial and fingerprint information as biometric identifiers. A 32-page Ordinary passport will cost Rf350, while a 64-page Ordinary passport will cost Rf600. Children under the age of 10 years and people applying for passports through diplomatic missions abroad will be issued with a 32-page non-electronic Ordinary passport, which will cost Rf250.

16.6.43 Sovereign Military Order of Malta

See also: Sovereign Military Order of Malta passport

Since 2005 the SMOM diplomatic and service passports include biometric features and are compliant with ICAO standards.

16.6.44 Moldova

See also: Moldovan passport

The Moldovan biometric passport is available from 1 January 2008. The new Moldovan biometric passport costs approximately 760 MDL (€45)*[95] and is obligatory from 1 January 2011. The passport of the Republic of Moldova with biometric data contains a chip which holds digital information, including the holder's signature, as well as the traditional information. It is valid for 7 years (for persons over 7) and 4 years (for persons less than 7) respectively. It was introduced as a request of European Union to safeguard the borders between the E.U. and Republic of Moldova.

16.6.45 Montenegro

See also: Montenegrin passport

The Montenegrin biometric passport was introduced in 2008. It costs approximately €40.

16.6.46 Mongolia

The Mongolian ministry of interior stated that first biometric passport will be issue at the end of 2016.

16.6.47 Mauritania

Cover of Mauritanian Biometric Passport

The issuance of the biometric passports was launched 6 May 2011.*[96]

It costs 115.68 US Dollar for issuance and is valid for five years only.

16.6.48 Morocco

See also: Moroccan passport

The Moroccan biometric passport was introduced in 2008. In December 2009, early limited trials have been extended, and the biometric passport is available from 25 September 2009 to all Moroccan citizens holders of an electronic identity card.*[97] It costs 300DH (approximately €27).

16.6.49 Mozambique

See also: Mozambican passport

Mozambique started to issue biometric passports in September 2014. The issuance of such passports was suspended 40 days later but was resumed in February 2015.*[98]

16.6.50 New Zealand

See also: New Zealand passport

Introduced in November 2005, like Australia and the USA, New Zealand is using the facial biometric identifier. There are two identifying factors: the small symbol on the front cover indicating that an electronic chip has been embedded in the passport, and the polycarbonate leaf in the front (version 2009) of the book inside which the chip is located. The cost is NZ$140 (when applying in person) or NZ$124.50 (when applying online—available only if already holding a passport) for adults, NZ$81.70 for children, valid for five years. However, in 2015 the New Zealand government approved for the reinstatement of a 10-year validity period for passports, which will come into effect on 30 November 2015.

16.6.51 Nigeria

See also: Nigerian passport

Nigeria is currently one of the few nations in Africa that issues biometric passports, and has done it since 2007.The harmonized ECOWAS Smart electronic passport issued by the Nigerian Immigrations Service is powered by biometric technology in tandem with the International Civil Aviation Organization (ICAO) specifications for international travels.

Travellers' data captured in the biometric passport can be accessed instantly and read by any security agent from any spot of the globe through an integrated network of systems configured and linked to a centrally-coordinated passport data bank managed by the Nigerian Immigrations Service.

16.6.52 Norway

See also: Norwegian passport

The introduction of biometric passports to Norway began in 2005 and supplied by Setec, costing NOK 450 for adults, or

c. €50, NOK 270 for children.

In 2007 the Norwegian government launched a 'multi-modal' biometric enrolment system supplied by Motorola. Motorola's new system enabled multiple public agencies to digitally capture and store fingerprints, 2D facial images and signatures for passports and visas.*[99]

The Norwegian biometrics company IDEX ASA has begun development of electronic ID cards (eID) with fingerprint security technology for use throughout the EU.*[100]

16.6.53 Pakistan

See also: Pakistani passport

In 2004, Pakistan became one of the first countries in the world to issue the biometric passports, which are compliant with ICAO standards and dubbed Multi-biometric e-Passports, however they do not carry the "chip inside" symbol (), which is mandatory for ICAO-standard electronic passports.

As of 2012, Pakistan has adopted the Multi-biometric e-Passport that is now compliant with ICAO standards.*[101]

In May 2016, Pakistan interior minister launched the project officially and promised that ePassports would be available fully in 2017 for citizens.*[102]

16.6.54 Panama

See also: Panamanian passport
 Panama has issued biometric passports since 2014. The cost of the passport went up from $50 to $100, and the inside contains images of recent government projects.

16.6.55 Peru

See also: Peruvian passport

On 21 February 2016, the *Superintendencia Nacional de Migraciones* announced that the first Peruvian biometric passports would be delivered by 26 February 2016.*[103]*[104] The first passport was issued for Peruvian opera singer Juan Diego Flórez. It will feature a new cover, along with several security improvements, in order to be exempted for visas for the Schengen Area.*[105] It will cost PEN98.50, approximately USD28, making it the cheapest passport in Latin America.*[106]

Cover of a Panamanian Biometric Passport

16.6.56 Philippines

See also: Philippine passport

On 11 August 2009, the first biometric passport was issued to then President Gloria Macapagal-Arroyo. The new e-passport has various security features, including a hidden encoded image; an ultra-thin, holographic laminate; and a tamper-proof electronic microchip and is priced at around ₱950.*[107]*[108]

16.6.57 Qatar

See also: Qatari passport

On 20 April 2008, Qatar started issuing biometric passports which are ICAO compliant. A Qatari passport costs QR200.*[109]

16.6.58 Russia

See also: Russian passport

Russian biometric passports were introduced in 2006. As of 2015, they cost 3500 rubles (approximately USD50) and use printed data, photo and fingerprints and are BAC-encrypted.*[110] Biometric passports issued after 1 March 2010 are valid for 10 years. Russian biometric passports are currently issued within Russia and in all of its consulates.

From 1 January 2015, the Government of Russia has issued passports which contain fingerprints.

16.6.59 Saudi Arabia

See also: Saudi Arabian passport

On 21 June 2006, Saudi Arabia started issuing biometric passports which are ICAO compliant. A Saudi Arabian passport costs SR150.

16.6.60 Serbia

See also: Serbian passport

Available since 7 July 2008, and cost 3.600 RSD or approximately €32.*[111] (Aged 3 or less a Serbian passport is valid for 3 years, aged 3 to 14 it is valid for 5 years, otherwise passport remain valid for 10 years.)

16.6.61 Singapore

See also: Singaporean passport

The Immigration and Checkpoints Authority (ICA)*[112] of Singapore introduced the Singapore biometric passport (BioPass) on 15 August 2006. With this, Singapore has met requirements under the US Visa Waiver Program which calls for countries to roll out their biometric passports before 26 October 2006.*[113]

The pricing for a Singaporean passport are, 70 SGD if applied via online, mail or ICA deposit box and 80 SGD if applied via Singapore Overseas Mission and Singaporean passports are valid for 5 years.

16.6.62 Somalia

See also: Somali passport

The new "e-passport" of Somalia was introduced and approved by the nation's Transitional Federal Government on 10 October 2006. It costs $100 USD to apply for Somalis living inside of Somalia, and $150 USD for Somalis living abroad. Somalia is now the first country on the African continent to have introduced the "e-passport" .*[114]

16.6.63 South Korea

See also: Republic of Korea passport

The Ministry of Foreign Affairs and Trade of South Korea started issuing biometric passports to its citizens on 25 August 2008. The cost is fixed to 55,000 Won or 55 US Dollars, and the validity of ordinary passport is 10 years.*[115]

16.6.64 South Sudan

See also: South Sudanese passport

The Republic of South Sudan started issuing internationally recognized electronic passports in January 2012.*[116] The passports were officially launched by the President Salva Kiir Mayardit on 3 January 2012 in a ceremony in Juba.*[117] The new passport will be valid for five years.*[118]

16.6.65 Slovakia

See also: Slovak passport

Biometric passports were first issued in Slovakia in 2008. The latest version was issued in 2014 and contains a contactless chip in the biodata card that meets ICAO specifications.*[119]

16.6.66 Sri Lanka

See also: Sri Lankan passport

From the 10 August 2015, the Department of Immigration and Emigration Sri Lanka has begun issuing ICAO compliant biometric passports to the public. As at October 2016, all new passports require both thumbs to be scanned and a digital facial mapping photograph be taken during the issuing process.*[120]

16.6.67 Sudan

See also: Sudanese passport

The Republic of the Sudan started issuing electronic passports to citizens in May 2009. The new electronic passport will be issued in three categories. The citizen's passport (ordinary passport) will be issued to ordinary citizens and will contain 48 pages. Business men/women who need to travel often will have a commercial passport that will contain 64 pages. Smaller passports that contain 32 pages only will be issued to children. The microprocessor chip will contain the holder's information. Cost to obtain a new passport will be SDG250 (approximately USD100), SDG200 for students and SDG100 for kids. The validity of the citizen's passport will be five years, or seven years for the commercial passport.*[121]

16.6.68 Switzerland

See also: Swiss passport

The Swiss biometric passport has been available since 4 September 2006. By a narrow majority of 50.14%, Swiss voters decided in May 2009 to accept the introduction of a biometric passport.*[122] Since 1 March 2010, all issued passports are biometric, containing a photograph and two fingerprints recorded electronically.*[123] The costs are CHF 140.00 for adults and CHF 60.00 for children (−18 years old).*[124]

16.6.69 Taiwan

See also: Taiwanese passport

The Taiwanese biometric passport has been available since 29 December 2008. It costs NT$1,600 for an ordinary passport with either 3, 5 or 10 years validity.*[125] Taiwanese Central Engraving and Printing Plant prints passports for the Ministry of Foreign Affairs of Taiwan for several decades. During this period, the passport has been redesigned various times. The current e-passport (or known as biometric passport) is fitted with RFID technology that facilitates Taiwanese passport immigration clearances worldwide.

16.6.70 Tajikistan

See also: Tajik passport

Biometric passports will be issued in Tajikistan from 1 February 2010. On 27 August 2009, Tajik Ministry of Foreign Affairs and German Muhlbauer signed a contract on purchase of blank biometric passports and appropriate equipment for Tajikistan.*[126]

16.6.71 Thailand

See also: Thai passport

The Ministry of Foreign Affairs of Thailand*[127] introduced the first biometric passport for Diplomats and Government officials on 26 May 2005. From 1 June 2005, a limited quantity of 100 passports a day was issued for Thai citizens, however, on 1 August 2005 a full operational service was installed and Thailand became the first country in Asia to issue an ICAO compliant biometric passport.*[128]

16.6.72 Togo

See also: Togolese passport

In August 2009, Togo became one of the first African countries to introduce biometric passports. The price of the passport was then set at 30,000 CFA Francs for Togolese residing in Togo. For Togolese residing abroad, the price varies.

16.6.73 Tunisia

See also: Tunisian passport

The Tunisia ministry of interior stated that it will start issuing biometric passports at the end of year 2016.

16.6.74 Turkey

See also: Turkish passport

Turkish passports which are compatible with European Union standards have been available since 1 June 2010.*[129] Colours of the new biometric passports have also been changed. Accordingly, regular passports; claret red, special passports; bottle green and diplomatic passports wrap black colours.*[130]

Most recently Turkish Minister of the State announced that the government is printing the new passports at government minting office since the private contractor failed to deliver.

The current cost of issuing a 10-year passport in Turkey is TRY620.60 (approximately US$215).[*][131][*][132]

16.6.75 Turkmenistan

See also: Turkmen passport

Turkmenistan became the first country in ex-USSR, in mid-Asia region to issue an ICAO-compliant biometric passport. The passport is available since 10 July 2008.[*][133]

16.6.76 Ukraine

See also: Ukrainian passport and Biometric passports in Ukraine

According to law, Ukraine was supposed to issue biometric passports and identity cards on 1 January 2013.[*][134] However, they did not become available until two years later in January 2015 and are fully compatible with European Union standards.

16.6.77 United Arab Emirates

See also: Emirati passport

The UAE ministry of interior stated that it would start issuing Emirati biometric passports at the end of 2010.[*][135]

16.6.78 United States

See also: United States passport

The biometric version of the U.S. passport (sometimes referred to as an electronic passport) has descriptive data and a digitized passport photo on its contactless chips, and does not have fingerprint information placed onto the contactless chip. However, the chip is large enough (64 kilobytes) for inclusion of biometric identifiers. The U.S. Department of State first issued these passports in 2006, and since August 2007 issues biometric passports only. Non-biometric passports are valid until their expiration dates.[*][136]

Although a system able to perform a facial-recognition match between the bearer and his or her image stored on the contactless chip is desired, it is unclear when such a system will be deployed by the U.S. Department of Homeland Security at its ports of entry.[*][137]

A high level of security became a priority for the United States after the attacks of 11 September 2001. High security required cracking down on counterfeit passports. In October 2004, the production stages of this high-tech passport commenced as the U.S. Government Printing Office (GPO) issued awards to the top bidders of the program. The awards totaled to roughly $1,000,000 for startup, development, and testing. The driving force of the initiative is the U.S. Enhanced Border Security and Visa Entry Reform Act of 2002 (also known as the "Border Security Act"), which states that such smartcard identity cards will be able to replace visas. As for foreigners travelling to the U.S., if they wish to enter U.S. visa-free under the Visa Waiver Program (VWP), they are now required to possess machine-readable passports that comply with international standards. Additionally, for travellers holding a valid passport issued on or after 26 October 2006, such a passport must be a biometric passport if used to enter the U.S. visa-free under the VWP.

16.6.79 Uruguay

See also: Uruguayan passport

The Uruguayan Ministry of the Interior started to issue biometric passports to Uruguayan citizens on 16 October 2015. The new passport complies with the standards set forth by the Visa Waiver Program of the United States.[*][138]

16.6.80 Uzbekistan

See also: Uzbekistani passport

In Uzbekistan, 23 June 2009 Islam Karimov issued a Presidential Decree[*][139] "On measures to further improve the passport system in the Republic of Uzbekistan." On 29 December 2009 the President of Uzbekistan signed a decree to change the dates for a phased exchange of populations existing passport to the biometric passport. In accordance with this decree, biometric passports will be phased in, beginning with 1 January 2011. In the first phase, the biometric passport will be issued to employees of ministries, departments and agencies of the republic, individuals who travel abroad or outside the country, as well as citizens who receive a passport in connection with the achievement of a certain age or for other grounds provided by law. The second phase will be for the rest of the population who will be able to get new passports for the period from 2012 to 2015.

16.6.81 Venezuela

See also: Venezuelan passport

Issued after July 2007, Venezuela was the first Latin American country issuing passports including RFID chips along other major security improvements. The chip has photo and fingerprints data.*[140]

16.6.82 Zimbabwe

Introduced in June 2017. *[141]

16.7 See also

- International Civil Aviation Organization Public Key Directory

16.8 References

[1] "Over 60+ countries now issuing ePassports - FindBiometrics". *FindBiometrics*. 2008-12-30. Retrieved 2017-04-05.

[2] "ICAO Document 9303, Part 1, Volume 1 (OCR machine-readable passports)" (PDF). *ICAO*. Retrieved 21 February 2017.

[3] Central Intelligence Agency (September 2011). "Surviving Secondary: An Identity Threat Assessment of Secondary Screening Procedures at International Airports" (PDF). Wikileaks.

[4] "ICAO PKD Participants". International Civil Aviation Organization. January 2017.

[5] "Metal shields and encryption for US passports". Newscientist.com. Retrieved 8 September 2010.

[6] Funke, Holger. "Overview of eMRTD test specifications". *blog.protocolbench.org*.

[7] Funke, Holger. "Automatic Border Control Systems (eGate)". *blog.protocolbench.org*.

[8] "Fingerprinting Passports" (PDF). Retrieved 8 September 2010.

[9] Goodin, Dan (26 January 2010). "Defects in e-passports allow real-time tracking, The Register, Dan Goodin, 26th Jan 2010". Theregister.co.uk. Retrieved 8 September 2010.

[10] "A Traceability Attack Against e-Passports, Tom Chothia and Vitaliy Smirnov, 14th International Conference on Financial Cryptography and Data Security 2010" (PDF). Retrieved 8 September 2010.

[11] "Attacks on Digital Passports" (PDF). Archived from the original (PDF) on 29 August 2008. Retrieved 8 September 2010.

[12] "Safest ever passport is not fit for purpose". Daily Mail. 5 March 2007. Retrieved 15 June 2012.

[13] "Belgian Biometric Passport does not get a pass". Dice.ucl.ac.be. Archived from the original on 6 April 2009. Retrieved 8 September 2010.

[14] Kim Zetter (3 August 2006). "Hackers clone E-Passports". *Wired*. Retrieved 18 October 2014.

[15] "Icao Pkd". .icao.int. Archived from the original on 18 October 2014. Retrieved 18 October 2014.

[16] Steve Boggan (6 August 2008). "Fakeproof e-passport is cloned in minutes". *The Sunday Times*. UK. Archived from the original on 8 July 2010. Retrieved 6 October 2010.

[17] "ePassport emulator". Dexlab.nl. Archived from the original on 12 April 2010. Retrieved 8 September 2010.

[18] "The Hacker's Choice ePassport tools". Freeworld.thc.org. Retrieved 8 September 2010.

[19] "The Hackers Choice (THC) ePassport RFID Vulnerability Demonstration". Google. Retrieved 8 September 2010.

[20] Lettice, John (30 September 2008). "Elvis has left the border: ePassport faking guide unleashed". Theregister.co.uk. Retrieved 8 September 2010.

[21] "ePassport reloaded goes mobile" (PDF). Retrieved 8 September 2010.

[22] (PDF) http://www2.icao.int/en/MRTD/Downloads/Supplements%20to%20Doc%209303/Supplement%20to%20ICAO%20Doc%209303%20-%20Release%207.pdf#page=35. Retrieved April 26, 2009. Missing or empty |title= (help)

[23] "PowerPoint Presentation" (PDF). Retrieved 8 September 2010.

[24] "BBC NEWS - Programmes - Click - ePassports 'at risk' from cloning". *bbc.co.uk*.

[25] Budapest Declaration on Machine Readable Travel Documents, FIDIS NoE, Budapest, September 2006

[26] "E-government: who controls the controllers?". Opendemocracy.net. Retrieved 9 February 2006.

[27] Jonathan P. Aus (25 September 2006). "Decision-making under Pressure: The Negotiation of the Biometric Passports Regulation in the Council" (in Norwegian). Arena.uio.no. Retrieved 8 September 2010.

[28] EC News article about the relevant regulations: Council Regulation (EC) 2252/2004, Commission Decision C(2005)409 adopted on 28 February 2005 and Commission Decision C(2006)2909 adopted on 28 June 2006

[29] Council Regulation (EC) No 2252/2004 of 13 December 2004, see preamble 10–14

[30] Opinion of the European Data Protection Supervisor on the proposal for a Regulation of the European Parliament and of the Council amending Council Regulation (EC) No 2252/2004 on standards for security features and biometrics in passports and travel documents issued by Member States, 6 August 2008

[31] "Dutch government to store fingerprints". Nrc.nl. 18 September 2009. Retrieved 8 September 2010.

[32] "COUNCIL REGULATION (EC) No 2252/2004 of 13 December 2004 on standards for security features and biometrics in passports and travel documents issued by Member States" (PDF). Official Journal of the European Union. 29 December 2004. Retrieved 6 October 2010.

[33] "Reisepass – Neuausstellung". *help.gv.at* (in German). Bundeskanzleramt Österreich. 11 June 2014. Retrieved 9 August 2014.

[34] "Reisepass – Minderjährige unter 18 Jahren". *help.gv.at* (in German). Bundeskanzleramt Österreich. 15 May 2014. Retrieved 9 August 2014.

[35] "Austria Adopts Fingerprint Passports". *english.cri.cn*. China Radio International. 30 March 2009. Retrieved 5 June 2010.

[36] "Bulgaria to Start Issuing Biometric IDs in March 2010". *Novinite.com*. Sofia News Agency. 1 October 2009. Retrieved 5 June 2010.

[37] "Priser på pas gældende fra 1. oktober 2004". *politi.dk* (in Danish). Danish National Police. 29 December 2009. Retrieved 5 June 2010.

[38] "Estonian Passport Application". Consulate General of Estonia in New York. Retrieved 5 June 2010.

[39] "Fingerprints to be included in new passports as from 29 June". Embassy of Finland, Washington. 29 June 2009. Retrieved 5 June 2010.

[40] "Passeport biométrique" (in French). Service-public.fr. Archived from the original on 22 July 2010. Retrieved 5 June 2010.

[41] "Elektronischer Reisepass" (in German). Bundesministerium des Innern. Archived from the original on 22 February 2010. Retrieved 5 June 2010.

[42] "Paßgesetz § 5 Gültigkeitsdauer" (in German). Bundesministerium der Justiz. Retrieved 5 June 2010.

[43] "Issuance of passports". Embassy of Greece Canberra. Archived from the original on 31 March 2012. Retrieved 10 October 2011.

[44] "Hellenic Ministry of Foreign Affairs". Archived from the original on 3 July 2010. Retrieved 5 June 2010.

[45] Molnár, Szilárd (6 July 2009). "HU: Fingerprint in Hungarian Passports". *ePractice*. Retrieved 5 June 2010.

[46] "Június végétől ujjlenyomat is lesz az új útlevelekben". *Kormányszóvivő.hu* (in Hungarian). 26 May 2009. Retrieved 5 June 2010.

[47] http://www.tata.hu/system/files/Utlevel.pdf

[48] "Passport". *esteri.it*.

[49] "Avvio emissione di passaporto elettronico con impronte digitali" (in Italian). Ambasciata d'Italia a La Valletta. 11 January 2010. Retrieved 5 June 2010.

[50] "Personal document issuing scheme". *Personalisation of Identity Documents Centre*. Ministry of the Interior. Retrieved 6 June 2013.

[51] "Passports". Embassy of Luxembourg in Washington. Archived from the original on July 5, 2010. Retrieved 5 June 2010.

[52] "Paspoort twee keer zo lang geldig, ID-kaart zonder vingerafdrukken". *rijksoverheid.nl*.

[53] "The new Polish passport with fingerprint". Polska Wytwórnia Papierów Wartościowych S.A. 22 June 2009. Retrieved 5 June 2010.

[54] "Archived copy". Archived from the original on 2013-11-07. Retrieved 2014-02-10.

[55] "Romanian Biometric Passport Prices". RCG. 7 January 2014. Retrieved 7 January 2014.

[56] "First biometric passport in Bucharest". RCG. 22 February 2010. Archived from the original on 31 May 2010. Retrieved 5 June 2010.

[57] "Slovenia to begin issuing second-generation biometric passports". *News*. Ministry of the Interior. 29 June 2009. Archived from the original on 29 March 2010. Retrieved 5 June 2010.

[58] "Passport and visa". *Swedavia*. Archived from the original on 5 July 2010. Retrieved 5 June 2010.

[59] "Passport fees". *www.gov.uk*.

[60] "Passport prices rise". *Directgov*. 7 July 2009. Archived from the original on 11 May 2011. Retrieved 5 June 2010.

[61] Ministerio del Interior y Transporte. "Ministerio del Interior y Transporte". *Ministerio del Interior y Transporte*.

[62] "Project of Biometric Passports in Armenia". *PLUS Journal*. 25 December 2008. Archived from the original on 21 July 2009. Retrieved 5 June 2010.

[63] Общество / Культура: У граждан Армении могут быть два вида паспортов. *Barev.NET* (in Russian). 20 October 2009. Archived from the original on 23 July 2011. Retrieved 5 June 2010.

[64] Армения. У граждан будет два типа удостоверений личности (in Russian). 21 October 2009. Retrieved 5 June 2010.

16.8. REFERENCES

[65] "SmartGate Frequently Asked Questions – What is an Australian ePassport?". Australian Customs Service.

[66] http://www.gov.bw/en/Ministries--Authorities/Ministries/Ministry-of-Labour--Home-Affairs-MLHA/Tools--Services/Services--Forms/Passport-Application-Form/

[67] http://www.sundaystandard.info/botswana-introduces-electronic-passport

[68] "Globo Vídeos – VIDEO – Polícia Federal apresenta novo modelo de passaporte". Video.globo.com. 10 December 2010. Retrieved 6 December 2011.

[69] "G1 - Polícia Federal amplia para dez anos prazo de validade de passaportes - notícias em Distrito Federal". *Distrito Federal*.

[70] "Brunei passport becomes Biometric passport". I4donline.net. Archived from the original on 8 December 2010. Retrieved 8 September 2010.

[71] http://www.camsinchew.com/node/37331 有效期 10 年・申辦費 100 美元・新版本護照啓用

[72] "The ePassport". Passport Canada. 6 December 2012. Retrieved 10 August 2011.

[73] "Passaporte eletrónico arranca em Cabo Verde". CEO Lusófono. 26 January 2016. Retrieved 10 March 2016.

[74] "Nueva cédula de identidad y pasaporte electrónicos" (in Spanish). Government of Chile. Retrieved 18 February 2014.

[75] "Foreign Minister Yang Jiechi Attends the Launch Ceremony for the Trial Issuance of E-Passports for Public Affairs". Ministry of Foreign Affairs. Retrieved 15 February 2011.

[76] "因公电子护照 31 日试点签发可使持照人快速通关". 中国网. Retrieved 15 February 2011.

[77] 中华人民共和国外交部公告 (in Chinese). Ministry of Foreign Affairs of the People's Republic of China. 1 June 2011. Retrieved 31 August 2011.

[78] "Chinese passports to get chipped". China Daily USA. Retrieved 5 May 2012.

[79] "(Spanish) Especial Pasaporte - Cancillería". *Cancillería*.

[80] Cancillería Colombia. "Costos".

[81] "Les nouveaux passeports de la République émergente du PDG en circulation". http://info241.com. 4 November 2014. Retrieved 14 November 2014. External link in |publisher= (help)

[82] "Electronic Passport System". Archived from the original on August 29, 2010. Retrieved March 28, 2010.

[83] "Fee Tables". Hong Kong Immigration Department. Retrieved 8 April 2016.

[84] After US tests, India to get first e-passport (16 May 2008). "After US tests, India to get first e-passport". NDTV.com. Archived from the original on 2 July 2008. Retrieved 8 September 2010.

[85] "Coming soon: Govt to roll out e-passports with electronic chip, bio-metric security features". indianexpress.com. 3 January 2017. Retrieved 6 August 2017.

[86] Mustaqim Adamrah (12 February 2010). "E-passport to premiere in January, government says". *The Jakarta Post*. Retrieved 18 January 2011.

[87] مدارک لازم جهت صدور گذرنامه (in Persian). 8 March 2014. Archived from the original on 9 February 2014. Retrieved 8 March 2014.

[88] IRQ103919.E (PDF). 23 December 2011. Retrieved 29 June 2016.

[89] http://www.haaretz.co.il/tmr/1.3151891

[90] Jewish News: The Yeshiva World, Israel Moving to Biometric Passport.

[91] "Gjatë verës nis lëshimi i pasaportave biometrike". Koha Ditore. 2 May 2011. Retrieved 2 May 2011.

[92] Lebanon to introduce biometric passports

[93] http://www.lesotho.gov.ls/gov_webportal/articles/2016/senate_endorses_passport_bill.html

[94] "Migration – Le passeport électronique obligatoire". L'Express de Madagascar. Archived from the original on 17 April 2016. Retrieved 8 April 2016.

[95] Centrul Resurselor Informaționale de Stat "Registru". "Centrul Resurselor Informaționale de Stat "Registru"". registru.md. Archived from the original on 7 September 2012. Retrieved 8 September 2010.

[96] "Mauritanians make biometric passports compulsory". panapress. Retrieved 2014-10-31.

[97] "Passport portal of the Kingdom of Morocco". Kingdom of Morocco. Retrieved 8 September 2010.

[98] "Moçambicanos necessitam de passaporte biométrico para entrar na África do Sul". Voice of America. Retrieved 8 April 2016.

[99] "Norway, Sweden push ahead and biometric passports" (PDF). Archived from the original (PDF) on 2013-01-02.

[100] Bernstein, Ralph (1 November 2011). "European Union". *Public Service Review* (22).

[101] http://unpan1.un.org/intradoc/groups/public/documents/un-dpadm/unpan048580.pdf

[102] http://tribune.com.pk/story/1103021/biometric-passports-to-be-introduced-in-2017/

[103] "Pasaporte electrónico se puede tramitar desde hoy" (in Spanish). El Comercio. 23 February 2016.

[104] "Todo lo que necesitas saber sobre el Pasaporte Electrónico" (in Spanish). Superintendencia Nacional de Migraciones. Retrieved 22 February 2016.

[105] "Visa Schengen: su eliminación estará lista para marzo del 2016" (in Spanish). El Comercio. 30 December 2015. Retrieved 23 February 2016.

[106] "Pasaporte peruano será el más barato de Latinoamérica, aseguran" (in Spanish). RPP. 21 February 2016.

[107] "Arroyo first to receive e-passport from DFA". 11 August 2009.

[108] "Frequently asked questions regarding the 'e-passport'". 11 August 2009.

[109] "Visa, Passport and Official Documents".

[110] Читаем биометрический паспорт (in Russian). 3 March 2009. Archived from the original on June 18, 2009. Retrieved 5 June 2010.

[111] "MUP RS – Putne isprave". Mup.gov.rs. Archived from the original on 31 August 2010. Retrieved 8 September 2010.

[112] "ICA - Immigration & Checkpoints Authority of Singapore". ica.gov.sg.

[113] "?". Archived from the original on October 1, 2006.

[114] "?". Archived from the original on January 21, 2008.

[115] "Passport Issuance Fees". Ministry of Foreign Affairs and Trade. Archived from the original on 6 March 2009. Retrieved 14 March 2011.

[116] "www.sudanradio.org". *sudanradio.org*. Archived from the original on 2012-12-31.

[117] http://www.sudantribune.com/South-Sudan-launches-passports-and,41183

[118] "gossmission.org". *gossmission.org*.

[119] "How to check a passport - Keesing Technologies". *Keesing Technologies*. Retrieved 2016-02-16.

[120] ICTA. "General Information on Passports". *immigration.gov.lk*.

[121] الجواز الالكتروني (in Arabic). Retrieved 5 June 2010.

[122] "50.1% Ja! Biometrischer Pass schafft es ganz knapp". *Blick.ch* (in German). May 17, 2009. Retrieved November 12, 2013.

[123] "Pass 10" (in German). 25 January 2010. Archived from the original on 7 June 2010. Retrieved 5 June 2010.

[124] "Gebühr/Preis und Gültigkeit" (in German). 6 May 2010. Archived from the original on 23 May 2010. Retrieved 5 June 2010.

[125] Accessed, October 6, 2010 Archived April 18, 2010, at the Wayback Machine.

[126] Yuldoshev, Avaz (27 October 2009). "First consignment of blank biometric passports for Tajikistan delivered to Dushanbe". Retrieved 5 June 2010.

[127] "The Minister of Foreign Affairs meets with senior executives of Italian private companies and chairs the meeting of Team Thailand and Honorary Consul Generals of Thailand in Italy". Ministry of Foreign Affairs, Kingdom of Thailand. Retrieved 6 October 2010.

[128] "Your partner in document verification". keesingfight-fraud.com. Archived from the original on September 30, 2007.

[129] "?". *Hurriyet Daily News* (in Turkish). Turkey.

[130] "Çağdaş pasaport yolda" (in Turkish). 18 June 2009. Retrieved 5 June 2010.

[131] "e-Pasaport". Retrieved 23 July 2016.

[132] "Umuma Mahsus (Lacivert) Pasaport Harçları" (in Turkish). Archived from the original on 16 March 2010. Retrieved 5 June 2010.

[133] "Turkmenistan introduces passports with biometric data of their holders". 10 July 2008. Retrieved 5 June 2010.

[134] "Українці зможуть отримати біометричні паспорти вже 1 січня 2015 року". unn.com.ua. Retrieved 3 November 2014.

[135] ""الداخلية": إصدار جواز السفر الالكتروني نهاية العام الجاري." Snrd.ae. Archived from the original on 11 April 2010. Retrieved 6 December 2011.

[136] "The U.S. Electronic Passport". Archived from the original on 27 August 2010. Retrieved 5 June 2010.

[137] Bronk, R. Christopher (May 2007). "Innovation by Policy: A Study of the Electronic Passport" (PDF). The James A. Baker III Institute for Public Policy. Archived from the original (PDF) on 2007-07-01. Retrieved 5 June 2010.

[138] "Emiten hoy el primer pasaporte electrónico" (in Spanish). 16 October 2015. Retrieved 10 December 2015.

[139] "Archived copy". Archived from the original on 2011-03-20. Retrieved 2011-01-21.

[140] "Nuevo pasaporte electrónico entrará en vigencia a finales de año" (in Spanish). 22 September 2006. Archived from the original on 16 May 2011. Retrieved 5 June 2010.

[141] http://www.technomag.co.zw/2017/06/13/government-introduces-electronic-chip-passports/

16.9 External links

ICAO passport standards and related materials:

- ICAO Doc 9303 Series. Machine Readable Travel Documents. Seventh Edition, 2015.
- International Civil Aviation Organization
- ICAO Document 9303, Part 2
- Supplement to ICAO Doc 9303 - Release_7
- LDS 1.7 PKI Maintenance

Open source and free tools:

- 'JMRTD' is an Open Source Java Implementation of Machine Readable Travel Documents
- 'RFIDIOt' is an open source python library for exploring RFID devices. The script mrpkey.py can read passport chips
- 'ePassport Viewer' is a GPL-friendly tool to read and checks ePassports
- 'epassport_emulator' is a freeware ePassport emulator for JavaCard and the Nokia Secure element

Chapter 17

Biometric voter registration

Biometric voter registration implicates using biometric technology (capturing unique physical features of an individual – fingerprinting is the most commonly used), most of the times in addition to demographics of the voter, for polling registration and/or authentication. The enrollment infrastructure allows collecting and maintaining a database of the biometric templates for all voters.

A biometric voting project might include introducing biometric registration kits for enrolment of voters; using electronic voter identification devices before and on Election Day; issuing of voter identification documents (i.e. biometric voter cards), among others. The chronological stages for adopting a biometric voting registration project usually include assessment; feasibility studies; securing funding; reviewing legislation; doing pilot projects and mock registration exercises; procurement; distribution of equipment, installation, and testing; recruitment and training of staff; voter information; deployment and, post-election audits.*[1]*[2]

The final aim of implementing biometric election technology is achieving de-duplication of the voting register,*[3] thus preventing multiple voter registration and multiple voting; improving identification of the voter at the polling station, and mitigating the incidence of voter fraud (e.g. buy/rent of voters IDs before an election).

However, it is vital that commissions carrying out these election projects first and foremost guarantee that the legal framework supports biometric voter identification,*[4]*[5] and then that the data captured during the registration process will be secured while maintaining two basic requirements: personalization and privacy. Likewise, it is imperative to have contingency mechanisms in place, in case biometric systems malfunction. One of the main challenges is to ensure that given the eventualities of technological hitches and failures, not a single voter is disenfranchised.*[6]

17.1 Countries with biometric voter registration

According to International IDEA's ICTs in Elections Database,*[7] as of 2016, the number of countries adopting biometrics in elections has steadily increased to over 50, with significant differences between regions. "While there are virtually no users in Europe, about half of the countries in Africa and Latin America use this technology in elections."*[8] 35 per cent of over 130 surveyed Electoral Management Bodies is capturing biometric data (such as fingerprints or photos) as part of their voter registration process (2016).*[1]

Countries which have used Biometric voting registration include Armenia,*[9]*[10] Angola,*[11]*[12] Bangladesh,*[13]*[14] Bhutan,*[15] Bolivia,*[16]*[17]*[18] Brazil,*[19]*[20] Burkina Faso,*[21] Cambodia,*[22]*[23] Cameroon,*[24] Chad,*[25]*[26] Colombia,*[27]*[28] Comoros,*[29]*[30] Congo (Democratic Republic of),*[31]*[32] Costa Rica,*[33] Cote d'Ivoire,*[34] Dominican Republic,*[35] Fiji, Gambia,*[36] Ghana,*[37] Guatemala, India,*[38]*[39] Iraq, Kenya,*[40]*[41] Lesotho, Liberia, Malawi, Mali, Mauritania, Mexico,*[42]*[43] Morocco, Mozambique, Namibia,*[44] Nepal,*[45] Nicaragua, Nigeria,*[46]*[47] Panama, Peru,*[42] The Philippines,*[48]*[49] Senegal, Sierra Leone,*[50]*[51] Solomon Islands, Somaliland,*[52] Swaziland, Tanzania,*[53] Uganda,*[54]*[55] Uruguay, Venezuela,*[42] Yemen,*[16]*[56] Zambia,*[57] and Zimbabwe.*[58]*[59]

17.2 Advocacy and criticism

Some promoters of biometric voting registration point out that this technology, if properly customised to the country's needs and well implemented, could offer better accessibility for citizens; help avoiding long queues and waiting times for registration and voting; add simplicity and speed to the election cycle (i.e. voter identification documents can make it easier for polling staff to verify voter details);*[5] make voters and Commissions feel confident about the quality of their registry (more accurate, reliable and complete data); improve e-voting security,*[60]*[61] considerably reduce or eliminate multiple registration and multiple voting, while mitigating the risk of impersonation, identity theft, the misuse of records of deceased voters, carousel voting and ballot-box stuffing.*[46]*[1]

On the other hand, those who criticise and disapprove the use of biometrics for voter identification maintain that using biometrics for election purposes raises concerns over voters' privacy, human dignity and governmental disclosure of personal information.*[4]*[62] Some critics go further to claim that biometrics in voting poses a serious threat to democracy, due to fears of violation of the secrecy of the vote (or correlation voter-vote).*[63]

The concerns as with any other application of biometrics for identification and authentication can be manifold.*[64]*[2] There is, for example, the possibility of voter disenfranchisement when characteristics typically used to identify or verify voters might become unavailable (e.g. bandaged finger, loss of a finger, failure to acquire due to dirt or fingerprint quality degradation).*[1]*[63] Similarly, there are also claims of potential logistical and procedural "new problems" that biometric identification devices can bring to an election cycle: increasing costs (initial purchase costs of biometric readers and infrastructure, costs linked to maintenance, storage and upgrades) and resources' allocation (human, time, material); additional training of commissions and polling staff, technological failures that could disenfranchise voters; and extra data storage that demands higher security.*[3]*[6]

17.3 See also

- Electoral roll
- Electronic identification
- Electronic pollbook
- Voter ID laws

17.4 References

[1] Wolf, Peter (2017). "Introducing Biometric Technology in Elections" (PDF). International Institute for Democracy and Electoral Assistance.

[2] "Biometric Voter Registration: Skills required and problems encountered?". aceproject.org. Retrieved 2017-07-19.

[3] Holtved, Ole (February 2011). "Biometrics in Elections" (PDF). International Foundation for Electoral Systems and USAID.

[4] "Biometrics: Friend or foe of privacy?" (PDF). Privacy International.

[5] Gelb, Alan; Clark, Julia (2013-02-11). "Using biometrics in development: lessons and challenges". *The Guardian*. ISSN 0261-3077. Retrieved 2017-07-18.

[6] "Biometric Voter Registration and Voter Identification". aceproject.org. Retrieved 2017-07-18.

[7] "ICTs in Elections Database". International IDEA. Retrieved 2017-07-18.

[8] "If the EMB uses technology to collect voter registration data, is biometric data captured and used during registration?". International IDEA. Retrieved 2017-07-18.

[9] "New Voter Authentication Devices (VADs) tested in Armenia". EC-UNDP Joint Task Force. Retrieved 2017-07-19.

[10] "Parliamentary Elections, 2 April 2017". OSCE. Retrieved 2017-07-19.

[11] BÖSL, ANTON. "Angola votes! FIRST PARLIAMENT ELECTIONS FOR 16 YEARS" (PDF).

[12] "ans_738_1109121176". International IDEA. Retrieved 2017-07-18.

[13] "EC starts distributing NID cards with biometric details". Business News 24 BD. Retrieved 2017-07-18.

[14] "Bangladesh Election Commission Begins Issuance of Biometric IDs". *FindBiometrics*. 2016-10-03. Retrieved 2017-07-18.

[15] "ans_738_591871763". International IDEA. Retrieved 2017-07-18.

[16] "The Biometric ID Grid: A Country-by-Country Guide". The Corbett Report. Retrieved 2017-07-18.

[17] "Carter Center Report on Biometric Registration in Bolivia".

[18] "Bolivia begins updating biometric voter registry".

[19] "MANDATORY, BIOMETRIC, AND ELECTRONIC: BRAZILIANS HEAD TO THE POLLS". *plus55*. Retrieved 2017-07-18.

[20] "In São Paulo, TSE's President presents overview of 2017–2018 Biometric Identification Program" (in Portuguese). Retrieved 2017-07-18.

[21] "ans_738_1545713035". International IDEA. Retrieved 2017-07-18.

[22] "Cambodia: Commune Elections Not Free or Fair". *Human Rights Watch*. 2017-06-12. Retrieved 2017-07-18.

[23] Vichea, Pang (2015-11-12). "Voter e-registration test run going 'smoothly'". *Phnom Penh Post*. Retrieved 2017-07-18.

[24] "Biometric Voter Registration in Cameroon". aceproject.org. Retrieved 2017-07-18.

[25] Debos, Marielle. "Biometric voting in Chad: new technology, same old political tricks". *The Conversation*. Retrieved 2017-07-18.

[26] "UNHCR, EU working with Chad on biometrics registration program for refugees". *BiometricUpdate*. 2017-01-25. Retrieved 2017-07-18.

[27] "Colombia". International IDEA. Retrieved 2017-07-18.

[28] "Registraduría Nacional del Estado Civil – Nuevas Tecnologías" (in Spanish). Registraduria Nacional del Estado Civil. Retrieved 2017-07-18.

[29] "Programme d'appui à la modernisation du fichier électoral de l'Union des Comores" (in French). Le PNUD en Union des Comores. Retrieved 2017-07-18.

[30] "Comoros". freedomhouse.org. Retrieved 2017-07-18.

[31] "ans_738_844038563". International IDEA. Retrieved 2017-07-18.

[32] "The Democratic Republic of Congo selects Gemalto mobile biometric enrollment solution to support fair elections". gemalto.com. Retrieved 2017-07-18.

[33] Rezazadeh, Reza (2003-05-12). *Electronic Electoral System: Simple, Abuse Free, Voter Friendly*. Xlibris Corporation. ISBN 978-1-4653-3194-6 – via Google Books.

[34] "Safran Tech Used in Two Recent Biometric Votes in Ivory Coast". *FindBiometrics*. 2017-02-01. Retrieved 2017-07-18.

[35] "Dominican Republic delegation is introduced to the Brazilian biometric registration system" (in Portuguese). Retrieved 2017-07-18.

[36] "Gambia Electoral Commission Reports on Biometric Registration Work". *FindBiometrics*. 2016-01-26. Retrieved 2017-07-18.

[37] Forson, Francisca Kakra. "Ghana Electoral Commission Assures Problem-free Second Round of Special Voting". *VOA*. Retrieved 2017-07-18.

[38] "EC to check bogus voting, link Aadhaar with electoral rolls". Hindustantimes.com. 2014-11-29. Retrieved 2017-07-18.

[39] "India's Election Commission to link biometric ID with electoral rolls". *BiometricUpdate*. 2014-12-02. Retrieved 2017-07-18.

[40] Gogineni, Roopa. "Kenya Introduces Biometrics for Voter Registration". *VOA*. Retrieved 2017-07-18.

[41] "Kenya implements biometric registration for all citizens". *BiometricUpdate*. 2015-02-05. Retrieved 2017-07-18.

[42] "If the EMB uses technology to collect voter registration data, is biometric data captured and used during registration?". International IDEA. Retrieved 2017-07-18.

[43] "Mexico Sticks with Safran for Voter Authentication". FindBiometrics. 2016-12-21. Retrieved 2017-07-18.

[44] "Biometric machine for voter registration unveiled". *Verified Voting*. 2013-12-23. Retrieved 2017-07-18.

[45] "Election Commission builds biometric voter database ahead of election". *Verified Voting*. 2013-01-11. Retrieved 2017-07-18.

[46] "Countries adopt biometrics for voter ID, fraud prevention – SecureIDNews". *SecureIDNews*. Retrieved 2017-07-18.

[47] "Nigeria's INEC Satisfied with New Voting Systems". *FindBiometrics*. 2015-03-12. Retrieved 2017-07-18.

[48] "Biometrics Now Mandatory For Voting In The Philippines". *FindBiometrics*. 2014-06-02. Retrieved 2017-07-18.

[49] Aurelio, Julie M. "Comelec earmarks P10-M for voter registration machines". newsinfo.inquirer.net. Retrieved 2017-07-18.

[50] "Biometric Election Tech Shipment Offers Reassurance in Sierra Leone". *FindBiometrics*. 2017-01-27. Retrieved 2017-07-18.

[51] "Sierra Leone News: 4000 Biometric registration kits arrive". *Awoko*. Retrieved 2017-07-18.

[52] "Daon Tech Supports Peacebuilding Work in Somaliland". *FindBiometrics*. 2016-01-14. Retrieved 2017-07-18.

[53] Clottey, Peter. "Tanzania Electoral Chief Promises Peaceful, Credible Election". *VOA*. Retrieved 2017-07-18.

[54] Clottey, Peter. "Uganda to Use Biometric Verification Machines for Elections". *VOA*. Retrieved 2017-07-18.

[55] "Uganda's Electoral Commission Trains Staff on Biometric Voter Identification". *FindBiometrics*. 2016-02-05. Retrieved 2017-07-18.

[56] "M2SYS Deploy Truevoter Yemen Biometric Voter Registration Solution – M2SYS". *M2SYS*. Retrieved 2017-07-18.

[57] Zulu, Delphine (2015-02-06). "Zambia: Biometric NRCs to Be Used in 2016 Elections". *The Times of Zambia (Ndola)*. Retrieved 2017-07-18.

[58] "Zimbabwe rules out biometric voting in 2018 elections, electronic system for registration only – Techzim". *Techzim*. 2017-01-25. Retrieved 2017-07-18.

[59] "BVR kits procurement gathers momentum". *The Herald*. Retrieved 2017-07-18.

17.4. REFERENCES

[60] "McAfee-Atlantic Council Joint Study Sees Important Role for Biometrics in e-Voting". *FindBiometrics*. 2014-10-08. Retrieved 2017-07-18.

[61] "Voting online made possible with selfie recognition technology". 2016-10-17. Retrieved 2017-07-18.

[62] "Biometrics Are Coming, Along With Serious Security Concerns". *WIRED*. Retrieved 2017-07-18.

[63] Vanguard, The Patriotic (2011-10-02). "Analysis of Biometric Voter Registration and Voting Process in Sierra Leone". *The Patriotic Vanguard*. Retrieved 2017-07-18.

[64] Das, Ravindra (2016-04-21). *Adopting Biometric Technology: Challenges and Solutions*. CRC Press. ISBN 978-1-4987-1745-8.

Chapter 18

Biometrics in schools

Biometrics in schools is a growing technology in the U.S. education sector. According to TechNavio, biometrics in the education market sector is expected to grow at a rate of 23.65% (CAGR) between 2014 and 2019.[*][1] Biometrics are unique physical or behavioral characteristics which can be used to automatically identify individuals. Biometric technologies capture, process and measure these characteristics electronically and compare them against existing records to create a highly accurate identity management capability.[*][2]

In 2015, the U.S. federal government spent $11.7 billion in reimbursements for the National School Lunch Program and $3.9 billion in reimbursements for the School Breakfast Program. That's 5 billion lunches and 2.3 billion breakfasts served annually[*][3]

18.1 Types of Biometrics Used in Schools

Fingerprint recognition technology in the biometric market has held the largest market size worldwide and has been widely adopted by many industries including schools. Fingerprint recognition is the most pervasive, oldest, simplest to install, and lowest cost biometric technology.[*][4] Although palm vein recognition, iris recognition and face recognition have been implemented in schools, finger scanning is by far the most commonly used technology in the U.S. education market.[*][5]

In the U.K., primarily the type of biometric employed is a fingerprint scan or thumbprint scan but vein and iris scanning systems are also in use.[*][6][*][7]

18.2 United Kingdom

Biometrics have been used in the UK since the early first decade of the 21st century, with some use of biometric technology in schools in Asia too. Biometric technology is used to address truancy, to replace library cards, or to charge for meals. School biometrics, typically electronic fingerprinting systems, have raised privacy concerns because of the creation of databases that would progressively include the entire population. The UK introduced legal duties on schools, if they wish to use biometric information about pupils, in the Protection of Freedoms Act 2012.[*][8]

Schools use pupils' biometric data for cashless catering, libraries, payment systems, registration and locker systems. In the UK biometric technology in schools was initially used for library book issue, approved for use by the UK's Information Commissioner's Office [*][9] in 2001 and the Department for Education [*][10] in 2002. Within a few years biometrics were being used for cashless catering systems, enabling relatives to deposit money into pupils' catering accounts, to be debited by a child's biometric fingerprint scan at the point of sale. In the USA biometrics systems are used for catering primarily, as mentioned above, with library and registration biometrics in use as well. Fingerprint locking systems are also used in the United Kingdom (fingerprint lock in the Holland Park School in London,[*][11]) databases, etc., in Belgium (Marie-José school in Liège [*][12][*][13]), in France, in Italy, etc.

When children use systems in which their biometric fingerprints are processed in school no image of the fingerprint is stored, although the fingerprint data stored can be potentially used in the same way as an image of a fingerprint. A series of digits (some 30) is created so the computer can recognise a child when he/she places their fingerprint on a scanner. The data stored can be interoperable[*][14] with Automated Fingerprint Identification Systems (AFIS) used by police and other agencies to store fingerprint data.

It is claimed to be impossible to reconstruct a finger print from biometric readers, although research in 2007 was undertaken and the paper 'From Template to Image: Reconstructing Fingerprints from Minutiae Points' [*][15] was published by the Institute of Electrical and Electronics Engineers.[*][16]

In 2002 the NGO Privacy International raised an alert that

tens of thousands of UK school children were being fingerprinted by schools, often without the knowledge or consent of their parents.[*][17] In 2002, the supplier Micro Librarian Systems, which use a technology similar to that used in US prisons and the German military, estimated that 350 schools throughout Britain were using such systems, to replace library cards.[*][17] In 2007, it was estimated that 3,500 schools (ten times more) are using such systems.[*][18] By 2009 the number of children fingerprinted was estimated to be two million.[*][19]

In the Protection of Freedoms Act; Part 1 "Regulation of Biometric Data", Chapter 2 schools and colleges are required to obtain the consent of one parent of a child under 18 for acquiring and processing the child's biometric information, and the Act gives the child rights to stop the processing of their biometric information regardless of any parental consent. It also states if any parent of the child objects to the processing of biometric information it must also be discontinued.

In addition to this schools are subject to the Data Protection Act and the Human Rights Act.

> Privacy International warned that the practice of finger printing for the purpose of library cards was in clear violation of the Human Rights Act and the Data Protection Act: The law states that privacy invasion must be proportionate to the threat. A few lost library cards do not warrant mass finger printing. It is also likely that the practice breaches Article 16 of the UN Convention on the Rights of the Child, that 'no child shall be subjected to arbitrary or unlawful interference with his or her privacy...'"[*][20]

Others claim that under the Data Protection Act (DPA), schools in the UK do not have to ask parental consent for such practices. Parents opposed to such practices may only bring individual complaints against schools.[*][21] Regardless of this the child's rights under the Protection of Freedoms Act remain unaffected.

Concerns have been raised about the civil liberties implications of fingerprinting children in schools.[*][22] In 2007 Early Day Motion 686, which called on the UK Government to conduct a full and open consultation with stakeholders about the use of biometrics in schools, secured the support of 85 Members of Parliament.[*][23]

In response to a complaint which they are continuing to pursue, in 2010 the European Commission expressed 'significant concerns' over the proportionality and necessity of the practice and the lack of judicial redress, indicating that the practice may break the European Union data protection directive.[*][24]

18.3 Belgium

The alleged use of taking children's fingerprints is to struggle against school truancy and/or to replace library cards or money for meals by fingerprint locks. In Belgium, this practice gave rise to a question in Parliament on 6 February 62007 by Michel de La Motte (Humanist Democratic Centre) to the Education Minister Marie Arena, who replied that they were legal insofar as the school did not use them for external purposes nor to survey the private life of children.[*][25] Such practices have also been used in France (Angers, Carqueiranne college in the Var —the latter won the Big Brother Award of 2005 for its hand geometry system, etc.) although the CNIL, official organism in charge of protection of privacy, has declared them "disproportionate." .[*][26] The CNIL, however, declared in 2002 hand geometry systems to be acceptable.

18.4 Early applications

The first reported use of biometric systems in U.S. schools was at Minnesota's Eagan High School in March 1997.[*][27] Eagan High School, a testing ground for education technology since it opened, allowed willing students to use fingerprint readers to speed up the borrowing of library books.

Penn Cambria School District in Cresson, PA was another earlier user of biometric technology.[*][28] In 2000, Food Service Solutions, a local software development company, designed and implemented a system where students bought lunch with just a fingerprint. The American Civil Liberties Union stated that this "could hasten the end of privacy rights" [*][29]

Biometric systems were first used in schools in the UK in 2001.[*][30] Use of this technology in schools has become wider spread, although there are currently no official figures for how many schools employ the technology.[*][31]

18.5 Applications

Biometric technologies in schools are used to borrow library books, for cashless canteen systems, vending machines, class attendance and payments into schools. Biometric technologies for home/school bus journeys are also under development.[*][32]

18.6 Ages

Biometric systems can be used by children as young as three years old.*[33]

18.7 Current usage

The two countries at the forefront employing biometric technology in schools are the UK and the United States. Biometric systems are also used in some schools in Belgium*[34] and Sweden*[35] but were withdrawn from China and Hong Kong schools due to privacy concerns.*[36] It was reported in August 2007 that Dubai are soon due to issue guidance to schools.*[37]

18.8 Security concerns

Concerns about the security implications of using conventional biometric templates in schools have been raised by a number of leading IT security experts, including Kim Cameron, architect of identity and access in the connected systems division at Microsoft, who cites research by Cavoukian and Stoianov to back up his assertion that "it is absolutely premature to begin using 'conventional biometrics' in schools".*[38]*[39]

18.9 Advantages

Biometric vendors claim benefits to schools such as improved reading skills, decreased wait times in lunch lines and increased revenues.*[40] They do not cite independent research to support this. Educationalist Dr. Sandra Leaton Gray of Homerton College, Cambridge stated in early 2007 that "I have not been able to find a single piece of published research which suggests that the use of biometrics in schools promotes healthy eating or improves reading skills amongst children... There is absolutely no evidence for such claims".*[41]

18.10 See also

- Biometrics
- Fingerprinting
- Big Brother
- Privacy International

18.11 References

[1] Perala, A (14 August 2015). "Biometrics in US education sector to see significant growth". *Findbiometrics Global Identity Management.* Retrieved 15 April 2017.

[2] "Identity is at the heart of the digital age". *www.ibia.org.* International Biometrics & identity Association (IBIA). Retrieved 15 April 2017.

[3] "School meal trends and stats". *www.schoolnutrition.org.* School Nutrition Association (SNA). Retrieved 15 April 2017.

[4] "Fingerprint biometrics: A global strategic business report". *www.strategyr.com.* Global Industry Analysts, Inc. Retrieved 15 April 2017.

[5] Perala, A. "Biometrics in US education sector to see significant growth". *www.findbiometrics.com.* Findbiometrics Global Identity Management. Retrieved 15 April 2017.

[6] Biometrics in Schools, Colleges and other Educational Institutions, Data Protection Commissioner, 2007

[7] Vein scanning in a primary school for food, in Scotland, Scotsman.com, October 2006

[8] "Protection of biometric information of children in schools". Department for Education. 13 December 2012. Retrieved 11 May 2015.

[9] http://biometricsinschools.files.wordpress.com/2013%5B%5D /08/ico-letter-2001.jpg

[10] http://biometricsinschools.files.wordpress.com/2013/08/dfes-letter-2002%5B%5D. jpg

[11] fingerprints for children of a London school (in French)

[12] fingerprints to secure the school? Archived 2007-07-01 at the Wayback Machine. (in French)

[13] the fingerprint reader in schools creates controversy Archived 2012-07-21 at Archive.is, 7 on 7, February 5, 2007 (in French)

[14] "Archived copy". Archived from the original on 2016-03-26. Retrieved 2016-07-28.

[15]

[16]

[17] Fingerprinting of UK school kids causes outcry, *The Register,* July 22, 2002

[18] Child fingerprint plan considered, *BBC*, March 4, 2007

[19] Singh, Y. "Why are we fingerprinting children?", The Guardian, March 7, 2009

[20] "Archived copy". Archived from the original on 2012-09-02. Retrieved 2012-10-03.

[21] Schools can fingerprint children without parental consent, *The Register*, September 7, 2006

[22] Porter, H. "Blindly fingerprinting children", The Guardian, November 18, 2009

[23] "EDM 686 – Biometric Data Collection In Schools". UK Parliament. 2007-01-19. Retrieved 2009-11-28.

[24] Europe tells Britain to justify itself over fingerprinting children in schools Telegraph, published 2010-12-14, accessed 2011-01-13

[25] Prises d'empreintes digitales dans un établissement scolaire Archived 2007-02-24 at the Wayback Machine., Question d'actualité à la Ministre-Présidente en charge de l'Enseignement obligatoire et de Promotion sociale (in French)

[26] Quand la biométrie s'installe dans les cantines au nez et à la barbe de la Cnil, *Zdnet*, September 9, 2003 (in French)

[27] "This Minnesota high school gives fingerprint scanning a whorl". eSchool News. 2000-09-01. Archived from the original on May 6, 2006. Retrieved 2006-11-20. External link in |publisher= (help)

[28] Fingerprints Pay For School Lunch

[29] "Fingerprints Pay For School Lunch". *CBS News*. January 24, 2001.

[30] "Biometrics And RFID Tracking In UK Education". The UK Column. 11 April 2013. Retrieved 11 May 2015.

[31] Knight, J. Parliamentary Written Answer 110750, Hansard, January 29, 2007.

[32] Grossman, W. "Is school fingerprinting out of bounds?", The Guardian, March 30, 2006.

[33] Devlin, K. "Nursery children to be fingerprinted", The Daily Telegraph, September 23, 2006.

[34] Fingerprint recognition in high schools used for registration. Archived May 16, 2007, at the Wayback Machine.

[35] Kvarnby School in Stockholm used to log into school computers. Archived November 25, 2006, at the Wayback Machine.

[36] China: Ballard, M. "Halt to school fingerprinting", The Register, November 9, 2006.

[37] Fingerprinting of pupils fails to score Archived September 29, 2007, at the Wayback Machine.

[38] Will biometrics grow up?

[39] Biometric Encryption: A Positive-Sum Technology that Achieves Strong Authentication, Security AND Privacy Archived 2007-06-14 at the Wayback Machine. Cavoukian,A and Stoianov,A March 2007

[40] Fingerprint Software Eliminates Privacy Concerns and Establishes Success (FindBiometrics) Archived 2009-03-14 at the Wayback Machine.

[41] LTKA – Experts warn of serious child fingerprinting risks (against schools fingerprinting our children)

General references

- https://web.archive.org/web/20110220075523/http://www.trust-e.co.uk/information-for-parents/page-14
- "Fingerprint Biometrics". *www.ibia.org*. International Biometrics & identity Association (IBIA). Retrieved 15 April 2017.

18.12 External links

- Biometrics in US Education Sector To See Significant Growth – Future for Biometrics in Schools, August 14, 2015.
- Biometrics in Schools – Latest news on the use and deployment of biometric systems in schools; particular emphasis on UK and US.
- School Biometrics: The Legal Conundrum – Patricia Deubel, Ph.D. / T.H.E. Journal, 10 April 2007.
- Biometrics in K-12: Ban or Buy? (Part 1) – Patricia Deubel, Ph.D. / T.H.E. Journal, 18 April 2007.
- Biometrics in K-12: Issues and Standardization (Part 2) – Patricia Deubel, Ph.D. / T.H.E. Journal, 25 April 2007.
- Biometrics in K-12: Vendor Claims and Your Business Plan (Part 3) – Patricia Deubel, Ph.D. / T.H.E. Journal, 2 May 2007.
- Index of relevant articles by Kim Cameron, architect of identity and access in the connected systems division at Microsoft.

18.12.1 Legislation

The following laws, legal opinions, or guidance are in place to regulate children's use of biometric technology. To date the practise of using biometrics in schools is only legally regulated in the USA:

- Legislation in Illinois, SB1702, 2007
- Legislation Illinois, SB2549, 2005–2006

18.12.2 Non statutory advice

USA

- Opinion of Jennifer M. Granholm, Attorney General, State of Michigan

Ireland

- Biometrics in Schools, Colleges and other Educational Institutions 2007 – Data Protection Commissioner

UK

- Portsmouth Finger Scanning Technology Guidance – June 2007. This, the first guidance issued in the UK, went to schools only in the Portsmouth area.
- The use of biometrics in schools – 23 July 2007. Advice from the UK Information Commissioner's Office.

Chapter 19

BioSlimDisk

BioSlimDisk is a biometric USB storage device.

19.1 Signature

The biometric USB hardware is encryption coupled with dual fingerprint authentication methodology to prevent crackers bypassing the encryption. Dual fingerprint authentication function increases the difficulty for crackers and prevents access from the outside into the secured nonvolatile internal memory storage. Signature is designed so that the encryption key is stored randomly in the flash memory. If it were cracked, a self destruct feature will trigger it to reformat and destroy all data stored in the device.*[1] *[2] *[3] *[4] *[5]

19.2 References

[1] "Blog Archive » BioSlimDisk biometric security token, Review, Comparison". Rohos. 2010-01-14. Retrieved 2010-02-08.

[2] "Review: BioSlimDisk Signature 512MB – New Tech Gadgets & Electronic Devices". Geek.com. Retrieved 2010-02-08.

[3] Peter de Boer (31 October 2007). "De BioSlimDisk Signature nader bekeken | Inleiding | Core" (in Dutch). Tweakers.net. Retrieved 2010-02-08.

[4] "Bioslimdisk Portable Encrypted Solutions | Product Review | Infosecurity Security Adviser". Infosecurityadviser.com. 2009-04-28. Archived from the original on 2009-11-28. Retrieved 2010-02-08.

[5] Kelly Jackson Higgins (January 4, 2010). "Secure USB Flaw Exposed". DarkReading. Retrieved 2010-02-08.

19.3 External links

- BioSlimDisk official website
- Rohos BioSlimDisk Signature Review
- Geek's BioSlimDisk Signature Review
- Tweaker's BioSlimDisk Signature Test Report (NL)
- Bioslimdisk Portable Encrypted Solutions, InfoSecurity Review
- The Limitations of Password Based Software Encryption World Leader

Chapter 20

Handwritten biometric recognition

Not to be confused with Optical character recognition.
Handwritten biometric recognition is the process of

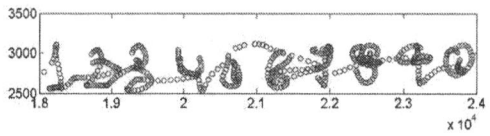

Example of handwritting of a sequence of digits. Its dynamic information is shown on the right. It is interesting to enphasize that movements in the air are also acquired by the digitizing tablet. These movements can be identified because pressure is equal to zero.

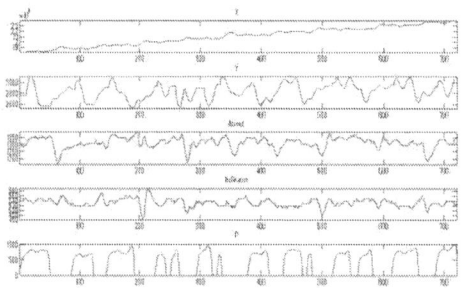

Example of dynamic information of handwritting.

identifying the author of a given text. Handwritten biometric recognition belongs to behavioural biometric systems because it is based on something that the user has learned to do.

20.1 Static and Dynamic recognition

Handwritten biometrics can be split into two main categories:

Static: In this mode, users writes on paper, digitize it through an optical scanner or a camera, and the biometric system recognizes the text analyzing its shape. This group is also known as "off-line".

Dynamic: In this mode, users writes in a digitizing tablet, which acquires the text in real time. Another possibility is the acquisition by means of stylus-operated PDAs. Dynamic recognition is also known as "on-line".Dynamic information usually consists of the following information:

- spatial coordinate x(t)
- spatial coordinate y(t)
- pressure p(t)
- azimuth az(t)
- inclination in(t)

Better accuracies are achieved by means of dynamic systems. Some technological approaches exist.[1][2][3][4][5]

20.2 Difference from OCR

Handwritten biometric recognition should not be confused with optical character recognition (OCR). While the goal of handwritten biometrics is to identify the author of a given text, the goal of an OCR is to recognize the content of the text, regardless of his author.

20.3 References

[1] Chapran, J. (2006). "Biometric Writer Identification: Feature Analysis and Classification". *International Journal of Pattern Recognition & Artificial Intelligence*: 483–503.

[2] Schomaker, L. (2007). "Advances in Writer Identification and Verification". *Ninth International Conference on Document Analysis and Recognition*. ICDAR: 1268–1273.

[3] Said, H. E. S.; TN Tan; KD Baker. "Personal identification based on handwriting". *Pattern Recognition*. **33** (2000): 149–160. doi:10.1016/S0031-3203(99)00006-0.

20.3. REFERENCES

[4] Schlapbach, A.; M Liwicki; H Bunke (2008). "A writer identification system for on-line whiteboard data". *Pattern recognition.* **41** (7): 2381–2397. doi:10.1016/j.patcog.2008.01.006.

[5] Sesa-Nogueras, Enric; Marcos Faundez-Zanuy (2012). "Biometric recognition using online uppercase handwritten text". *Pattern recognition.* **45** (1): 128–144. doi:10.1016/j.patcog.2011.06.002.

Chapter 21

Private biometrics

A form of biometrics, also called Biometric Encryption or BioCryptics, in which the prover is protected against the misuse of template data by a dishonest verifier.

Biometric identification requires that a verifier searches for matches in a data base that contains data about the entire population. This introduces the security and privacy threat that the verifier who steals biometric templates from some (or even all) persons in the data base can perform impersonation attacks. When a private verification system is used on a large scale, the reference data base has to be made available to many different verifiers, who, in general, cannot be trusted. Information stolen from a data base can be misused to construct artificial biometrics to impersonate people. Creation of artificial biometrics is possible even if only part of the template is available.

To develop an insight in the security aspects of biometrics, one can distinguish between verification and private verification. In a typical verification situation, access to the reference template allows a malicious verifier to artificially construct measurement data that will pass the verification test, even if the prover has never exposed herself to a biometric measurement after the enrollment.

In **private verification**, the reference data should not leak relevant information to allow the verifier to (effectively) construct valid measurement data. Such protection is common practice for storage of computer passwords. When a computer verifies a password, it does not compare the password typed by the user with a stored reference copy. Instead, the password is processed by a cryptographic one-way function F and the outcome is compared against a locally stored reference string $F(y)$. So y is only temporarily available on the system hardware, and no stored data allows calculation of y. This prevents attacks from the inside by stealing unencrypted or decryptable secrets.

21.1 Comparison with handling computer passwords

The main difference between password checking and biometric private verification is that during biometric measurements it is unavoidable that noise or other aberrations occur. Noisy measurement data are quantized into discrete values before these can be processed by any cryptographic function. Due to external noise, the outcome of the quantization may differ from experiment to experiment. In particular if one of the biometric parameters has a value close to a quantization threshold, minor amounts of noise can change the outcome. Minor changes at the input of a cryptographic function are amplified and the outcome will bear no resemblance to the expected outcome. This property, commonly referred to as 'confusion' and 'diffusion', makes it less trivial to use biometric data as input to a cryptographic function. The notion of near matches or distance between enrollment and operational measurements vanishes after encryption or any other cryptographically strong operation. Hence, the comparison of measured data with reference data can not be executed in the encrypted domain without prior precautions to contain the effect of noise.

Meanwhile, it is important to realize that protection of the reference data stored in a database is not a complete solution to the above-mentioned threats. After having had an opportunity to measure operational biometric data, a dishonest verifier uses these measurement data. This can happen without anyone noticing it: Victor grabs the fingerprint image left behind on a sensor. This corresponds to grabbing all keystrokes including the plain passwords typed by a user.

21.2 References

- Jeroen Breebaart, Christoph Busch, Justine Grave, Els Kindt: *A Reference Architecture for Biometric Template Protection based on Pseudo Identities.* In Arslan Brömme, Christoph Busch, Detlef Hühnlein (Eds.): *BIOSIG 2008*, 2008, pages 25–37, Lecture Notes in Informatics 137, Gesellschaft für Informatik, http://www.jeroenbreebaart.com/papers/biosig/biosig2008.pdf.

- Ileana Buhan, Emile Kelkboom, Koen Simoens: *A Survey of the Security and Privacy Measures for Anonymous Biometric Authentication Systems*. International Conference on Intelligent Information Hiding and Multimedia Signal Processing (IIH-MSP 2010), 2010, IEEE Computer Society, http://www.cosic.esat.kuleuven.be/publications/article-1462.pdf.

- Ann Cavoukian, Alex Stoianov: *Biometric Encryption: A Positive-Sum Technology that Achieves Strong Authentication, Security and Privacy*. Discussion paper of the Office of the Information and Privacy Commissioner of Ontario, 2007, http://www.ipc.on.ca/images/Resources/bio-encryp.pdf.

- Ann Cavoukian, Alex Stoianov: *Biometric Encryption: The New Breed of Untraceable Biometrics*. In:

- Nikolaos V. Boulgouris, Konstantinos N. Plataniotis, Evangelia Micheli-Tzanakou (Eds.): *Biometrics: Theory, Methods, and Applications*, 2009, John Wiley & Sons, Inc., Hoboken, NJ, USA, pages 655-710, ISBN 978-0-470-24782-2.

- Ari Juels and Martin Wattenberg. *A fuzzy commitment scheme*. In ACM Conference on Computer and Communications Security, pages

28–36, 1999.

- Pim Tuyls, Boris Skoric, Tom Kevenaar (Editors), *Security with Noisy Data: Private Biometrics, Secure Key Storage and Anti-Counterfeiting (Hardcover)*, Springer, 2007, ISBN 978-1-84628-983-5.

- Jean-Paul Linnartz and Pim Tuyls, *New Shielding functions to enhance privacy and prevent misuse of biometric templates*, 4th International Conference on Audio and Video Based Biometric Person Authentication, Guildford, United Kingdom, 9–11 June 2003.

- White paper *Private Identity Matching*, http://www.priv-id.com/images/Technology-primer.pdf.

Chapter 22

Signature recognition

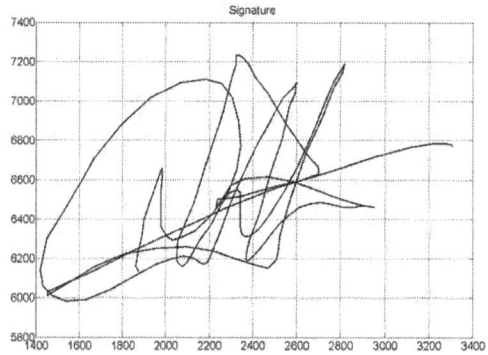

Example of signature shape.

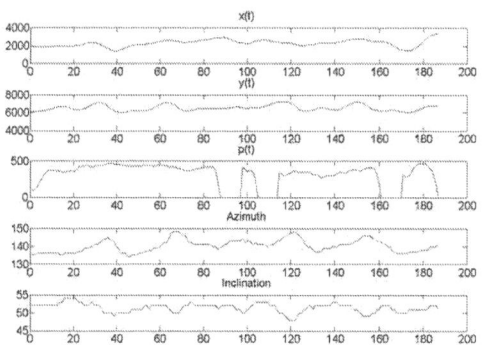

Example of dynamic information of a signature. Looking at the pressure information it can be seen that the user has lift the pen 3 times in the middle of the signature (areas with pressure equal to zero).

Signature recognition is a behavioural biometric. It can be operated in two different ways:

Static: In this mode, users write their signature on paper, digitize it through an optical scanner or a camera, and the biometric system recognizes the signature analyzing its shape. This group is also known as "off-line".

Dynamic: In this mode, users write their signature in a digitizing tablet, which acquires the signature in real time. Another possibility is the acquisition by means of stylus-operated PDAs. Some systems also operate on smart-phones or tablets with a capacitive screen, where users can sign using a finger or an appropriate pen. Dynamic recognition is also known as "on-line". Dynamic information usually consists of the following information:

- spatial coordinate x(t)
- spatial coordinate y(t)
- pressure p(t)
- azimuth az(t)
- inclination in(t)
- pen up/down

The state-of-the-art in signature recognition can be found in the last major international competition.[*][1]

The most popular pattern recognition techniques applied for signature recognition are dynamic time warping, hidden Markov models and vector quantization. Combinations of different techniques also exist.[*][2]

22.1 Related techniques

Recently, a handwriten biometric approach has also been proposed.[*][3] In this case, the user is recognized analyzing his handwritten text (see also Handwritten biometric recognition).

22.2 Databases

Several public databases exist, being the most popular ones SVC,[*][4] and MCYT.[*][5]

22.3 References

[1] Houmani, Nesmaa; A. Mayoue; S. Garcia-Salicetti; B. Dorizzi; M.I. Khalil; M. Mostafa; H. Abbas; Z.T. Kardkovàcs; D. Muramatsu; B. Yanikoglu; A. Kholmatov; M. Martinez-Diaz; J. Fierrez; J. Ortega-Garcia; J. Roure Alcobé; J. Fabregas; M. Faundez-Zanuy; J. M. Pascual-Gaspar; V. Cardeñoso-Payo; C. Vivaracho-Pascual (March 2012). "BioSecure signature evaluation campaign (BSEC'2009): Evaluating online signature algorithms depending on the quality of signatures". *Pattern Recognition.* **45** (3): 993–1003. doi:10.1016/j.patcog.2011.08.008.

[2] Faundez-Zanuy, Marcos (2007). "On-line signature recognition based on VQ-DTW". *Pattern recognition.* **40** (3): 981–992. doi:10.1016/j.patcog.2006.06.007.

[3] Chapran, J. (2006). "Biometric Writer Identification: Feature Analysis and Classification". *International Journal of Pattern Recognition & Artificial Intelligence.* **20**: 483–503. doi:10.1142/s0218001406004831.

[4] Yeung, D. H.; Xiong, Y.; George, S.; Kashi, R.; Matsumoto, T.; Rigoll, G. (2004). "SVC2004: First international signature verification competition". *Lecture Notes in Computer Science.* LNCS-3072: 16–22.

[5] Ortega-Garcia, Javier; J. Fierrez; D. Simon; J. Gonzalez; M. Faúndez-Zanuy; V. Espinosa; A. Satue; I. Hernaez; J.-J. Igarza; C. Vivaracho; D. Escudero; Q.-I. Moro. "MCYT Baseline Corpus: A Multimodal Biometric Database". *IEE Proceedings - Vision, Image and Signal Processing.* **150**: 395–401. doi:10.1049/ip-vis:20031078.

Chapter 23

Vein matching

Vein matching, also called **vascular technology**,[*][1] is a technique of biometric identification through the analysis of the patterns of blood vessels visible from the surface of the skin.[*][2] Though used by the Federal Bureau of Investigation and the Central Intelligence Agency,[*][3] this method of identification is still in development and has not yet been universally adopted by crime labs as it is not considered as reliable as more established techniques, such as fingerprinting. However, it can be used in conjunction with existing forensic data in support of a conclusion.[*][2][*][4]

While other types of biometric scanners are more popular for security systems, Vascular scanners are growing in popularity. Fingerprint scanners are more frequently used, but Naito says they generally do not provide enough data points for critical verification decisions. Since fingerprint scanners require direct contact of the finger with the scanner, dry or abraded skin can interfere with the reliability of the system. Skin diseases, such as psoriasis can also limit the accuracy of the scanner, not to mention direct contact with the scanner can result in need for more frequent cleaning and higher risk of equipment damage. Vascular scanners do not require contact with the scanner, and since the information they read is on the inside of the body, skin conditions do not affect the accuracy of the reading. Vascular scanners also work with extreme speed, scanning in less than a second. As they scan, they capture the unique pattern veins take as they branch through the hand. Compared to the Retinal Scanner, which is more accurate than the vascular scanner, the retinal scanner has much lower popularity, because of its intrusive nature. People generally are uncomfortable exposing their eyes to an unknown light, not to mention retinal scanners are more difficult to install, since variances in height and face angle must be accounted for.[*][5]

23.1 History of Vein Matching

Joe Rice, an automation controls engineer at Kodak's Annesley Factory, invented vein pattern recognition in the early 1980s in response to his bank cards and identity being stolen. He developed essentially a barcode reader for people and assigned the rights to the UK's NRDC (National Research Development Corporation) See his website from 1993 onwards at: https://sites.google.com/site/veinpatternhome/

His original patent on vein matching Google "WO1985004088A1"

His US Patent on Vein Matching issued 1987 Google "US patent 4699149"

Details of his 1990 Smart door handle

His biometric Summit Talk Google " A thirdway for Biometrics"

Joe's predictions for the future of Vein pattern technology: http://biometrics.mainguet.org/types/vein_JoeRice.htm

The NRDC/ BTG (Thatcher privatised NRDC into BTG) made little headway in licensing vein pattern technology. The world was wedded to fingerprints and Iris patterns and Governments (the main buyers of biometric solutions) wanted open view biometrics for surveillance purposes, not a hidden, personal biometric solution!

In the late 1990s BTG said they were dropping vein patterns through no commercial interest. Joe was unhappy with the BTGs decision and their implementation of vein pattern technology so he gave the following talk to the Biometric Summit in Washington DC Google " A Third Way for Biometrics" on how he would develop vein pattern recognition. This view was countered by a following speaker from IBG (The US based international Biometric Group) who said there was insufficient information content in vein patterns for them to be used as a viable biometric.

In 2002 Hitachi and Fujitsu launched vein biometric products and veins have turned out to be one of the most consistent, discriminatory and accurate biometric traits.

Fast forward to the mid 2000s and Joe started to get a few approaches from people and firms who wanted to pick his brains on vein biometrics and the biowatch proposal, eventually Joe received an invitation from Matthias Vanoni to partner in a Swiss company Biowatch SA to develop and commercialise the biowatch.

23.2 Commercial applications

Vascular/vein pattern recognition (VPR) technology has been developed commercially by Hitachi since 1997, in which infrared light absorbed by the hemoglobin in a subject's blood vessels is recorded (as dark patterns) by a CCD camera behind a transparent surface.[6] The data patterns are processed, compressed, and digitized for future biometric authentication of the subject. Finger scanning devices have been deployed for use in Japanese financial institutions, kiosks, and turnstiles.[7] Mantra Softech marketed a device in India that scans vein patterns in palms for attendance recording.[8] Fujitsu developed a version that does not require direct physical contact with the vein scanner for improved hygiene in the use of electronic point of sale devices.[9]

Computer security expert Bruce Schneier stated that a key advantage of vein patterns for biometric identification is the lack of a known method of forging a usable "dummy", as is possible with fingerprints.[10]

23.3 Forensic identification

According to a 31,000-word investigative report published in January 2011 by Georgetown University faculty and students,[11][12][13][14][15] U.S. federal investigators used photos from the video recording of the beheading of American journalist Daniel Pearl to match the veins on the visible areas of the perpetrator to that of captured al-Qaeda operative Khalid Sheikh Mohammed, notably a "bulging vein" running across his hand.[4] The FBI and the CIA used the matching technique on Mohammed in 2004 and again in 2007.[3] Officials were concerned that his confession, which had been obtained through torture (namely waterboarding), would not hold up in court and used vein matching evidence to bolster their case.[2]

23.4 Other applications

Some US hospitals, such as NYU Langone Medical Center, use a vein matching system called Imprivata PatientSecure, primarily to reduce errors. Additional benefits include identifying unconscious or uncommunicative patients, and saving time and paperwork.[16] Dr. Bernard A. Birnbaum, chief of hospital operations at Langone, says "vein patterns are 100 times more unique than fingerprints".[17]

However, the newspaper reports of the use of vein matching to determine whether Khalid Sheikh Mohammed was Mr Pearl's murderer quoted FBI officials who described the technique as "less reliable" than fingerprints.

23.5 See also

- Finger Vein recognition
- Iris recognition
- Retinal scan

23.6 References

[1] Finn, Peter (20 January 2011). "Report: Top al-Qaeda figure killed Pearl". *The Washington Post*. Retrieved 21 January 2011.

[2] Blackburn, Bradley (20 January 2011). "Report Says Justice Not Served in Murder of Daniel Pearl, Wall Street Journal Reporter". ABC News. pp. 1–2. Retrieved 20 January 2011.

[3] Cratty, Carol (20 January 2011). "Photos of hands backed up Pearl slaying confession, report finds". CNN. Retrieved 21 January 2011.

[4] Ackerman, Spencer (20 January 2011). "Qaeda Killer's Veins Implicate Him In Journo's Murder". *Wired*. Retrieved 21 January 2011.

[5] http://www.nbcnews.com/id/25378726/ns/technology_and_science-innovation/

[6] HRSID Vein Recognition overview

[7] "Finger Vein Authentication Technology". Hitachi America, Ltd. Retrieved 21 January 2011.

[8] "PV2000". India: Mantra Softech Pvt. Ltd. Retrieved 21 January 2011.

[9] "Your hand is the key: The world's first contactless palm vein authentication technology". *PalmSecure*. Fujitsu. Retrieved 21 January 2011.

[10] Schneier, Bruce (8 August 2007). "Another Biometric: Vein Patterns". *Schneier on Security*. Retrieved 21 January 2011.

[11] Stanglin, Douglas (20 January 2011). "Report: Forensic evidence ties 9/11 plotter to Pearl's killing". *USA Today*. Retrieved 21 January 2011.

[12] Benjamin Wittes (2011-01-20). "So KSM Really Did Kill Daniel Pearl". Lawfare. Retrieved 2013-10-10. The investigation produced a lengthy report concluding, among other things, that Khalid Sheikh Mohammed was telling the truth when he boasted at his CSRT hearing of "decapitat[ing] with my blessed right hand the head of the American Jew, Daniel Pearl."

[13] Asra Q. Nomani; et al. (2011-01-20). "The Pearl Project". The Center for Public Integrity. Retrieved 2013-10-10.

[14] Peter Finn (2011-01-20). "Khalid Sheik Mohammed killed U.S. journalist Daniel Pearl, report finds". Washington Post. Retrieved 2013-10-10. A recently completed investigation of the killing of Daniel Pearl in Pakistan nine years ago makes public new evidence that a senior al-Qaeda operative executed the Wall Street Journal reporter.

[15] Ben Farmer (2011-01-20). "Daniel Pearl was beheaded by 9/11 mastermind". The Telegraph (UK). Retrieved 2013-10-10. They replied: 'The photo you sent me and the hand of our friend inside the cage seem identical to me.' Both the CIA and FBI use the mathematical modelling technique, though it is not considered as reliable as fingerprinting.

[16] Allen, Jonathan (28 July 2011). "New York hospital using palm-scanners". *Stuff.co.nz*. Retrieved 30 July 2011.

[17] By Plasencia, Amanda (28 July 2011). "Hospital Scans Patient Hands to Pull Medical Info". *NBC New York*. Retrieved 30 July 2011.

23.7 Further reading

- Prasanalakshmi, Balaji; Sampath, Kannammal (2009). "A secure cryptosystem from palm vein biometrics". *Proceedings of the 2nd International Conference on Interaction Sciences: Information Technology, Culture and Human*. ACM, New York, NY, USA.

- Watanabe, Masaki; Shiohara, Morito; Sasaki, Shigeru (September 2005). "Palm vein authentication technology and its applications" (PDF). *Proceedings of the Biometric Consortium Conference*. Fujitsu Laboratories.

- Zhang, Yi-Bo; Li, Qin; You, Jane; Bhattacharya, Prabir (2007). "Palm Vein Extraction and Matching for Personal Authentication". *Lecture Notes in Computer Science*. Concordia University. **4781**: 154–164. doi:10.1007/978-3-540-76414-4_16.

- Chen, Liukui; Zheng, Hong (May 2009). "Finger Vein Image Recognition Based on Tri-value Template Fuzzy Matching" (PDF). *Proceedings of the 9th WSEAS International Conference on Multimedia Systems & Signal Processing*. Wuhan University: 206–211. ISBN 978-960-474-077-2. ISSN 1790-5117.

- Kumar, A.; Prathyusha, K.V. (September 2009). "Personal Authentication Using Hand Vein Triangulation and Knuckle Shape" (PDF). *IEEE Transactions on Image Processing*. IEEE Signal Processing Society. **18** (9): 2127–2136. ISSN 1057-7149. PMID 19447728. doi:10.1109/TIP.2009.2023153.

- Chen, Haifen; Lu, Guangming; Wang, Rui (December 2009). "A new palm vein matching method based on ICP algorithm". *International Conference on Information Systems*. Harbin Institute of Technology: 1207–1211. ISBN 978-1-60558-710-3. doi:10.1145/1655925.1656145.

- Sarkar, Ishani; Alisherov, Farkhod; Kim, Tai-hoon; Bhattacharyya, Debnath (March 2010). "Palm Vein Authentication System: A Review" (PDF). *International Journal of Control and Automation*. **3** (1).

23.8 External links

- Finger-vein image recognition combining modified hausdorff distance with minutiae feature matching at The Free Library

- *The Truth Left Behind: Inside the Kidnapping and Murder of Daniel Pearl* at the Center for Public Integrity

Chapter 24

Voice analysis

Voice analysis is the study of speech sounds for purposes other than linguistic content, such as in speech recognition. Such studies include mostly medical analysis of the voice (phoniatrics), but also speaker identification. More controversially, some believe that the truthfulness or emotional state of speakers can be determined using Voice Stress Analysis or Layered Voice Analysis.

24.1 Typical voice problems

A medical study of the voice can be, for instance, analysis of the voice of patients who have had a polyp removed from their vocal cords through an operation. Objective evaluation of voice quality improvement requires some way to measure of voice quality. An experienced voice therapist can quite reliably evaluate the voice, but this requires extensive training and is still subjective.

Another active research topic in medical voice analysis is vocal loading evaluation. The vocal cords of a person who speaks for an extended time suffer from tiring—that is, the process of speaking exerts a load on the vocal cords and tires the tissue. Among professional voice users (e.g., teachers, sales people) this tiring can cause voice failures and sick leaves. To evaluate these problems, vocal loading must be objectively measured.

24.2 Analysis methods

Voice problems that require voice analysis most commonly originate from the vocal folds or the laryngeal musculature that controls them, since the folds are subject to collision forces with each vibratory cycle and to drying from the air being forced through the small gap between them, and the laryngeal musclature is intensely active during speech or singing and is subject to tiring. However, dynamic analysis of the vocal folds and their movement is physically difficult. The location of the vocal folds effectively prohibits direct, invasive measurement of movement. Less invasive imaging methods such as x-rays or ultrasounds do not work because the vocal cords are surrounded by cartilage, which distorts image quality. Movements in the vocal cords are rapid, fundamental frequencies are usually between 80 and 300 Hz, thus preventing usage of ordinary video. Stroboscopic, and high-speed videos provide an option but to see the vocal folds, a fiberoptic probe leading to the camera must be positioned in the throat, which makes speaking difficult. In addition, placing objects in the pharynx usually triggers a gag reflex that stops voicing and closes the larynx. In addition, stroboscopic imaging is only useful when the vocal fold vibratory pattern is closely periodic.

The most important indirect methods are currently inverse filtering of either microphone or oral airflow recordings and electroglottography (EGG). In inverse filtering, the speech sound (the radiated acoustic pressure waveform, as obtained from a microphone) or the oral airflow waveform from a circumferentially vented (CV) mask is recorded outside the mouth and then filtered by a mathematical method to remove the effects of the vocal tract. This method produces an estimate of the waveform of the glottal airflow pulses, which in turn reflect the movements of the vocal folds. The other kind of noninvasive indirect indication of vocal fold motion is the electroglottography, in which electrodes placed on either side of the subject's throat at the level of the vocal folds record the changes in the conductivity of the throat according to how large a portion of the vocal folds are touching each other. It thus yields one-dimensional information of the contact area. Neither inverse filtering nor EGG are sufficient to completely describe the complex 3-dimensional pattern of vocal fold movement, but can provide useful indirect evidence of that movement.

24.3 See also

- Voice biometrics
- Speech processing

- Audio signal processing
- Digital signal processing
- Stuttering

24.4 External links

- Voice Problems and Vocal Disorder Online Community (VoiceMatters.net)

Chapter 25

Identity Cards Act 2006

The Identity Cards Act 2006 (c 15) was an Act of the Parliament of the United Kingdom that has since been repealed. It created national identity cards, a personal identification document and European Union travel document, linked to a database known as the **National Identity Register** (NIR), which has since been destroyed.

The introduction of the scheme was much debated, and various degrees of concern about the scheme were expressed by human rights lawyers, activists, security professionals and IT experts, as well as politicians. Many of the concerns focused on the databases underlying the identity cards rather than the cards themselves. The Act specified fifty categories of information that the National Identity Register could hold on each citizen,*[1] including up to 10 fingerprints, digitised facial scan and iris scan, current and past UK and overseas places of residence of all residents of the UK throughout their lives and indexes to other Government databases (including National Insurance Number*[2]) – which would allow them to be connected. The legislation on this resident register also said that any further information could be added.*[3]

The legislation further said that those renewing or applying for passports must be entered on to the NIR. It was expected that this would happen soon after the Identity and Passport Service (IPS), which was formerly the UK Passport Service, started interviewing passport applicants to verify their identity.*[4]

The Conservative/Liberal Democrat Coalition formed after the 2010 general election announced that the ID card scheme would be scrapped.*[5]*[6] The Identity Cards Act was repealed by the Identity Documents Act 2010 on 21 January 2011, and the cards were invalidated with no refunds to purchasers.*[7] Foreign nationals from outside the European Union, however, continue to require an ID card for use as a biometric residence permit under the provisions of the UK Borders Act 2007 and the Borders, Citizenship and Immigration Act 2009.*[8]*[9]

Only workers in certain high-security professions, such as airport workers, were required to have an identity card in 2009, and this general lack of compulsory ID remains the case today. Therefore, driving licences, particularly the photocard driving licence introduced in 1998, along with passports, are now the most widely used ID documents in the United Kingdom. Nobody in the UK is required to carry any form of ID. In everyday situations most authorities, such as the police, do not make spot checks of identification for individuals, although they may do in instances of arrest. Some banks will accept a provisional driving licence only from young people, the upper age limit for which varies from bank to bank, while others will accept it from all ages.*[10]*[11]*[12]

25.1 Development

25.1.1 Reasons for introduction

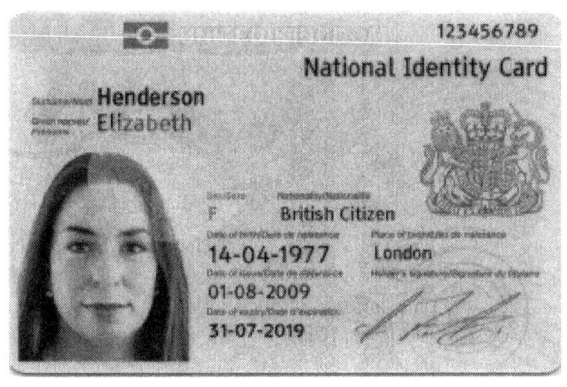

A promotional image of a UK National Identity Card released by the Home Office.

Initial attempts to introduce a voluntary identity card were made under the Conservative administration of John Major, under the then Home Secretary Michael Howard. At the Labour party conference in 1995, Tony Blair demanded that "instead of wasting hundreds of millions of pounds on compulsory ID cards as the Tory Right demand, let that money provide thousands more police officers on the beat in our local communities."*[13] It was included in the Conservative election manifesto for the 1997 general election, but Labour won that election.

A proposal for ID cards, to be called "entitlement cards", was initially revived by the Home Secretary at the time, David Blunkett, following the terrorist attacks of 11 September 2001,*[14] but was reportedly opposed by Cabinet colleagues. However, rising concerns about identity theft and the misuse of public services led to a proposal in February 2002 for the introduction of entitlement cards to be used to obtain social security services, and a consultation paper, *Entitlement Cards and Identity Fraud*, was published by the Home Office on 3 July 2002.*[15] A public consultation process followed, which resulted in a majority of submission by organisations being in favour of a scheme to verify a person's identity accurately. However, it was clear that the ability to properly identify a person to their true identity was central to the proposal's operation, with wider implications for operations against crime and terrorism.

In 2003, Blunkett announced that the government intended to introduce a "British national identity card" linked to a national identity database, the National Identity Register. The proposals were included in the November 2003 Queen's Speech, despite doubts over the ability of the scheme to prevent terrorism. Feedback from the consultation exercise indicated that the term "entitlement card" was superficially softer and warmer, but less familiar and "weaselly", and consequently the euphemism was dropped in favour of "identity card".*[16]

During a private seminar for the Fabian Society in August 2005, Tony McNulty, the minister in charge of the scheme, stated "perhaps in the past the government, in its enthusiasm, oversold the advantages of identity cards", and that they "did suggest, or at least implied, that they might well be a panacea for identity fraud, for benefit fraud, terrorism, entitlement and access to public services". He suggested that they should be seen as "a gold standard in proving your identity".*[17] Documentation released by the Home Office demonstrated analysis conducted with the private and public sector showed the benefits of the proposed identity card scheme could be quantified at £650m to £1.1bn a year, with a number of other, less quantifiable, strategic benefits —such as disrupting the activities of organised crime and terrorist groups.*[18]

25.1.2 Legislative progress

The Identity Cards Bill was included in the Queen's Speech on 23 November 2004, and introduced to the House of Commons on 29 November.

It was first voted on by Members of Parliament following the second reading of the bill on 20 December 2004, where it passed by 385 votes to 93. The bill was opposed by 19 Labour MPs, 10 Conservative MPs, and the Liberal Democrats, while a number of Labour and Conservative members abstained, in defiance of party policies. A separate vote on a proposal to reject the Bill was defeated by 306 votes to 93. Charles Clarke, the new Home Secretary, had earlier rejected calls to postpone the reading of the Bill following his recent appointment.

The third reading of the bill in the Commons was approved on 11 February 2005 by 224 votes to 64; a majority of 160. Although being in favour in principle, the Conservatives officially abstained, but 11 of their MPs joined 19 Labour MPs in voting against the Government. The Bill then passed to the House of Lords, but there was insufficient time to debate the matter, and Labour were unable to do a deal with the Conservatives in the short time available in the days before Parliament was dissolved on 11 April, following the announcement of the 2005 general election.*[19]

Labour's manifesto for the 2005 election stated that, if returned to power, they would "introduce ID cards, including biometric data like fingerprints, backed up by a national register and rolling out initially on a voluntary basis as people renew their passports". In public speeches and on the campaign trail, Labour made clear that they would bring the same Bill back to Parliament. In contrast, the Liberal Democrat manifesto opposed the idea because, they claimed, ID cards "don't work", while the Conservatives made no mention of the issue.

After the 2005 election

Following their 2005 election victory, the Labour Government introduced a new Identity Cards Bill, substantially the same as the previous Bill, into the Commons on 25 May. The Conservatives joined the Liberal Democrats in opposing the Bill, saying that it did not pass their "five tests". These tests included confidence that the scheme could be made to work, and its impact on civil liberties. In December 2005 the Conservative party elected a new leader, David Cameron, who opposes ID cards in principle.

The second reading of the Bill on 28 June was passed, 314 votes to 283, a majority of 31.

At its third reading in the Commons on 18 October, the majority in favour fell to 25, with 309 votes in favour to 284 against.*[20] In the report stage between the readings, the Bill was amended to prevent the National Identity Register database being linked to the Police National Computer.

In early 2006, the Bill was passed through the House of Lords committee stage, where 279 amendments were considered. One outcome of this was a vote demanding that the Government instruct the National Audit Office to provide a full costing of the scheme over its first ten years, and another demanding that a "secure and reliable method" of recording and storing the data should be found. A third defeat limited the potential for ID cards to be required before people could access public services.[*][21] On 23 January the House of Lords defeated the government by backing a fully voluntary scheme.[*][22]

The committee stage ended on 30 January, and the third reading of the Bill took place on 6 February, after which it returned to the Commons. There, on 18 February, the legislation was carried by a majority of 25, with 25 Labour MPs joining those opposing it. Following the defeats in the Lords, the government changed the Bill in order to require separate legislation to make the cards compulsory; however, an amendment to make it possible to apply for a biometric passport without having to register on the National Identity Register database was defeated, overturning the Lords' changes to make the Bill fully voluntary. The Lords' amendment requiring a National Audit Office report was rejected.

The Bill returned to the Lords on 6 March, where the Commons amendments were reversed by a majority of 61.[*][23] The defeat came despite ministers warning that the Lords should follow the Salisbury Convention by refraining from blocking a manifesto commitment. Both Conservatives and Liberal Democrats stated generally in 2005 that they no longer felt bound to abide by the convention, while in this specific case several Lords stated that it would not apply as the manifesto commitment was for implementation on a "voluntary basis" as passports are renewed, rather than being compulsory as passports are renewed.

Subsequent votes:

- 13 March: House of Commons —majority of 33 for Government (310 to 277)[*][24]

- 15 March: House of Lords —majority of 35 against Government (218 to 183)[*][25]

- 16 March: House of Commons —majority of 51 for Government (292 to 241)[*][26]

- 20 March: House of Lords —majority of 36 against Government (211 to 175)[*][27]

- 21 March: House of Commons —majority of 43 for Government (284 to 241)[*][28]

On 29 March, the House of Lords voted in favour of a new plan with a majority of 227 (287 to 60).[*][29] Under this scheme, everyone renewing a passport from 2008 would be issued an ID card and have their details placed on the national ID card database. The Government said that until 2010, people could choose not to be issued a card, though they would still have to pay for one, and still be placed on the database.

The Bill received Royal Assent on 30 March 2006.

25.1.3 Timescale and implementation progress

On 11 October 2006, the Labour government announced a timescale described as "highly ambitious" by computer experts.[*][30] The Home Office announced that it would publish an ID management action plan in the months from November 2006, followed by agreements with departments on their uses for the system. There was to be a report on potential private sector uses for the scheme before 2007 Budget.

On 25 September 2006, Home Office Minister Liam Byrne said that "There are opportunities which give me optimism to think that actually there is a way of exploiting systems already in place in a way which brings down the costs quite substantially" .[*][31]

Emails leaked in June 2006[*][32] indicated that the plan was already in difficulty, with plans for the early introduction of a limited register and ID card with reduced biometrics known as the "early variant" described as a "huge risk".

Due to the costs of developing a new system from scratch, in 2007 the Government approved an alternative plan to use the Department for Work and Pensions' Customer Information System to store the biographical information, linked to a new database to store biometrics, despite concerns over issues of inter-departmental governance, funding and accountability which were never resolved.[*][33]

The schedule for putting passport applicants' and renewers' details on the National Identity Register (NIR) was never announced. A nationwide network of 68 interview offices for first-time passport applicants started opening in June 2007 and is now complete. The interview consisted mainly of asking applicants to confirm facts about themselves, which someone attempting to steal their identity may not know. The government has stated that all personal information used in the interview not required for the application was destroyed shortly after the passport was issued.[*][34] Fingerprints were not taken. Plans to take iris scans were dropped, although the Government had not ruled them out as a future option.[*][35]

In March 2008, the Home Secretary announced that people could choose to have an identity card, a passport, or both when they become available (although they could not opt out of having their details recorded on the NIR). On 25 November 2008 people making applications to remain in the United Kingdom as a student or based on marriage were required to have an identity card. Under those plans it was estimated that by the end of 2014–15 about 90% of all foreign nationals would have been issued with one. On 22 January 2008, the Home Office confirmed that large volumes of cards would not be issued until 2012; however, ID cards were issued to workers in critical locations, starting with airside workers in Manchester and London City airports in 2009, and young people were being offered cards in 2010.

A leaked document, published on 29 January 2008, suggested that "universal compulsion should not be used unless absolutely necessary ... due the need for inevitably controversial and time-consuming primary legislation" but that "various forms of coercion, such as designation of the application process for identity documents issued by UK ministers (e.g. passports) were an option to stimulate applications in a manageable way" .*[36]

In January 2008 the *Financial Times* reported that Accenture and BAE Systems had withdrawn from the procurement process. Fujitsu Services, CSC, EDS, IBM, Steria and Thales Group were still negotiating framework agreements with the government.*[37]

On 1 August 2008 it was confirmed that Thales Group was awarded a 4-year contract to work on the design, building, testing and operation of the National Identity Scheme.*[38]

On 25 September 2008 Jacqui Smith unveiled replicas of the first actual cards to be issued as residence permits to foreign nationals.*[39]*[40]

The first to receive ID cards were foreign nationals, from 25 November 2008 until the programme's cancellation. National Identity Cards for UK nationals became available to people resident in the Greater Manchester area on 30 November 2009.*[41] Ordinary British citizens were then meant to be offered (on a voluntary basis at first, but later in larger volumes) ID cards from 2011 to 2012.*[39] A Home Office minister, Meg Hillier, said that they would be a "convenient" way for young people to prove their age when going to bars and that at £30 they are cheaper than purchasing passports,*[42] although the total cost including processing fees was expected to be up to £60,*[43] more expensive than a passport cost before the introduction of the ID card and database scheme – the Conservatives and Liberal Democrats criticised the increase in passport costs as being needed for the ID card scheme.*[44]*[45] In December 2009, while on a trip to promote identity cards, Meg Hillier had to admit she had forgotten hers and was left unable to display one for photographers.*[46]*[47]

25.1.4 Pilot schemes and partial rollouts

- **non-EU foreign nationals on student or marriage/civil partnership visas (compulsory)** – from November 2008 until the programme's cancellation, non-European Union foreign nationals with permission to stay in the UK on the basis of a student visa or a marriage/civil partnership visa would, when applying to extend their stay, be required to apply for an ID card.*[48]

- **Air industry staff (compulsory)** (cancelled) – a pilot scheme involving compulsory IDs for 30,000 air industry staff, planned to start in September 2009 at Manchester and London City airports, was cancelled in June 2009, after substantial opposition from unions.*[49]

- **Greater Manchester residents (voluntary)** – a pilot scheme open to all residents of Greater Manchester, from October 2009; which was expanded to Merseyside, then the rest of the North-West in early 2010.*[50] 13,200 people signed up. The Manchester Evening News revealed in 2010 that senior Whitehall officials were urged to email friends and relatives encouraging them to buy cards, because of fears about the level of demand.*[51]

- **Air industry staff (voluntary)** – a pilot scheme involving free, voluntary ID cards for airside workers, began in November 2009 until the programme's cancellation at Manchester and London City airports.*[52]

- **Young people opening bank accounts (voluntary)** – in 2010 young people would have been encouraged to get ID cards when they opened bank accounts.*[53]

- **London residents (voluntary)** – was a planned pilot scheme in 2010 open to all residents of London.*[53]

- **over the age of 16 if registered for IPS newsletter updates (voluntary)**, begun in 2010*[54]

- **over the age of 16 applying for a passport** intended in 2011–2012, optional, but applicants' details would have been entered into the National Identity Register*[53]

25.1.5 2010 general election

During the 2010 general election campaign, the published manifestos of the various parties revealed that the Labour Party planned to continue the introduction of the identity card scheme, while all other parties pledged to discontinue plans to issue ID cards. The Conservative party also explicitly pledged to scrap the National Identity Register, while the wording of several other manifestos implied that this may have been the position of certain other parties too.*[55]

25.1.6 Ending of the scheme

Main article: Identity Documents Act 2010

In the Conservative – Liberal Democrat Coalition Agreement that followed the 2010 general election, the new government announced that they planned to scrap the ID card scheme, including the National Identity Register (as well as the next generation of biometric passports and the ContactPoint database), as part of their measures "to reverse the substantial erosion of civil liberties under the Labour Government and roll back state intrusion." *[5]*[6]

In a document published in May 2010 at the time of the Queen's Speech, the new Government announced that the scrapping of the scheme would save approximately £86 million over the following 4 years, and avoid a further £800 million in maintenance costs over the decade which were to have been recovered through fees.*[56]

On 27 May 2010, the draft Identity Documents Act 2010 was published with the aim of having it passed into law by August 2010.*[57] The government missed this target but expected the bill to become law before the new year.*[58] The Bill was passed by the House of Commons on 15 September 2010 and received Royal Assent on 21 December 2010. Section 1(1) of the Identity Documents Act repealed the Identity Cards Act 2006 on 21 January 2011 (making all ID cards invalid) and mandated the destruction of all data on the National Identity Register by 21 February 2011.*[59]

The register was officially destroyed on Thursday, 10 February when the final 500 hard drives containing the register were shredded at RDC in Witham, Essex.*[60]

Home Office Minister Damian Green said: "This marks the final end of the identity card scheme: dead, buried and crushed ... What we are destroying today is the last elements of the national identity register, which was always the most objectionable part of the scheme." *[61]

A banker from Germany with joint British and Swiss nationality was arguably the last person to officially use the ID card on a flight from Düsseldorf to Manchester on 21 January 2011, landing 90 minutes before the scheme was officially scrapped at midnight.*[62]

25.2 Historical and international comparisons

25.2.1 ID cards during the World Wars

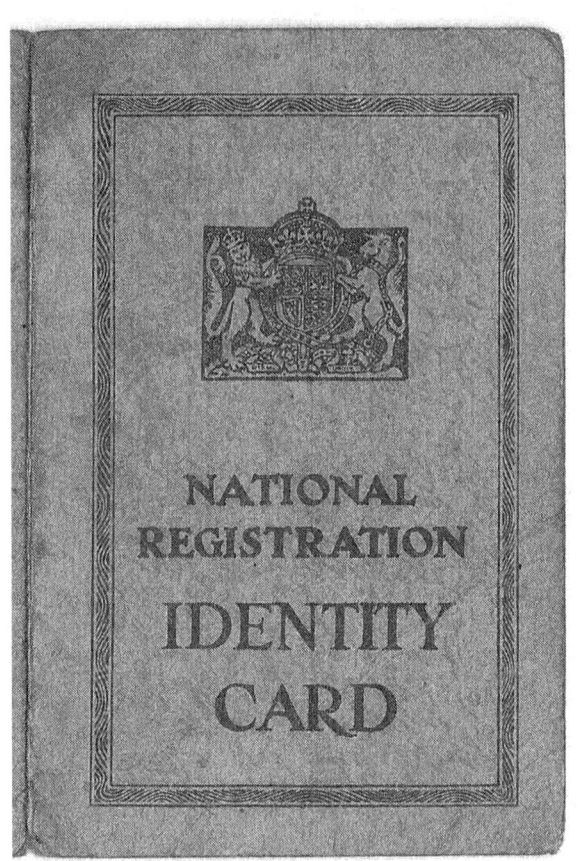

A mid-20th century ID card

Compulsory identity cards were first issued in the United Kingdom during World War I, and abandoned in 1919. Cards were re-introduced during World War II under the National Registration Act 1939, but were abandoned seven years after the end of that war, in 1952, amid widespread public resentment. The National Register however, became the National Health Service Register and is maintained to this day.

The World War I identity card scheme was highly unpopular, though accepted in the light of the prevailing national emergency. It is possible to take a small measure of how the national identity scheme was received from remarks by the historian A. J. P. Taylor in his *English History, 1914–1945*, where he describes the whole thing as an "indignity" and talks of the Home Guard "harassing" people for their cards.*[63]

After the Second World War the government of Clement

Attlee decided to continue the scheme in the face of the Cold War and the perceived Soviet threat, though it grew ever less popular. In the mind of the public it was more and more associated with bureaucratic interference and regulation, reflected, most particularly, in the 1949 comedy film *Passport to Pimlico*. Identity cards also became the subject of a celebrated civil liberties case in 1950. Clarence Henry Willcock, a member of the Liberal Party, refused to produce his after being stopped by the police. During his subsequent trial he argued that identity cards had no place in peace time, a defence rejected by the magistrate's court. In his subsequent appeal, *Willcock v Muckle*, the judgment of the lower court was upheld.

Protest reached Parliament, where the Conservative and Liberal peers voiced their anger over what they saw as "Socialist card-indexing". After the defeat of the Labour Government in the general election of October 1951 the incoming Conservative administration of Winston Churchill was pledged to get rid of the scheme, "to set the people free", in the words of one minister. Cheers rang out when on 21 February 1952 the Minister for Health, Harry Crookshank, announced in the House of Commons that national identity cards were to be scrapped. This was a popular move, adopted against the wishes of the police and the security services, though the decision to repeal the 1939 legislation was, in significant part, driven by the need for economies. By 1952 national registration was costing £500,000 per annum (equivalent to £13,000,000 in 2015) and required 1500 civil servants to administer it.

25.2.2 International comparisons

Identity cards

During the UK Presidency of the EU in 2005 a decision was made to "agree common standards for security features and secure issuing procedures for ID cards (December 2005), with detailed standards agreed as soon as possible thereafter. In this respect, the UK Presidency has put forward a proposal for EU-wide use of biometrics in national ID cards." *[64]

Australia started work on a health and social services access card, but the government elected in the 2007 federal election cancelled it.

Belgium has introduced the Electronic identity card or eID card from 2004 and by 2012 every citizen in Belgium must have an e-ID card for identity purposes. A variant exists for children, but that is not compulsory.

Biometrics in identity and travel documents

There has been an international move towards the introduction of biometrics into identity and travel documents. The ICAO has recommended that all countries adopt biometric passports, and the United States has made it a requirement for entering the US under the visa waiver programme. Biometric border control systems have been established in the United States and the United Arab Emirates, and the EU is introducing biometric visas. However, it should be noted that, internationally, the only requirement for biometric passports is a digital photograph.

25.3 System

25.3.1 Legal requirements

Under the NIS UK Residents who wanted or were required to apply for an ID card would have been required to fulfil certain functions:

- Attend in person to have their fingerprints recorded at one of the Identity & Passport Service's High Street partners.

- Promptly inform the police or Identity & Passport Service if a card is lost or damaged, and apply for a new card.*[65]

- Promptly inform the Identity & Passport Service of any change of address.

- Promptly inform the Identity & Passport Service of any prescribed change of circumstances affecting the information recorded about them in the Register.

Failure to do so would have meant a penalty of up to £1,000 or a shortened permission to stay.*[66]

25.3.2 National Identity Register

Key to the ID Card scheme was a centralised computer database, the National Identity Register (NIR). To identify someone it would not have been necessary to check their card, since identity could be determined by a taking a biometric scan and matching it against a database entry.

ID cards for foreign nationals were produced by the Driver and Vehicle Licensing Agency (DVLA) in Swansea on behalf of the Home Office.

25.3.3 Identity Registration Number

One entry on the NIR was the Identity Registration Number. The Home Office had recognised that a unique identifier was needed as a primary key for the database.

The Home Office's Identity Cards Benefits Overview document[*][67] described how the IRN would have enabled data sharing amongst police databases (including the Police DNA database), legal databases, and even corporate databases (including bank and travel operators).

25.3.4 Types of cards

Three types of identity cards were issued:[*][68]

- The *National Identity Card*, which was lilac and salmon in colour, was issued to British citizens only. It contained the text "British Citizen" and was a valid travel document for entry into any EEA state and Switzerland until its invalidation in 2010.

- The *Identification Card* was turquoise and green in colour and did not mention the holder's nationality. It was issued to EU, EEA and Swiss citizens living in the UK (including Irish citizens living in Northern Ireland[*][69]). It was also issued to certain family members of EU/EEA citizens, to British citizens to whom certain conditions or restrictions apply, and as an additional card to a person living in two gender roles.[*][70]

- The *Identity Card for Foreign Nationals* was blue and pink in colour[*][71] and was issued to certain categories of immigrants from non-EU/EEA countries.

25.4 Use as travel document

Until midnight on 21 January 2011, the National Identity Card was officially recognised as a valid travel document by the EEA and Switzerland, following which the United Kingdom instructed immigration authorities therein to cease accepting it as a valid travel document. It also became accepted voluntarily by a number of other European countries but its current validity in these additional countries remains unclear, given that its acceptance and subsequent denial by these countries was never mandated by the United Kingdom through EU or EEA channels. It was the only travel document valid for use by UK nationals throughout the EEA and Switzerland, other than a valid British citizen passport or a pink Gibraltar identity card. The exception to this was for travel to Ireland. All British citizens are entitled to enter Ireland without the need to carry a valid travel document, on account of the Common Travel Area agreement.

- European Union[*][72]
- Iceland (EEA) [*][73]
- Liechtenstein (EEA) [*][73]
- Norway (EEA) [*][73]
- Switzerland [*][73]

It became accepted also by:

- Albania [*][74][*][75]
- Andorra [*][76] "Any travel document recognised by France or Spain"
- Bosnia and Herzegovina [*][77]
- Croatia [*][78]
- Faroe Islands [*][79]
- Macedonia [*][80]
- Monaco [*][81]
- Montenegro [*][82]
- Morocco (only for tours organised by a travel agency for groups of more than three people)[*][83]
- San Marino [*][84]
- Serbia [*][85]
- Vatican City [*][86]

It was also accepted as a travel document to enter the British Crown Dependencies and the British Overseas Territories:

- Gibraltar Part of EU
- Guernsey Part of Common Travel Area —no travel document required to enter from the UK. (NB: Air travellers require photo-ID for airline security purposes.)
- Isle of Man Part of Common Travel Area —no travel document required to enter from the UK. (NB: Air travellers require photo-ID for airline security purposes.)
- Jersey Part of Common Travel Area —no travel document required to enter from the UK. (NB: Air travellers require photo-ID for airline security purposes.)

All other overseas territories require a fully valid passport. Of the two countries closest to the UK not to accept UK ID cards, Ukraine and Belarus, the latter requires not only a passport but also for British citizens to obtain a visa in advance (except if entering and exiting through Minsk airport and staying for max 5 days).

Controversially, some travel companies initially refused to carry passengers with UK National Identity Cards.*[87]

25.5 Reaction

The announcement of the scheme had seen a mixed reaction from both the public and from figures connected to terrorism and law enforcement.

25.5.1 Public reaction

For a detailed account of opinion polls concerning the National Identity Card, see Opinion polls on the British national identity card.

Over a period of time, public opinion, as measured by opinion polls, appears to have shifted away from support for the scheme towards opposition. This appeared to have become more of a concern since the disclosure of the loss of 25 million records by Her Majesty's Revenue and Customs.

In 2003, the announcement of the scheme was followed by a public consultation exercise, particularly among 'stakeholder groups'. At March 2003 the government stated that the overall results were:

> in favour: 2606 responses (61%)
>
> against: 1587 responses (38%)
>
> neutral: 48 responses (1%)

By July 2006, an ICM poll*[88] indicated that public support had fallen to 46%, with opposition at 51%.

A further poll by YouGov/Daily Telegraph, published on 4 December 2006, indicated support for the identity card element of the scheme at 50%, with 39% opposed. Support for the national database was weaker, with 22% happy and 78% unhappy with the prospect of having their data recorded. Only 11% trusted the government to keep the data confidential. 3.12% of the sample were prepared to undergo long prison sentences rather than have a card.*[89]

25.5.2 Terrorism and crime

Eliza Manningham-Buller, the former head of Britain's counter-intelligence and security agency MI5 was on record as supporting the introduction of identity cards, as was Sir Ian Blair, former Commissioner of the Metropolitan Police and his predecessor, Sir John (now Lord) Stevens. The Association of Chief Police Officers was also supportive.

However, in November 2005 Dame Stella Rimington, who was Director General of MI5 before Eliza Manningham-Buller, questioned the usefulness of the proposed scheme.*[90] This intervention caused a good deal of controversy amongst supporters and opponents of the scheme, especially as Manningham-Buller stated that ID cards would in fact disrupt the activities of terrorists, noting that significant numbers of terrorists take advantage of the weaknesses of current identification methods to assist their activities.

Lord Carlile was appointed after 11 September attacks on New York and Washington in 2001 to independently review the working of the Terrorism Act 2000 and subsequent anti-terrorist laws.*[91] Talking on GMTV on 29 January 2006, he expressed his views on the proposed legislation, saying*[92] that ID cards could be of limited value in the fight against terrorism but that Parliament had to judge that value against the curtailment of civil liberties. Speaking on the same programme, Lord Stevens of Kirkwhelpington, former Met Police Commissioner, argued in favour for the need for identity cards, saying they had benefits in tackling serious crimes, such as money laundering and identity theft.

25.6 Objections to the scheme

25.6.1 Costs

Independent studies including one by the London School of Economics had suggested that costs could be as much as £12 billion to £18 billion.*[93] The reliability of this study was challenged by the Labour Government which disputed some of the assumptions used in the calculations, such as the need to retake biometric information every 5 years. The government argued that this assumption had not been supported by any research in the London School of Economics report, and that biometric experts quoted in the LSE reports had sought to distance themselves from its findings. The Government also claimed that the authors of these estimates were established opponents to the scheme and could not be considered unbiased academic sources.

Tony McNulty, Home Office minister who was responsible for the scheme, responded by saying a "ceiling" on costs would be announced in October 2005.*[94] There were indications that the Labour Government was looking at ways of subsidising the scheme by charging other Government

Departments, with the implication that this would result in increased charges for other Government services to individuals or businesses.[*][95]

After the 2005 general election the Home Office stated that it would cost £584 million a year to run the scheme. In October 2006, the Government declared it would cost £5.4bn to run the ID cards scheme for the next 10 years.[*][96] In May 2007 the Home Office forecast a cost rise of £400m to £5.3 billion,[*][97] a figure revised in November 2007 to £5.612bn.[*][98]

The Labour Government abandoned plans for a giant new computer system to run the national identity card scheme. Instead of a single multibillion-pound system, information was held on three existing, separate databases.[*][99]

An estimate from the Home Office placed the cost of a 10-year passport and ID card package at £85, while after the 2005 general election in May 2005 they issued a revised figure of over £93,[*][100] and announced that a "standalone" ID card would cost £30.[*][101] In 2009, it was announced that retailers would be collecting fingerprints and photographs, and that they would be able to charge for this, meaning that the total cost for a standalone ID card was expected to be up to £60.[*][43]

25.6.2 Effectiveness

The then Home Secretary David Blunkett stated in 2004 said the cards would stop people using multiple identities and boost the fight against terrorism and organised crime. However, human rights group Liberty disputed this, pointing out that the existence of another form of ID cards in Spain did not prevent the Madrid train bombings.[*][102]

However, Blunkett subsequently made a significant U-turn. At his opening speech for Infosecurity Europe on 27 April 2009, he stepped back from the concept of a full National Identity Database for every citizen, saying it would be sufficient to improve the verification of passports.[*][103][*][104][*][105]

His successor, Charles Clarke, said that ID cards "cannot stop attacks", in the aftermath of the 7 July 2005 London bombings, and added that he doubted it would have prevented the atrocities. However, he felt that on the balance between protecting civil liberties and preventing crime, ID cards would help rather than hinder.[*][106]

25.6.3 Ethnic minorities

The Government's Race Equality Impact Assessment[*][107] indicated significant concern among ethnic groups over how the Police would use their powers under an Identity Cards Act 2006, with 64% of black and 53% of Indian respondents having expressed concern, particularly about the potential for abuse and discrimination. In their January 2005 report[*][108] on the Bill, the Commission for Racial Equality stated that the fear of discrimination is neither misconceived nor exaggerated, and note that this is also an ongoing issue in Germany, the Netherlands and France.

The CRE were also concerned that disproportionate requirements by employers and the authorities for ethnic minorities to identify themselves may lead to a two-tiered structure amongst racial groups, with foreign nationals and British ethnic minorities feeling compelled to register while British white people do not.[*][109][*][110]

According to the CRE, certain groups who move location frequently and who tend to live on low incomes (such as Gypsies, travellers, asylum-seekers and refugees) would risk being criminalised under the legislation through failing to update their registration each time they moved due to lack of funds to pay the fee that may be charged.

25.6.4 Concerns raised by the Information Commissioner

In a press release on 30 July 2004,[*][111] Richard Thomas the Information Commissioner's Office stated that a NIR raised substantial data protection and personal privacy concerns. He sought clarification of why so much personal information needed to be kept as part of establishing an individual's identity and indicated concern about the wide range of bodies who would view the records of services individuals have used. The Commissioner also pointed out that those who renew or apply for a driving licence or passport were to be automatically added to the National Identity Register, and so would lose the option of not registering. He subsequently stated: "My anxiety is that we don't sleepwalk into a surveillance society." [*][112] In February 2003, on BBC Radio 4's *Today* programme, he warned that ID cards could become a target for identity theft by organised crime.

25.6.5 Human rights

On 2 February 2005, the UK Parliament's Joint Committee on Human Rights questioned the compatibility of the Bill with Article 8 of the European Convention on Human Rights (the right to respect for private life) and Article 14 (the right to non-discrimination),[*][113] both of which are encapsulated in the Human Rights Act 1998.

25.6.6 Feature creep

Even without new primary legislation, the Identity Cards Act 2006 allowed the potential scope of the scheme to be much greater than that usually publicised by the Government.

For example, Gordon Brown was reported to be "planning a massive expansion of the ID cards project that would widen surveillance of everyday life by allowing high-street businesses to share confidential information with police databases." *[114] He apparently described how "police could be alerted as soon as a wanted person used a biometric-enabled cash card or even entered a building via an iris-scan door." *[115]

The wartime National Registration ID card expanded from 3 functions to 39 by the time it was abolished.*[116]

Concerns had also been raised following Tony Blair's response to an ID card petition stating that the fingerprint register would be used to compare the fingerprints of the population at large against the records of 900,000 unsolved crimes. Opposition MPs claimed that the use of the biometric data in this way would directly breach promises given during the Commons debate that there would be adequate safeguards preventing the use of ID card data for "fishing expeditions" .*[117]*[118]

25.6.7 Database extent and access

Home Office forecasts envisaged that "265 government departments and as many as 48,000 accredited private sector organisations" would have had access to the database, and that 163 million identity verifications or more would take place each year.*[119] However, the IPS had stated that only the data needed for the passport would have been kept*[120] and that organisations that have permission to access the data held on the Register could only have done so with the individual's permission, unless to prevent or investigate a crime.

25.6.8 Vulnerable individuals

The CRE had also recommended that more work was required to protect the interests of vulnerable individuals. For example, people escaping from domestic violence or a forced marriage may have been at risk if their previous names or addresses were disclosed. Minister Meg Hillier, in a letter to *The Spectator* magazine, claimed that as the ID card would not have someone's address on it, it would protect such a person's privacy in a way currently unavailable.*[121]

25.6.9 Identity theft

In May 2005 Tony Blair said "ID cards are needed to stop the soaring costs of identity theft".*[122] However, security experts claimed that placing trust in a single document may make identity theft easier, since only this document needs to be targeted.*[123]

25.6.10 Technology

Elsewhere, doubts remained concerning the practicability of the scheme. Existing government systems were not appropriate for the issuing to UK citizens from 2009.*[30]

Tests of facial recognition software dating from 2006 showed error rates of up to 52 percent for the disabled.*[124]

The cards could stop some credit cards from working properly, when kept in the same wallet.*[51]

25.7 Opposition campaigns

In May 2006, NO2ID launched the "Renew for Freedom" campaign,*[125] urging passport holders to renew their passports in the summer of 2006 to delay being entered on the National Identity Register. This followed the comment made by Charles Clarke in the House of Commons that "anyone who feels strongly enough about the linkage [between passports and the ID scheme] not to want to be issued with an ID card in the initial phase will be free to surrender their existing passport and apply for a new passport before the designation order takes effect" .*[126]

In response, the Home Office said that it was "hard to see what would be achieved, other than incurring unnecessary expense" by renewing passports early.*[127] However, the cost of a passport was £51 at the time, then increased in 2006 and 2007 to £72 and was due to rise to £93 after the introduction of ID cards.

On 14 November 2007, the NO2ID opposition group called for financial donations from the 11,360 people who had pledged to contribute to a fighting fund opposing the legislation.*[128] The organisation planned to challenge the statutory instruments that were planned to be brought in to enable the ID card scheme.*[129]

Baroness Williams and Nick Clegg said in 2007 that they would take part in civil disobedience campaigns by refusing to register for an ID card, or to attend photographic sittings.*[130]

25.7.1 Scotland

Although policy on passports and the National Identity Scheme was not an area devolved to the Scottish Government, on 19 November 2008 the Scottish Parliament voted*[131] to reject the ID card scheme, with no votes against the government motion, and only the Scottish Labour MSPs abstaining. In 2005 the previous Labour-Liberal Democrat coalition government had stated*[132] that "the proposals for an identity card scheme confine themselves to reserved policy areas only", and that ID cards will not be needed to access devolved services in Scotland, e.g. health, education, the legal system and transport.

However, the similar Scottish National Entitlement Card has been introduced.

25.7.2 Northern Ireland

The introduction of compulsory ID cards to Northern Ireland would likely have provoked serious opposition given the large Nationalist community who regard themselves as Irish and not British.*[133] In an effort to counter this, the British Government decided not to include the Union Flag on the card, and had stated that a separate card will be issued to Northern Irish people who identify their nationality as Irish. The separate card would not have included any statement of nationality, and as such, could not have been used as a travel document as only the Irish Government may issue travel documents for Irish citizens. Home Secretary Alan Johnson had also stated that the inclusion of Northern Irish people on the National Identity Register of British citizens would not have prevented such people from claiming full Irish citizenship rights.*[134]

25.8 See also

- Opinion polls on the British national identity card
- "Kiss Ya Lips (No I.D.)", a protest song by Ian Brown
- National Insurance number
- Human rights in the United Kingdom
- Common Travel Area
- Schengen Information System
- Mass surveillance in the United Kingdom
- Work card
- Scottish National Entitlement Card

25.9 Notes

[1] The citation of this Act by this short title was authorised by section 44(1) of this Act. Due to the repeal of this Act, it is now authorised by section 19(2) of the Interpretation Act 1978.

25.10 References

[1] "Identity Cards Act 2006 (c. 15)". Opsi.gov.uk. Retrieved 8 May 2010.

[2] Kable (7 January 2010). "Johnson reveals ID register linked to NI numbers". Theregister.co.uk. Retrieved 8 May 2010.

[3] "Identity Cards Act 2006 (c. 15)". Opsi.gov.uk. Retrieved 8 May 2010.

[4] "Webcam interview essential for new passports". Stornoway Today. Archived from the original on 24 March 2007. Retrieved 8 May 2010.

[5] Conservative Liberal Democrat Coalition Agreement, Conservative Party, Published 12 May 2010. Retrieved 13 May 2010

[6] Conservative Liberal Democrat Coalition Agreement, Liberal Democrats, Published 12 May 2010. Retrieved 13 May 2010

[7] Porter, Andrew; Kirkup, James (24 May 2010). "ID card scheme will be scrapped with no refund to holders". *Daily Telegraph*. London.

[8] Comment: ID cards by the backdoor? politics.co.uk, published 6 June 2010. Retrieved 7 June 2010

[9] Cancellation of identity cards: FAQs Immigration and Passport Service

[10] "HSBC Customer Support - Proof of Identity - HSBC Bank UK".

[11] https://www.halifax.co.uk/SecurityandPrivacy/pdf/personal-information-and-identity.pdf

[12] "Page Not Found".

[13] Hansard text for 19 May 2005

[14] BBC News: A question of identity, 25 September 2001

[15] Entitlement Cards and Identity Fraud – A Consultation Paper Archived 1 August 2012 at Archive.is, 3 July 2003

[16] A Summary of Findings from the Consultation Exercise on Entitlement Cards and Identity Fraud Archived 27 November 2007 at the Wayback Machine., page 45, November 2002

[17] Labour admits ID card 'oversell', 4 August 2005

[18] http://www.identitycards.gov.uk Archived 28 September 2007 at the Wayback Machine.

[19] Patrick Wintour (6 April 2005). "Casino and ID card bills hit in deal on legislation | Special Reports | Guardian Unlimited Politics". London: Politics.guardian.co.uk. Retrieved 8 May 2010.

[20] "Politics | Labour survives ID card rebellion". BBC News. 18 October 2005. Retrieved 8 May 2010.

[21] Russell, Ben (17 January 2006). "ID Cards Bill in crisis after peers inflict defeat —UK Politics, UK —Independent.co.uk". London: News.independent.co.uk. Archived from the original on 4 July 2008. Retrieved 8 May 2010.

[22] "Politics | ID cards scheme in Lords defeats". BBC News. 23 January 2006. Retrieved 8 May 2010.

[23] "Clarke vows to overturn ID cards defeat | Special Reports | Guardian Unlimited Politics". London: Politics.guardian.co.uk. 7 March 2006. Retrieved 8 May 2010.

[24] Craig Hoy (13 March 2006). "Clarke warns peers on ID cards". ePolitix.com. Retrieved 8 May 2010.

[25] "ID cards remain in parliamentary limbo". ePolitix.com. 15 March 2006. Retrieved 8 May 2010.

[26] "ID card law sent back to the Lords". ePolitix.com. 16 March 2006. Retrieved 8 May 2010.

[27] Daniel Forman (20 March 2006). "Ministers lose another ID cards vote". ePolitix.com. Retrieved 8 May 2010.

[28] "Politics | MPs stand firm on identity cards". BBC News. 21 March 2006. Retrieved 8 May 2010.

[29] "Politics | Deal paves the way for ID cards". BBC News. 30 March 2006. Retrieved 8 May 2010.

[30] Sarah Arnott. "Cost of ID card technology pencilled in at £800m". vnunet.com. Archived from the original on 30 September 2007. Retrieved 8 May 2010.

[31] Stone, Ollie (25 September 2006). "Politics | Identity card cost 'may be cut'". BBC News. Retrieved 8 May 2010.

[32] Emails from Whitehall officials in charge of ID cards – Sunday Times – Times Online published 2006-06-09. Retrieved 3 February 2011

[33] Analysis: How the IT behind ID cards was never going to work Computer Weekly, published 2011-01-31. Retrieved 3 February 2011

[34] Churcher, Joe. "Passport interviews 'next step to ID cards' – Scotland on Sunday". Scotlandonsunday.scotsman.com. Retrieved 8 May 2010.

[35] "Government drops iris scan plan | Pinsent Masons LLP". Out-law.com. Retrieved 8 May 2010.

[36] Francis Elliott, ID cards may be issued by coercion, says leaked memo, The Times, 28 January 2008

[37] Companies abandon ID card project, Financial Times, 23 January 2008

[38] Thales awarded National Identity Scheme contract. Ips.gov.uk. Retrieved on 13 June 2011.

[39] "First ID card unveiled by Home Secretary as scheme builds momentum" (Press release). Identity and Passport Service. 25 September 2008.

[40] Christopher Hope (17 December 2008). "Jacqui Smith unveils the UK's new identity card —with no sign of Britain". London: The Daily Telegraph.

[41] "National identity card launched in Manchester". IPS. 30 November 2009. Retrieved 8 May 2010.

[42] "UK | UK Politics | ID cards 'good for going to bars'". BBC News. 16 November 2009. Retrieved 8 May 2010.

[43] "UK | UK Politics | Retailers reject ID security fear". BBC News. 6 May 2009. Retrieved 8 May 2010.

[44] "UK | Passport cost to increase by 29%". BBC News. 24 July 2006. Retrieved 8 May 2010.

[45] "UK | UK Politics | Cost of new passports to increase". BBC News. 7 July 2009. Retrieved 8 May 2010.

[46] ID card minister forgets ID card. The Register. 16 December 2009

[47] "Identity minister forgets ID card". BBC News. 15 December 2009. Retrieved 8 May 2010.

[48] *BBC*, 25 September 2008, Foreign national ID card unveiled

[49] *The Guardian*, 30 June 2009, Identity card trial for air industry staff dropped

[50] *Rochdale Online*, 21 August 2009, ID cards rolled out in Greater Manchester

[51] Revealed: The full shambles of the ID card trial in Greater Manchester | Manchester Evening News Archived 1 January 2011 at the Wayback Machine.. menmedia.co.uk (30 December 2010). Retrieved on 13 June 2011.

[52] *IPS Website*, 30 November 2009, Press release

[53] *BBC*, 2 July 2009, Q&A: Identity cards

[54] "Who can get the card : ID Cards : Directgov". Idsmart.direct.gov.uk. 1 January 2009. Archived from the original on 5 February 2010. Retrieved 8 May 2010.

[55] "Opposition policies on identity cards, London School of Economics & Political Science, April 15, 2010". Blogs.lse.ac.uk. Archived from the original on 17 April 2010. Retrieved 8 May 2010.

[56] Queen's Speech – Identity Documents Bill Archived 28 May 2010 at the Wayback Machine., Number10.gov.uk, published 25 May 2010. Retrieved 27 May 2010

25.10. REFERENCES

[57] "Identity cards scheme will be axed 'within 100 days'". *BBC News Online*. 27 May 2010.

[58] "Home Office fails to axe ID cards in 100 days". *BBC News Online*. 23 August 2010.

[59] "Home office press release". *Home Office Online*. 21 January 2011.

[60] Espiner, Tom. (11 February 2011) Government destroys final ID cards data | ZDNet UK Archived 13 February 2011 at the Wayback Machine.. Zdnet.co.uk. Retrieved on 13 June 2011.

[61] Mathieson, SA (10 February 2011). "Minister destroys national identity register". *The Guardian*. London.

[62] "Final stop for ID cards as trial bites the dust". *Manchester Evening News*. 22 January 2011. Archived from the original on 25 January 2011.

[63] A. J. P. Taylor, *English History, 1914–1945*, pp. 563, 599.

[64] "UK Presidency advances EU-wide ID card standards, data retention and intelligence sharing to fight terrorism, 14 July 2005". Retrieved 8 May 2010.

[65] Alan Travis, home affairs editor (6 July 2009). "£1,000 fines to bolster ID cards as Tories pledge to scrap scheme | Politics". London: The Guardian. Retrieved 8 May 2010.

[66] "UK Border Agency | Identity card holders' responsibilities". Ukba.homeoffice.gov.uk. Archived from the original on 5 August 2012. Retrieved 8 May 2010.

[67] Identity Cards Scheme – Benefits Overview Archived 18 February 2006 at the Wayback Machine., Home Office

[68] "Types of identity card". IPS. Retrieved 8 May 2010.

[69] Alan Travis, home affairs editor (30 July 2009). "No room for union flag as Alan Johnson unveils the British identity card | Politics". London: The Guardian. Retrieved 8 May 2010.

[70] The Identity Cards Act 2006 (Prescribed Information) Regulations 2009 (No. 2794), 1 (2) (d-j)

[71] "UK Border Agency | Identity cards for foreign nationals". Ukba.homeoffice.gov.uk. Archived from the original on 18 March 2009. Retrieved 8 May 2010.

[72] "EEA countries". IPS. Retrieved 8 May 2010.

[73] "HM Passport Office - GOV.UK".

[74] "Passport and visa requirements". *Timatic*. International Air Transport Association (IATA) through Olympic Air.

[75] Albania. Moveoneinc.com. Retrieved on 13 June 2011.

[76] "Passport and visa requirements". *Timatic*. International Air Transport Association (IATA) through Olympic Air.

[77] "Passport and visa requirements". *Timatic*. International Air Transport Association (IATA) through Olympic Air.

[78] http://www.mvpei.hr/MVP.asp?pcpid=1615&dmid=187#pocdrz

[79] "Home".

[80] "Passport and visa requirements". *Timatic*. International Air Transport Association (IATA) through Olympic Air.

[81] "Passport and visa requirements". *Timatic*. International Air Transport Association (IATA) through Olympic Air.

[82] "Archived copy". Archived from the original on 9 October 2012. Retrieved 2010-12-12.

[83] Formalities For Foreigners Archived 16 July 2011 at the Wayback Machine.. Maec.gov.ma (11 November 2003). Retrieved on 13 June 2011.

[84] "Passport and visa requirements". *Timatic*. International Air Transport Association (IATA) through Olympic Air.

[85] "Serbia's EU membership path - News - European Parliament".

[86] "Passport and visa requirements". *Timatic*. International Air Transport Association (IATA) through Olympic Air.

[87] Multiple travel firms refuse ID cards as passport alternative. The Register. 23 December 2009

[88] "Media Centre —Polls". ICM Research. 4 April 2010. Archived from the original on 11 February 2007. Retrieved 8 May 2010.

[89] Survey Report. Web.archive.org (27 February 2008). Retrieved on 13 June 2011.

[90] "Politics | Ex-MI5 chief sparks ID card row". BBC News. 17 November 2005. Retrieved 8 May 2010.

[91] Jurisprudence Review Launch 2007 Archived 25 July 2008 at the Wayback Machine., ICLR. Retrieved 15 October 2010

[92] "UK | UK Politics | ID cards are of 'limited value'". BBC News. 29 January 2006. Retrieved 8 May 2010.

[93] "ID Cards —UK's high tech scheme is high risk". LSE. Archived from the original on 29 January 2009. Retrieved 8 May 2010.

[94] "(BBC)". BBC News. 4 August 2005. Retrieved 8 May 2010.

[95] Clarke's ID card cost laundry starts to break surface. The Register. 5 July 2005

[96] "Politics | ID card scheme cost put at £5.4bn". BBC News. 9 October 2006. Retrieved 8 May 2010.

[97] "ID card cost rises above £5bn". BBC News. 10 May 2007. Retrieved 8 May 2010.

[98] "ID card scheme 'to cost £5.6bn'". BBC News. 8 November 2007. Retrieved 8 May 2010.

[99] "Giant ID computer plan scrapped". *BBC News*. British Broadcasting Corporation. 19 December 2006. Retrieved 22 December 2007.

[100] Matthew Tempest and agencies (25 May 2005). "ID card cost soars as new bill published | Politics | guardian.co.uk". London: Guardian. Retrieved 8 May 2010.

[101] "UK | UK Politics | Q&A: Identity cards". BBC News. 2 July 2009. Retrieved 8 May 2010.

[102] "UK | Politics | ID cards 'cannot stop terrorism'". BBC News. 24 April 2004. Retrieved 8 May 2010.

[103] Wakefield, Jane (28 April 2009). "Technology | Blunkett seeks 'end to ID cards'". BBC News. Retrieved 8 May 2010.

[104] "Scrap ID cards plan, says David Blunkett | Politics | guardian.co.uk". London: Guardian. 28 April 2009. Retrieved 8 May 2010.

[105] "Infosecurity Europe: Opening Address presented by the RT Hon David Blunkett MP, Member of Parliament for Sheffield, Brightside". Yada-yada.co.uk. Archived from the original on 27 November 2009. Retrieved 8 May 2010.

[106] "Politics | ID cards 'wouldn't stop attacks'". BBC News. 8 July 2005. Retrieved 8 May 2010.

[107] http://www.identitycards.gov.uk/downloads/Identity_cards_Bil_Race_Equality.pdf Archived 28 September 2007 at the Wayback Machine.

[108] http://83.137.212.42/sitearchive/cre/downloads/id_cards.doc Archived 11 September 2008 at the Wayback Machine.

[109] "Politics | Race watchdog warns on ID cards". BBC News. 15 June 2004. Retrieved 8 May 2010.

[110] Identity Cards Bill, House of Lords report stage

[111] "(.doc file)". Informationcommissioner.gov.uk. 30 April 2010. Archived from the original on 25 March 2006. Retrieved 8 May 2010.

[112] Rachel Sylvester. "Beware rise of Big Brother state, warns data watchdog —Britain". London: Times Online. Retrieved 8 May 2010.

[113] "IDABC —UK: UK's ID Cards Bill wins parliamentary vote despite Human". Ec.europa.eu. Retrieved 8 May 2010.

[114] Gaby Hinsliff, political editor (6 August 2006). "Brown to let shops share ID card data | Politics | The Observer". London: Observer.guardian.co.uk. Retrieved 8 May 2010.

[115] Elliott, Francis (6 August 2006). "ID plans: powers set to widen —UK Politics, UK —Independent.co.uk". London: News.independent.co.uk. Archived from the original on 7 January 2009. Retrieved 8 May 2010.

[116] Wakefield, Jane (20 May 2004). "Technology | Opponents take on ID card plans". BBC News. Retrieved 8 May 2010.

[117] Jones, George (20 February 2007). "ID cards 'will allow crime fingerprint checks'". London: Telegraph. Retrieved 8 May 2010.

[118] Press Association (20 February 2007). "Blair under fire over police access to ID card database | Special Reports | Guardian Unlimited Politics". London: Politics.guardian.co.uk. Retrieved 8 May 2010.

[119] http://www.identitycards.gov.uk/library/procurement_strategy_market_soundings.pdf Archived 12 December 2006 at the Wayback Machine.

[120] Identity and Passport Service | Home Office. Ips.gov.uk. Retrieved on 13 June 2011.

[121] Hillier, Meg (7 January 2009). "ID cards are attractive". *The Spectator*.

[122] "Politics | Blair defends identity card plan". BBC News. 25 May 2005. Retrieved 8 May 2010.

[123] "Bloomberg.com: U.K". Quote.bloomberg.com. 26 April 2004. Archived from the original on 30 September 2007. Retrieved 8 May 2010.

[124] Moss, David (14 August 2009). "The biometric delusion". Theregister.co.uk. Retrieved 8 May 2010.

[125] "renew your passport". renew for freedom. Retrieved 8 May 2010.

[126] "I beg to move, That this House does...: 21 Mar 2006: House of Commons debates". TheyWorkForYou.com. 20 December 2004. Retrieved 8 May 2010.

[127] "Politics | Home Office questions ID protest". BBC News. 25 May 2006. Retrieved 8 May 2010.

[128] Calling in NO2ID's pledge "refuse" Archived 18 November 2007 at the Wayback Machine., 14 November 2007

[129] No2ID calls in pledge cash to 'probe' ID Act's enabling laws, The Register, 15 November 2007

[130] "Peer 'ready to defy ID card law'". *BBC News Online*. 10 November 2007.

[131] Holyrood rejects identity cards, BBC News, 19 November 2008

[132] ceu@scotland.gsi.gov.uk, Scottish Government, St. Andrew's House, Regent Road, Edinburgh EH1 3DG Tel:0131 556 8400 (16 June 2005). "Scottish government's position on ID cards".

[133] "Microsoft Word —More than just a card FINAL.doc" (PDF). Archived from the original (PDF) on 20 February 2010. Retrieved 8 May 2010.

[134] "UK | Northern Ireland | ID card 'recognises Irish rights'". BBC News. 30 July 2009. Retrieved 8 May 2010.

25.11 External links

- Map of centres September 2006
- Online Debate – ID cards and the National Identity Register should be scrapped August 2007
- UK Borders Agency website information on ID cards for foreign nationals
- Types of identity card IPS page describing what the different cards look like

25.11.1 News stories

- 22 November 2007, *BBC*, Is Brown re-thinking ID cards?
- 22 November 2007, *BBC*, Cameron calls for ID cards halt
- 14 September 2006, *epolitix.com*, Minister defends data-sharing scheme
- 7 August 2006, *Guardian*, Hackers crack new biometric passports
- 14 June 2006, *epolitix.com*, Amnesty for illegal immigrants once identity cards in place?
- 24 May 2006, *BBC*, Lib Dems back the "Renew for Freedom" campaign
- 4 April 2006, *The Register*, Passport rule change anticipates ID refusenik sabotage efforts
- 30 March 2006, *BBC*, Identity cards will be made compulsory if Labour wins the next election
- 13 February 2006, *The Daily Mirror*, Motorists could soon be forced to carry an ID card
- 30 January 2006, *BBC*, Transsexuals 'to get 2 ID cards'
- 17 January 2006, *The Times*, Peers deal Blair blow over '£19bn cost of ID cards'
- 8 December 2005, *Guardian*, A pan-European ID card will make a bad idea even worse
- 17 November 2005, *BBC*, Ex-MI5 chief sparks ID card row
- 18 October 2005, *The Scotsman*, Microsoft security officer: ID cards will lead to 'massive fraud'
- 12 October 2005, *The Register*. support for ID cards collapses
- 4 August 2005, *BBC*, Labour admits ID card 'oversell'
- 8 July 2005, *BBC*, Chales Clarke: ID cards wouldn't stop bomb attacks
- 5 July 2005, *The Register*, "Clarke's ID card cost laundry starts to break surface"
- 3 July 2005, *The Observer*, Rebels ready to face prison over ID cards: Refuseniks will copy Australian tactics to foil scheme
- 2 July 2005, Muriel Gray in *The Guardian*, Nobody has nothing to hide: Identity cards will deprive the innocent of one of their most basic rights
- "ID cards 'will reveal detail of daily life'" – Information commissioner warns of surveillance society
- December 2004 *Guardian* Legal advice on ID cards kept secret
- December 2004 *Guardian* If you value your freedom, reject this sinister ID card
- April 2004 *Times* It is right to experiment with identity cards
- April 2004 *Bloomberg* Identity Card Will Make Fraud Easier
- July 2003 *Statewatch* Identity cards in the UK – a lesson from history
- July 2002 *Daily Telegraph* £70 ID card to combine passport and car licence
- September 2001 *Guardian* ID cards might solve asylum crisis

25.11.2 Guides

- March 2005 *London School of Economics* An Assessment of the UK Identity Cards Bill and its Implications
- February 2005 *Bow Group report* The Case Against ID Cards, by Rt Hon Peter Lilley MP
- December 2004 *BBC* Identity card Q&A
- June 2004 *Electricinca* An analysis of the British national identity card
- May 2004 *The Register* Everything you never wanted to know about the UK ID card
- November 2003 *Guardian* Q&A
- September 2001 *Telegraph* The case for and against identity cards
- *Trevor Mendham* UK ID Cards – the case against
- *A map of the debate on UK ID cards and the related ID databases"*, Debategraph

25.11.3 Opposition groups

- No2ID: UK campaign against ID Cards and mass surveillance
- Defy-ID
- Liberty – human rights concerns
- Reform
- idFolly
- Haringey Against Identity Cards

Chapter 26

International Identity Federation

The **International Identity Federation**[*][1] was formed in Scotland during 2006. IDFed provides a web-based solution to the problem of identification of injured or distressed travellers. Members are issued with a unique number embossed onto an identity tag or bracelet. They use a forum to log their movements and this relates back to a secure data base controlled by a 24-hour monitoring service. In the event of a member being found helpless, the 18-digit unique number (created using an algorithm) is used to inform next of kin or employers and aid in the treatment and/or repatriation of the member.[*][2] For those undertaking particularly hazardous or dangerous journeys, DNA profiling is also offered through a UK Department of Justice approved specialist DNA laboratory based at the Southern General Hospital in Glasgow.

26.1 Sources

- The Press and Journal (http://www.pressandjournal.co.uk). Wednesday April 2, 2008. "Identity Tags Provide Reassurance."

- *Daily Record* (http://www.dailyrecord.co.uk) Wednesday April 2, 2008. "Tagging Along."

- Courier and Advertiser (http://www.thecourier.co.uk Wednesday April 2, 2008. "New Service for Travellers."

- Scottish Television (stv.tv) Evening News, April 1, 2008.

26.2 References

[1] http://www.idfed.org/

[2] Courier and Advertiser (thecourier.co.uk) Wednesday April 2nd 2008. Page 4; "New Service for Travellers." written by Liz Fowler

26.3 Text and image sources, contributors, and licenses

26.3.1 Text

- **Biometrics** *Source:* https://en.wikipedia.org/wiki/Biometrics?oldid=802257948 *Contributors:* Derek Ross, Malcolm Farmer, Andre Engels, Youssefsan, Ant, Heron, Michael Hardy, Pnm, Kku, Tregoweth, Ronz, J-Wiki, Snoyes, Den fjättrade ankan~enwiki, Kingturtle, Nerd~enwiki, Julesd, Cimon Avaro, Cherkash, Feedmecereal, M0mms, Dcoetzee, RickK, Daniel Quinlan, Dbabbitt, Raul654, Pakaran, Bwefler, Robbot, Chealer, Fredrik, Kizor, Chris 73, Pingveno, Auric, Hadal, Giftlite, DocWatson42, Dinomite, Oberiko, Mintleaf~enwiki, Jonathan O'Donnell, Dmmaus, Khalid hassani, Neilc, Utcursch, Antandrus, Stonor, Starx, Micahbales, Coderanger, Anirvan, Neale Monks, Ukexpat, Mike Rosoft, Mormegil, Econrad, Rich Farmbrough, ArnoldReinhold, Fleung, Notinasnaid, Xezbeth, Bender235, Mateo SA, Loren36, Shanes, Cohesion, ToastieIL, Zr40, Sanmartin, Alansohn, Neonumbers, Adequate~enwiki, Gregmcpherson, Sjschen, InShaneee, Velella, Tocksin, Ceyockey, Inarius, Demiurg, Stemonitis, Simetrical, Mindmatrix, Lgallindo, SDC, Elenthalion, Waldir, Bruno Unna, Rjwilmsi, Kinu, Vegaswikian, Haya shiloh, XLerate, Kjhebert, Vickatespence, Peripatetic, Ttwaring, Alejo2083, FlaBot, SchuminWeb, Old Moonraker, Michael.R.Crusoe, Slant, TheDJ, Jrtayloriv, Intgr, Shaile, Peterl, Whosasking, UkPaolo, Albanaco, Wavelength, SpikeJones, RussBot, Hede2000, Mark Ironie, Bergsten, Stephenb, Gaius Cornelius, Shaddack, Chaos, Standarshy, Anomalocaris, NawlinWiki, Szenti, Welsh, Matticus78, Larry laptop, Gujamin, DeadEyeArrow, Everyguy, Phaedrus86, AramPerez, FF2010, Bobstopper, Open2universe, Nikkimaria, Theda, Closedmouth, Toddgee, Nothlit, JLaTondre, GoodSirJava, M.A.Dabbah, JDspeeder1, Pschearer, Xiaojeng~enwiki, DVD R W, Luk, SmackBot, Elonka, IddoGenuth, EJ-Vargas, Hftf, Melchoir, Unyoyega, VarunRajendran, KocjoBot~enwiki, Ultramandk, PJM, Canthusus, ApersOn, Evanoff, Valley2city, Bluebot, Lord of the Left Hand, Iain.dalton, EncMstr, Deli nk, WETaylor, Octahedron80, Reaper X, Chendy, Mike hayes, Kotra, Jacob Poon, OrphanBot, KevM, LouScheffer, UU, Digitize, Krich, Teehee123, MichaelBillington, Suvablee0506, John wesley, G716, Takowl, Ronaz, Ohconfucius, JGF Wilks, JaeRae, SashatoBot, Calvados~enwiki, Khazar, Tazmaniacs, Natarajuab, Epingchris, Shadowlynk, Minna Sora no Shita, Ben Moore, QDE-can, Martinp23, Dicklyon, Ryulong, Locutus, Nabeth, Kvng, Dl2000, BananaFiend, Shoeofdeath, IvanLanin, Igoldste, Hikui87~enwiki, Lucy-marie, Chickenmonkey, Nhgaudreau, KimChee, Dan1679, Wikifarzin, Little Grabbi, Tedmarynicz, CmdrObot, Zshortman, ShelfSkewed, Neelix, Bmk, Equendil, Phatom87, Cahk, Miguel.lima, Cyhawk, Jose.canedo, Kevin23, Dancter, Tawkerbot4, David Moss, Hontogaichiban, Synetech, Ixchup, SummonerMarc, Gvrdoc, Thijs!bot, ConceptExp, Melissa123~enwiki, Mojo Hand, Lotte Monz, Marek69, Risingjsun, Reswobslc, SusanLesch, Dawnseeker2000, Escarbot, BuffaloChip97, AntiVandalBot, Gioto, MelissaRichard, Djechelon, AWoodside-LElliott, Alphachimpbot, Spartaz, AndreasWittenstein, JAnDbot, Husond, TomChatt, BenjaminGittins, MER-C, Instinct, Merkwuerdiglichlieber, Steveprutz, Z22, Magioladitis, Bongwarrior, VoABot II, T@nn, JamesBWatson, Eugene Cuprin, Crabbi, Recurring dreams, JaGa, Estesce, CliffC, The Almighty Cornholio, Jim.henderson, Rettetast, DrRisk13, PrestonH, J.delanoy, Littledaniel 93, Herbythyme, Javawizard, Pgilmon, Ncmvocalist, NewEnglandYankee, DadaNeem, KylieTastic, STBotD, Homer Landskirty, FloydRTurbo, VolkovBot, ABF, Jeff G., Linnartz, Philip Trueman, TXiKiBoT, Oshwah, Kiranwashkindar, Vipinhari, Liko81, Salvar, Wiikipedian, JWinTX, LeaveSleaves, Everything counts, Wasted Sapience, Falcon8765, A Raider Like Indiana, Vid2vid, Monty845, Nagy, Ltka, Andypdavis, Hypersubtext, Double Dickel, SieBot, Jsc83, Caltas, Triwbe, Nitelken, Joe3600, Sicarii3, Vodnokon4e, NiteSensor23, Free Software Knight, Brankow, Nboavida, Manway, Sanya3, Torinfo, Hariva, Abihana, Precious Roy, Loren.wilton, Westmannn, ClueBot, Anjith2006, Kamalswami, Tabsys, The Thing That Should Not Be, EoGuy, Unbuttered Parsnip, Niceguyedc, Tmazhindu, Sravee294, Heshwa, Anon lynx, Repat, Iner22, Leonard^Bloom, Southern Forester, Predator47, Mcado2dm, SomeAssemblyRequired, Iohannes Animosus, Amsaim, Thingg, Pcheesman, MEGAC1pher, Qwfp, SoxBot III, DumZiBoT, XLinkBot, SHAH7424, Funya445, Chazmoney3, Izzy67, TFOWR, WikHead, ZooFari, Toshtrent, Michaeljthieme, Viper10003, Addbot, Some jerk on the Internet, JayMangat, Acolin f, DaughterofSun, Nikomis, Gizziiusa, Download, Subverted, Favonian, West.andrew.g, Tassedethe, 84user, Eragon22125, Mr.Xp, Chidicon, Xario, Tide rolls, Bfigura's puppy, Verbal, OlEnglish, הגמל התימני, Michael.micha.shafir, आशीष भटनागर, Luckas-bot, Yobot, Themfromspace, Aljpeters, THEN WHO WAS PHONE?, Greyrobe, Kamikaze-Bot, AnomieBOT, DemocraticLuntz, Cmcginni, 1exec1, Exsybaser, Jim1138, JackieBot, Piano non troppo, Gompa~enwiki, Flewis, Materialscientist, Wikibob777, Citation bot, Ap aravind, Zaasaa, Dithridge, Addihockey10, Capricorn42, TheCuriousGnome, Jgeorge60, GrouchoBot, Andriusval, Daivesh, SCΛRECROW, Philshinn~enwiki, MerlLinkBot, Shadowjams, Jenolds, Dougofborg, Frozenevolution, FrescoBot, P.amitsingh, Anthillharry, HJ Mitchell, PeterEastern, Eccso, Citation bot 1, ANDROBETA, Pinethicket, Jonesey95, A8UDI, Tri7megi7tu7, Barenholtz, Reconsider the static, Trappist the monk, Amandallene3, Fama Clamosa, Saponniah, Biospecialist, DARTH SIDIOUS 2, Mean as custard, RjwilmsiBot, Ripchip Bot, Noommos, Saftorangen, EmausBot, WikitanvirBot, Fandraltastic, Princesspeople123, Paulguckian, Primefac, GoingBatty, Tommy2010, VillemVillemVillem, Wikipelli, Dcirovic, K6ka, Sciprecision, SentinelAlpha, Abhinavcambridge, Ingjaimehdz, Fæ, Denytrendy, Jahub, Rpalanisenthi, Msraia, AManWithNoPlan, Ocaasi, Photojack50, BrownRock, Milostewart, Donner60, SBaker43, Sven Manguard, DASHBotAV, Bundawda, Spicemix, Jrsage, ClueBot NG, Jack Greenmaven, MelbourneStar, Raghith, Gilderien, Movses-bot, Architypist2, Awareinc, YuMaNuMa, Cntras, J349, Widr, WikiMSL, Sephalon1, Ngilchr, Helpful Pixie Bot, Magtei, Brabantois, BG19bot, Northamerica1000, Mcic2011, MusikAnimal, Konullu, AwamerT, RoshanJadhav, Vatr5, Nawalani, Chmarkine, Meatsgains, Alyxzandrea, Snow Blizzard, Klilidiplomus, Rutviksutrave, Crticialthought74, AjitaR, David.moreno72, Octavius SV, Brie721, Lwam99, Total-MAdMaN, Deberhardt1988, JYBot, Txhoney, Enterprisey, EagerToddler39, Biernat2, CuriousMind01, NNWikipedia, Lugia2453, MFZBCN, FaradayLupin, Mennowitteveen, Hyphod76, Spencer.mccormick, Ams3nf, Crsini, DavidLeighEllis, Lgnlint, Quenhitran, KathleenErickson80004, Mdfahimshaikh, TheArtMaster, Ravishyam Bangalore, Melcous, Monkbot, Scarlettail, AMResearchNext1, Bingomadangle, CherryCoke027, Pblowry, KasparBot, Ibernesi, Blue rose red, BushelCandle, Allwi.in.oz, Wwjd12x, InternetArchiveBot, Minecraftpsyco, GreenC bot, Windowwiper96, Zcarrot, NoToleranceForIntolerance, Frederick Ehiagwina, E-DemSnoopy, Lacypaperclip, KolbertBot and Anonymous: 732

- **Access control** *Source:* https://en.wikipedia.org/wiki/Access_control?oldid=800465911 *Contributors:* Timo Honkasalo, The Anome, Edward, Patrick, BAxelrod, Wikiborg, Vaceituno, Texture, Auric, HaeB, Tea2min, Alan Liefting, Andycjp, Omassey, Bishonen, Bderidder, CHoltje, El C, Sietse Snel, BrokenSegue, SPUI, Espoo, Hu, Wtmitchell, Dave.Dunford, Neetij, H2g2bob, Ringbang, Woohookitty, Mindmatrix, Camw, Daira Hopwood, Jeff3000, PeregrineAY, BD2412, Rjwilmsi, Amire80, Syced, Eubot, Old Moonraker, Nihiltres, Gurch, Intgr, Chobot, DVdm, Bgwhite, Borgx, RussBot, Stephenb, Gaius Cornelius, NawlinWiki, Grafen, Welsh, PrologFan, Auminski, Galar71, Ka-Ping Yee, Back ache, Guinness man, DEng, NeilN, Carlosguitar, Exit2DOS2000, SmackBot, McGeddon, Zzm7000, Xaosflux, Yamaguchi 先生, Delfeye, Silly rabbit, DHN-bot~enwiki, Frap, RoyalBlueStuey, Andreiij, Trbdavies, Luís Felipe Braga, DMacks, Chris0334, ArielGlenn, Kuru, Yan Kuligin, Twredfish, Slakr, Optakeover, Waggers, E-Kartoffel, Thatcher, Kvng, Hu12, Hetar, MikeHobday, Linkspamremover, CmdrObot, Cydebot, Gogo Dodo, Jedonnelley, Soifranc, RickinBaltimore, Dawnseeker2000, QuiteUnusual, MER-C, NE2, Roleplayer, Knokej, Magioladitis, McGov1258, George A. M., Web-Crawling Stickler, Americanhero, Billbl, FisherQueen, Bostonvaulter, CliffC, CommonsDelinker, Fmjohnson, KylieTastic,

26.3. TEXT AND IMAGE SOURCES, CONTRIBUTORS, AND LICENSES

Memoroid~enwiki, Hersfold, Philip Trueman, TXiKiBoT, SecurityEditor, Ziounclesi, Mazarin07, Andy Dingley, Scouttle, EverGreg, Nickbernon, Wikiscottcha, SieBot, Therepguy, Happysailor, Yerpo, Brankow, Lightmouse, Sfan00 IMG, ClueBot, Josang, Binksternet, Apacheguru, DragonBot, LadyAngel89, Eldub1999, The Founders Intent, Dekisugi, Aitias, Talsetrocks, Wsimonsen, XLinkBot, Pichpich, Tonypdmtr, Mitch Ames, MystBot, RyanCross, Addbot, Actatek, Leszek Jańczuk, Fluffernutter, Rickfray, MrOllie, Subverted, Stantry, م‌اين, Gail, Jarble, Ben Ben, Luckas-bot, Yobot, Bunnyhop11, KamikazeBot, Timothyhouse1, AnomieBOT, Advancesafes55, Piano non troppo, Willowrock, Materialscientist, Brightgalrs, StewartNetAddict, Securitywiki, Sionk, Jgeorge60, Andriusval, SassoBot, Kernel.package, Jray123, Ruuddekeijzer, Indyaedave, Nageh, Mark Renier, Sae1962, DrilBot, Winterst, Swamy.narasimha, Testplt75, Serols, Jandalhandler, Lotje, Vrenator, Clarkcj12, Zeeshankhuhro, Lingliu07, Mean as custard, EmausBot, John of Reading, WikitanvirBot, Timtempleton, Immunize, Feptel, Iancbend, Dcirovic, Gagandeeps117, Abhinavcambridge, Josve05a, Krd, Iwatchwebmaster, Staszek Lem, RISCO Group, SBaker43, Ssbabudilip, Dineshkumar Ponnusamy, ChuispastonBot, ClueBot NG, Sesha Sayee K V, KunjanKshetri, Catlemur, Chester Markel, Secguru1, Animusnovo, Mesoderm, Stewartjohnson229900, Widr, Ellerose, BG19bot, Thekillerpenguin, Mathnerd314159, David.moreno72, Cyberbot II, The Illusive Man, Cfeltus, Dexbot, Codename Lisa, Vinjadhav, TwoTwoHello, TheSadnessOfBeing, Epicgenius, David.brossard, Dewoller, Fan Zhang-IHC, Luesand, Mattgavenda, Kahtar, TotaliTech, Mumbui, Weikrx, Mgavenda, Rsschomburg, Fixture, Virgo hariom, Clark Steph, Cmontgomery11, Cybersecurity101, Filedelinkerbot, KaiserKIS, SCHAPPY, Ggotero, Hphaikuku, Thetechgirl, KH-1, Politikuserna, Zubairul, KasparBot, JJMC89, Fabtron, Yasuo Miyakawa, DatGuy, InternetArchiveBot, Cyber Awareness Crew, Serge10duke, Stevedaikaads, Wiki Ankit parikh, Locksmithatlanta, Gluons12, Karolina Kielar, Redwebraising, Pyrrhonist05, MaxsmithPL, Anonume, Lockbox, Jerryyipsk, Avalkov, Nephigrant, Webmastersilverhost, Murkl, DoctorDake, Inner Focus, KolbertBot and Anonymous: 294

- **Fingerprint** *Source:* https://en.wikipedia.org/wiki/Fingerprint?oldid=803185169 *Contributors:* Eloquence, The Anome, Arvindn, Ant, Heron, Jdlh, Patrick, Zocky, Liftarn, Menchi, Ixfd64, Delirium, Ellywa, Ronz, J-Wiki, Bueller 007, Julesd, Tristanb, Rl, Lukobe, Etaoin, Emperorbma, Dcoetzee, Ww, E23~enwiki, Topbanana, HarryHenryGebel, Ank329, Rossumcapek, Paul W, Robbot, Alouden, Romanm, Calmypal, Koba-chan, Pingveno, Academic Challenger, Meelar, Auric, Diderot, Dodger~enwiki, Lupo, Robartin, Matt Gies, DocWatson42, Paul Richter, Dinomite, Seabhcan, Inter, Tom harrison, Fastfission, Monedula, Everyking, Mcapdevila, Zumbo, Finn-Zoltan, Solipsist, Mckaysalisbury, Edcolins, Neilc, Alexf, Geni, CryptoDerk, Antandrus, MisfitToys, MacGyverMagic, Oneiros, Ruzulo, OwenBlacker, DragonflySixtyseven, Secfan, FrozenUmbrella, Necrothesp, MasterSlowPoke, Rgrg, Jh51681, Syvanen, Canterbury Tail, Mike Rosoft, Discospinster, Rich Farmbrough, Rhobite, Rama, Bender235, ESkog, JoeSmack, RJHall, Rick MILLER~enwiki, J-Star, Aude, Tom, Art LaPella, Orlady, Jpgordon, Bobo192, Vervin, Sortior, Smalljim, Reinyday, Adrian~enwiki, Ctrl build, La goutte de pluie, Eritain, Larryv, John Fader, Pearle, Hooperbloob, Methegreat, Jumbuck, Honeycake, Alansohn, Etxrge, Anthony Appleyard, Duffman~enwiki, Javier Jelovcan, Riana, AzaToth, Comrade009, Hohum, Zsero, Beakerboy, Wtmitchell, Evil Monkey, Josh3736, Sciurinæ, Tom.k, Ekedolphin, Alkarex, Richard Arthur Norton (1958-), Jeffrey O. Gustafson, Mel Etitis, OwenX, Mindmatrix, Spettro9, Brentdax, Tabletop, Dmol, GregorB, Isnow, Xiong Chiamiov, Gimboid13, Tokek, Palica, Lusitana, MassGalactusUniversum, Graham87, Ipsissimus~enwiki, BD2412, FreplySpang, Ductapedaredevil, Jshadias, Drbogdan, Rjwilmsi, Nightscream, Virtualphtn, JHMM13, Syced, Rangek, SchuminWeb, Old Moonraker, Margosbot~enwiki, EmielMols, RexNL, Nimur, Quuxplusone, Celendin, Intgr, Alphachimp, Chobot, Bornhj, DVdm, Bgwhite, The Rambling Man, YurikBot, Borgx, Kafziel, Kymacpherson, Icarus3, Me and, GLaDOS, Mythsearcher, Stephenb, Gaius Cornelius, Shaddack, Cryptic, VanGoldfish, NawlinWiki, UCaetano, Grafen, Irishguy, FreelanceWizard, Azazell0, Blufflimbo, RL0919, Vivaldi, Misza13, Mad Max, Bucketsofg, Mysid, Kkmurray, Crisco 1492, Jkelly, Sandstein, Malekhanif, Liucougar, Theda, Closedmouth, TBadger, Le sacre, Mais oui!, Samuel Blanning, NiTenIchiRyu, Flamma, DVD R W, Krótki, Cappelle, Crystallina, Mohsens, Ozzmosis, SmackBot, JMPerez~enwiki, Reedy, InverseHypercube, AnonUser, Well, girl, look at you!, Thunderboltz, Chairman S., Delldot, Eskimbot, Stimpy, WildElf, Jtwang, Kevin Maloney, Swerdnaneb, Edgar181, Apers0n, Septegram, Gilliam, Ohnoitsjamie, Psiphiorg, Master Jay, SlimJim, Persian Poet Gal, Master of Puppets, Deli nk, Baa, NYKevin, Tamfang, Frap, JRPG, Rrburke, Kittybrewster, Addshore, ClairSamoht, COMPFUNK2, Jmlk17, Jumping cheese, Decltype, Sokolesq, BinaryTed, Smerus, HYC, SashatoBot, Lambiam, Serein (renamed because of SUL), Arodb, BrownHairedGirl, Kuru, John, Microchip08, Euchiasmus, Hbachus, Geppyp, Scientizzle, Tazmaniacs, Tom08, Peterlewis, JohnWittle, Rodrigue, Yashkochar, Werdan7, KentWA, Doczilla, Asyndeton, Andreworkney, BranStark, BananaFiend, Iridescent, Phase4, Paul venter, J Di, Courcelles, Thebigone45, Tawkerbot2, Yashgaroth, Daniel5127, Nydas, IronChris, Amniarix, Fvasconcellos, Switchercat, JForget, CmdrObot, Earthlyreason, Cyrus XIII, THINMAN, Runningonbrains, Colombiano21, Anil1956, Casper2k3, Mohiuddinahmed, Pi Guy 31415, Tim1988, MaxEnt, Pewwer42, Nilfanion, The Photographer, Arrataz, Cydebot, Steel, Gogo Dodo, Red Director, CLPEandFFS, David Santos, Jose.canedo, Marssociety, Tawkerbot4, Doug Weller, ErrantX, GangstaEB, Omicronpersei8, Zalgo, Daniel Olsen, Almodabber, BetacommandBot, JamesAM, Epbr123, Wikid77, PabloP, Ucanlookitup, Headbomb, Marek69, James086, Doyley, Cool Blue, Randall00, Escarbot, AntiVandalBot, CobraWiki, Quintote, Dalgspleh, Dylan Lake, Tillman, David Shankbone, AubreyEllenShomo, M-hwang, ClassicSC, JAnDbot, Empath, Andonic, Hut 8.5, TurkishHorde, Underleger, Bongwarrior, VoABot II, Fusionmix, Kuyabribri, Hasek is the best, Think outside the box, Keith H., WhatamIdoing, Loonymonkey, Josephcn, DerHexer, Pax:Vobiscum, DGG, MartinBot, Mandaglione, Tholly, Anaxial, CommonsDelinker, J.delanoy, ISC PB, Mokailleet, Ntalamai, Maurice Carbonaro, OttoMäkelä, Nothingofwater, Paris1127, Rod57, Elkost, Gustronico, Aervanath, Whitebro, Heyitspeter, Nick Sbr, Daniel olcorn, Xavier Giró, Prhartcom, Wa3pxx, Lithis7, Cometstyles, Justarandomgeek, Treisijs, Ken Saladin, Qqzzccdd, CardinalDan, PACKRATDC3, Deor, Nunziapr~enwiki, VolkovBot, CWii, Mrh30, Bry9000, LokiClock, PJHaseldine, Dreadatticus, Rostowicz, Philip Trueman, Martinevans123, Oshwah, Sroc, WilliamSommerwerck, Remindmelater, Tom239, Dchall1, Someguy1221, Olly150, Corvus cornix, Frank.ZKH, LeaveSleaves, Tsob, Raymondwinn, Seb az86556, KC Panchal, Justinfr, Snowbot, Christopher Connor, Jeeny, Madhero88, Cassandra B~enwiki, MP 12, Purplepong, Cmcnicoll, Falcon8765, Unused0030, Brianga, Ltka, Deconstructhis, CMBJ, Keegan brown, Xgllo, SieBot, Jack Merridew, Winchelsea, Cwkmail, RJaguar3, BeckyAnne, Sendura, Mandsford, EditorInTheRye, Brankow, Lisatwo, Tombomp, Hobartimus, Banandar123, Brandonm0720, Janfri, HighInBC, Crazycasta, Tegrenath, Wee Curry Monster, Salaribio, Mx. Granger, ClueBot, The Thing That Should Not Be, Cliff, Jepmatt, Ajdonnison, CyrilThePig4, Maulsull, Niceguyedc, Hegvald, Excirial, Alexbot, Jusdafax, Xobster, Snookumz, Sun Creator, Mattyl34, Arjayay, Jotterbot, Martijn van Mensvoort, Danzach2, JasonAQuest, Rui Gabriel Correia, Chaosdruid, Thingg, Aitias, Versus22, Ubardak, HundredManSlayer, Omer88f, Remote2, Nhekman, Fastily, Spitfire, Jack008~enwiki, Jovianeye, Ost316, Skarebo, SilvonenBot, PL290, Vianello, Aunt Entropy, Ridgididgi, WPjcm, Belfunk, Metodicar, Kajabla, Addbot, Mr0t1633, Kelly, Milominderbinder2, Ronny corral, Barsoomian, Jojhutton, Tcncv, AnnaJGrant, Nohomers48, TutterMouse, 4insix, AlphOmegGuy, Vishnava, Leszek Jańczuk, Diptanshu Das, Glane23, Sixxgrand, Roux, Favonian, LinkFA-Bot, West.andrew.g, 5 albert square, Mobit, Jimlaw512, Tide rolls, Lightbot, OlEnglish, Samuel Pepys, Wireless friend, Thebiggnome, Luckas-bot, Yobot, 2D, Kartano, II MusLiM HyBRiD II, MrDude23, Carleas, THEN WHO WAS PHONE?, Rachmaninoff, AnomieBOT, Kristen Eriksen, Etan J. Tal, AdjustShift, Kingpin13, Ulric1313, Flewis, Goodtimber, ImperatorExercitus, 90 Auto, Citation bot, ArthurBot, LovesMacs, Xqbot, Capricorn42, Jeffrey Mall, Acebulf, Bitf05a015, DustFormsWords, Roy.samarendra, Twirligig, Zumalabe, Leffanie, GainLine, Shadowjams, Jiggajigga52, Fingersx12, FrescoBot, Fortdj33, Paine Ellsworth, Tobby72, Bulbtree, Diddy88888, JMS

Old Al, Ibmua, Qwell the pell, Lee Flailmarch, Sangeethsajilal97, Qwerty671, Alpha plus (a+), Cannolis, HamburgerRadio, Citation bot 1, SuaveArt, Shshme, Pinethicket, I dream of horses, DGZxYo, Jivee Blau, Spidey104, Codwiki, RedBot, Fumitol, SIbuff, FoxBot, Trappist the monk, Fama Clamosa, Lotje, Overagainst, Fjianjiang, Vrenator, Hefiz, Aminor315, TheGrimReaper NS, Wamelculi, Diannaa, EyeKnows, DARTH SIDIOUS 2, Difu Wu, RjwilmsiBot, Strawfair, DASHBot, EmausBot, John of Reading, Jurjenb, Noonaj, RenamedUser01302013, Tisane, Pudding30, Dcirovic, Doug757, Freethron, Richard asr, Chuck Baggett, Ida Shaw, Schemel, Karunyans, Halfvind, Balln1, Reikat705, Muslim Editor, Wayne Slam, L Kensington, Deutschgirl, Philafrenzy, Satheesan.vn, DASHBotAV, Georgy90, Special Cases, ClueBot NG, Horoporo, Robertwest100, Jeffreybarke, Kmchanw, Frietjes, Metrónomo, CopperSquare, Waramat, 6Anesthesia, Helpful Pixie Bot, Bibcode Bot, BG19bot, Aeroplanepics0112, Sanglorian, TheGoodBadWorst, BattyBot, Jimw338, Khazar2, Countered, Soulparadox, Dobie80, Cup o' Java, Dexbot, DriveByWire, Mysterious Whisper, Sidsandyy, Drajay1976, Me, Myself, and I are Here, FamAD123, Iztwoz, Perfect Purple Pyramid, Comp.arch, Marigold100, Sam Sailor, Anrnusna, Melanie gaillet, Drsoumyadeepb, Monkbot, Darius robin, Henning1973, Daß Wölf, Adûnâi, DangerousJXD, Vorpzn, VexorAbVikipædia, Benjaminikuta, HelpUsStopSpam, Nøkkenbuer, KasparBot, Lemondoge, Charlotte135, InternetArchiveBot, Jwred, Bender the Bot, FrancisF23, Jon Kolbert, Magic links bot, Bcus, Raghvendra99674010, KolbertBot and Anonymous: 727

- **Facial recognition system** Source: https://en.wikipedia.org/wiki/Facial_recognition_system?oldid=802621451 Contributors: Michael Hardy, DopefishJustin, Kku, Ahoerstemeier, Ronz, Kingturtle, BAxelrod, Dysprosia, Hyacinth, Lumos3, Pakcw, RedWolf, Altenmann, Yosri, Ancheta Wis, DocWatson42, Seabhcan, Ds13, Foobar, Neilc, Beland, Stonor, OwenBlacker, Trevor MacInnis, Discospinster, Bender235, Zy26, Violetriga, Bobo192, Remuel, Nsaa, Gary, TheParanoidOne, Guy Harris, Diego Moya, Jtalledo, Wtmitchell, Forderud, Velho, TheGoblin, Apokrif, Waldir, Rjwilmsi, Charliex, MZMcBride, OKtosiTe, Old Moonraker, Jrtayloriv, Srleffler, Bgwhite, Wavelength, ChantillyToyota, Pi Delport, Victordk13, Gaius Cornelius, Bovineone, Alynna Kasmira, Ahills60, IceCreamAntisocial, Phaedrus86, Hirak 99, Dfinkel, ColinMcMillen, NickelShoe, SmackBot, Gilliam, Skizzik, Octahedron80, Aaron Solomon Adelman, Frap, KerathFreeman, RoyalBlueStuey, Otis182, Bollinger, Erwin, TastyPoutine, JoeBot, KimChee, Dan1679, Raysonho, BeenAroundAWhile, JohnCD, Tschel, Hukkinen, Penbat, Miguel.lima, CLPEandFFS, Tomseidler, Dancter, Red7jon, Thijs!bot, Rusl, Sylenius, SusanLesch, Yvetuy, Escarbot, Gioto, AnAj, Mlapierre-jheisler, Spartaz, MER-C, The Transhumanist, Pcubbage, Magioladitis, Michi.bo, Ab quake, Jimjamjak, Indon, A3nm, Calltech, War wizard90, Emmaleah, MartinBot, Mandaglione, GeorgHH, Tbone55, Hans Dunkelberg, Maurice Carbonaro, Reedy Bot, Plasticup, KylieTastic, Jamesontai, Idiomabot, VolkovBot, DMB71387, WOSlinker, Philip Trueman, Oshwah, Erik the Red 2, Shaftesbury, Wikidemon, BotKung, Curb Safe Charmer, Altermike, Spinningspark, Kycook, Random1973, Marcochen, Malcolmxl5, ToePeu.bot, Jaxo tm, WRK, NiteSensor23, Fcummins, TubularWorld, HairyWombat, Caspiax, ClueBot, Tabsys, Zybler, Enthusiast01, CaNNoNFoDDa, Gwguffey, 1ForTheMoney, Maraparacc, Wyatt915, Shervinemami, D0762, MrOllie, Newfraferz87, Dwwest, Verbal, Lightbot, Alexkonoplev, Jarble, Legobot, Yobot, Bunnyhop11, Fraggle81, Librsh, Carleas, SeederBOT, Jim1138, Flewis, Materialscientist, Clickey, Anthony Ratkov, LilHelpa, PhDOnPoint, Samuel lo hk, Solphusion~enwiki, Seeder122, Andriusval, Miriska, Doulos Christos, Kavitachate12, Richbham, Oliphant104, Mycotoxin, HamburgerRadio, Pinethicket, PlyrStar93, Full-date unlinking bot, Lotje, Dalba, RjwilmsiBot, John of Reading, Nork007, T3dkjn89q00vl02Cxp1kqs3x7, Rockin291, Klbrain, Wikipelli, Dcirovic, AvicBot, Abhinavcambridge, Ida Shaw, Thargor Orlando, SporkBot, Oktie, SBaker43, Tdietterich, Tot12, Tstefanick, Scottclowe, BhargavSushant, ClueBot NG, Cris.palmero, Masssly, Widr, WikiMSL, Maybenot2day, BG19bot, Northamerica1000, Mifter Public, Mark Arsten, Eidenberger, Dannygg, Jimw338, IjonTichyIjonTichy, TwoTwoHello, SFK2, RatiborNN, 069952497a, Me, Myself, and I are Here, Epicgenius, Magnolia677, Sajjad farokhi, Teaminthemix, Mahende1, CensoredScribe, George8211, FockeWulf FW 190, Golopotw, Acschell, Zergeist2, Fixuture, Wheetabixxxxxxxxx, Kylo Ren, Tony85poon, Moonshoeislies, Darius robin, Pgrobison, Lepro2, Zhangj1079, Hunenmensch, Ira Leviton, -Ergo, Klokkeslett, Carlosrez, Sergiokgutierrez, InternetArchiveBot, Fasadi, Dnava005, GreenC bot, Smoodie1120, Hyksandra, Udelledo, Bender the Bot, Magic links bot, Usperson1, Patbeppo, KolbertBot, ManuBN88 and Anonymous: 303

- **DNA** Source: https://en.wikipedia.org/wiki/DNA?oldid=802476633 Contributors: AxelBoldt, Magnus Manske, Peter Winnberg, Marj Tiefert, Lee Daniel Crocker, Eloquence, Mav, Bryan Derksen, Slrubenstein, RK, Andre Engels, Fnielsen, Ted Longstaffe, LA2, Danny, Aldie, Unukorno, Toby Bartels, PierreAbbat, Ortolan88, Ben-Zin~enwiki, Anthere, Ellmist, Graft, Heron, Hephaestos, Someone else, Stevertigo, Spiff~enwiki, Lir, Erik Zachte, Lexor, Wwwwolf, Ixfd64, Fruge~enwiki, TakuyaMurata, (, Pde, Pcb21, Goatasaur, Egil, 168..., Looxix~enwiki, Ellywa, Mortene, Ahoerstemeier, Fcrick, Mac, Docu, Snoyes, CatherineMunro, JWSchmidt, Kingturtle, Glenn, Cyan, Poor Yorick, Kwekubo, Rotem Dan, Llull, Samuel~enwiki, Mxn, Raven in Orbit, Quickbeam, Hashar, Mulad, Crusadeonilliteracy, Adam Bishop, Timwi, Dcoetzee, Nohat, Wikiborg, Reddi, Lfh, Jwrosenzweig, Gutza, Rednblu, Wik, Zoicon5, Steinsky, Patrick0Moran, Tpbradbury, Tom Allen, Samsara, Thue, Bevo, Shizhao, Dbabbitt, Raul654, Gakrivas, Bcorr, Pakaran, Jerzy, Lumos3, Donarreiskoffer, Robbot, Astronautics~enwiki, Chris 73, Schutz, Vyasa, Peak, Stewartadcock, Sverdrup, Academic Challenger, Timrollpickering, Bkell, Factual, Moink, Hadal, Jstech, Anthony, Neckro, Pifactorial, Diberri, Cyrius, Dmn, Dina, Ancheta Wis, Giftlite, JamesMLane, Graeme Bartlett, DocWatson42, Nunh-huh, Kapow, Netoholic, Fastfission, MSGJ, Obli, Everyking, No Guru, Dratman, Curps, Michael Devore, Bensaccount, Guanaco, Jorge Stolfi, Pascal666, Horatio, Luigi30, Solipsist, Ojl, Avala, SWAdair, Bobblewik, Alan Au, Delta G, Wmahan, Stevietheman, Adenosine, Utcursch, Pgan002, Andycjp, Dullhunk, J. 'mach' wust, CryptoDerk, Gazibara, Kums, Antandrus, Onco p53, MisfitToys, G3pro, PDH, Jossi, Rdsmith4, Mikko Paananen, OwenBlacker, Kevin B12, PFHLai, Magadan, Icairns, Troels Arvin, Figure, Asbestos, Neutrality, Golnazfotohabadi, JohnArmagh, Jh51681, Sonett72, Adashiel, Thorwald, Mike Rosoft, Alkivar, D6, Freakofnurture, Rdb, A-giau, William Pietri, ElTyrant, Rich Farmbrough, Guanabot, Ffirehorse, Cacycle, Qutezuce, Vsmith, EliasAlucard, Mgtoohey, Mjpieters, Zazou, Bender235, ESkog, Kbh3rd, Swid, Loren36, Danny B-), Brian0918, Charm, Ben Webber, Kwamikagami, Mwanner, Phoenix Hacker, Aude, Shanes, Susvolans, RoyBoy, EurekaLott, Andreww, Causa sui, Bobo192, Kghose, Infocidal, R. S. Shaw, Brim, Dungodung, Arcadian, Redquark, Timl, Tomgally, La goutte de pluie, Jojit fb, Malcolm rowe, Vanished user 19794758563875, Kierano, Hagerman, Bijee~enwiki, HasharBot~enwiki, Sam Burne James, Emoticon, Jumbuck, Danski14, Mithent, Gary, Anthony Appleyard, Chino, Borisblue, Atlant, SemperBlotto, Ricky81682, Loris, Benjah-bmm27, Wouterstomp, AzaToth, Yamla, Water Bottle, Echuck215, Seans Potato Business, Kocio, InShaneee, Hu, Malo, VladimirKorablin, Snowolf, PaePae, Wtmitchell, Melaen, Schapel, BaronLarf, ClockworkSoul, Unconventional, KingTT, Knowledge Seeker, Cburnett, Evil Monkey, Cal 1234, Max Naylor, CloudNine, TenOfAllTrades, Sciurinæ, Inge-Lyubov, Lerdsuwa, LFaraone, Gene Nygaard, Alai, Mattbrundage, Netkinetic, Johntex, Kitch, Adrian.benko, RyanGerbil10, Falcorian, Tariqabjotu, Ashujo, DeAceShooter, Stemonitis, Natarajanganesan, Gmaxwell, Zanaq, Kelly Martin, OwenX, Woohookitty, Mindmatrix, Katyare, TigerShark, Yansa, Rocastelo, Carcharoth, Nitecrawler, Lincher, Colorajo, JeremyA, Duncan.france, MONGO, Nirajrm, Astrowob, Eleassar777, Bbatsell, GregorB, SCEhardt, TheAlphaWolf, Junes, Palica, Turnstep, Marudubshinki, Mandarax, Frostyservant, RichardWeiss, Yasha~enwiki, Graham87, Deltabeignet, Magister Mathematicae, 168.., 169, GoldRingChip, Buxtehude, BD2412, Patrick2480, FreplySpang, Lion Wilson, Yurik, RxS, Effeietsanders, Whoutz, Crzrussian, Drbogdan, Rjwilmsi, Biriwilg, Koavf, Virtualphtn, Collins.mc, Vary, JHMM13, Bruce1ee, Jmcc150, Salix alba, JonMoulton, Oblivious, DonSiano, Ligulem, Gjuggler, Sean-

26.3. TEXT AND IMAGE SOURCES, CONTRIBUTORS, AND LICENSES

Mack, Bubba73, Brighterorange, Bhadani, Ucucha, AySz88, Sango123, Yamamoto Ichiro, Kevmitch, Wobble, Ravidreams, FlaBot, Tufflaw, RobertG, Latka, Nihiltres, Crazycomputers, Chanting Fox, Chill Pill Bill, RexNL, Takometer, Karrmann, Jrtayloriv, DevastatorIIC, Wikipedia Administration, ZScout370, Alphachimp, McDogm, Daycd, BMF81, GordonWatts, Smithbrenon, Mstroeck, King of Hearts, Chobot, Nauseam, Antilived, Bornhj, Bjwebb, Mhking, Bgwhite, Digitalme, Whosasking, The Rambling Man, Wavelength, Cathalgarvey, Sceptre, DarkAvenger, WAvegetarian, Jumbo Snails, Spaully, DNA EDIT WAR, Editing DNA, Anarchy on DNA, DNA is shyt, I hate DNA, Zell Miller's DNA, Trent Lott's DNA, RJC, Mark Ironie, Bernie Sanders' DNA, Orrin Hatch's DNA, Bill Nelson's DNA, Tom Harkin's DNA, Chuck Grassley's DNA, Richard Durbin's DNA, Max Baucus' DNA, Splette, SpuriousQ, Stephenb, Wimt, GeeJo, Tavilis, Anomalocaris, Canadaduane, Alcides, Sentausa, Fabhcún, EWS23, Alzhaid~enwiki, Dysmorodrepanis~enwiki, Wiki alf, Chick Bowen, Jaxl, Tmueck, Psora, Taco325i, Yoninah, Djm1279, Ragesoss, Retired username, Sangwine, Robdurbar, Bert Macklin, PhilipO, Misza13, Grafikm fr, Fs, Tony1, Dbfirs, Lockesdonkey, Roy Brumback, DeadEyeArrow, Bota47, Biolinker, Cosmotron, Supspirit, Scope creep, Caerwine, Chris84~enwiki, Somoza, Wknight94, Leptictidium, WAS 4.250, FF2010, Alarob, Calaschysm, 2over0, Theodolite, BenBildstein, Theda, Spondoolicks, Hurricanehink, Jolt76, GraemeL, Acer, JoanneB, Heathhunnicutt, Zahiri, ArielGold, Curpsbot-unicodify, Michigan user, RunOrDie, Stuhacking, Kungfuadam, Banus, Aironswitch, Samuel Blanning, BiH, DVD R W, Jer ome, Luk, RichG, Blastwizard, Itub, Twilight Realm, A bit iffy, SmackBot, Eperotao, Spongebobsqpants, Tarret, Chodges, Slashme, Prodego, KnowledgeOfSelf, TestPilot, Royalguard11, Martin.Budden, Melchoir, Nitramrekcap, Shoy, Joconnol, Jacek Kendysz, Pkirlin, Delldot, Blondtraillite, Zephyris, Yamaguchi 先生, Cool3, Aksi great, Peter Isotalo, Gilliam, Julian Diamond, Chaojoker, Jimwong, ERcheck, Carbon-16, Yaser al-Nabriss, Kazkaskazkasako, Izehar, Unint, Keegan, Persian Poet Gal, RDBrown, MidgleyDJ, Uthbrian, Adamstevenson, Dlohcierekim's sock, DHN-bot~enwiki, Terraguy, Zven, Darth Panda, Firetrap9254, Gracenotes, Lightspeedchick, Mikker, Hgrosser, Zsinj, Can't sleep, clown will eat me, Jorvik, Danielkueh, Snowmanradio, OOODDD, Roadnottaken, TheKMan, Anita1988, GVnayR, Grover cleveland, Aldaron, NoIdeaNick, Spectrogram, Iapetus, Nakon, Savidan, Ne0Freedom, Jiddisch~enwiki, Rezecib, RandomP, Smokefoot, SteveHopson, Drphilharmonic, Portugue6927, Fagstein, DMacks, Kendrick7, Where, Jls043, Pilotguy, Madeleine Price Ball, SashatoBot, Nishkid64, Visium, Rory096, Zahid Abdassabur, Kuru, General Ization, J 1982, Tazmaniacs, Kahlfin~enwiki, Epingchris, Statsone, Shadowlynk, AstroChemist, JoshuaZ, Edwy, Coredesat, Mgiganteus1, MichaelHa, IronGargoyle, Fernando S. Aldado~enwiki, The Man in Question, Loadmaster, Digger3000, Alexandremas~enwiki, Hvn0413, Mantissa128, Echo park00, Munita Prasad, Mr Stephen, Dicklyon, Serephine, 4u1e, SandyGeorgia, Sir.Loin, Midnightblueowl, Ryulong, Elb2000, Carlo.milanesi, Squirepants101, EggBurger, Autonova, Glen Hunt's DNA, Keith-264, PaulGS, SimonD, Vanished user, Paul venter, Ariel Pontes, Younusporteous, Paul Foxworthy, ScottBS, Esurnir, Adambiswanger1, Tawkerbot2, Alegoo92, Ouishoebean, Davidbspalding, Redneckjimmy, Fvasconcellos, Switchercat, JForget, RSido, Friendly Neighbour, Gatortpk, CmdrObot, Mattbr, Wafulz, Zarex, Lavateraguy, SupaStarGirl, Leevanjackson, CWY2190, GHe, STAN SWANSON, Mintman16, Outriggr (2006-2009), Cerberus lord, Logical2u, Nthornberry, Moorice~enwiki, RoddyYoung, GeoMor, Sopoforic, A D 13, WillowW, Bryan, Steel, YanWong~enwiki, Michaelas10, Gogo Dodo, Travelbird, Red Director, Trd300gt, Hughdbrown, Sloth monkey, Studerby, Shekharsuman, B, Tawkerbot4, Carstensen, Doug Weller, DumbBOT, Narayanese, ErrantX, Omicronpersei8, Daniel Olsen, Gimmetrow, Casliber, Sabbre, Konrad Foerstner, Thijs!bot, Bloger, Opabinia regalis, Ante Aikio, Russ47025, Headbomb, West Brom 4ever, NorwegianBlue, Tellyaddict, Peter K., GAThrawn22, EdJohnston, Eddycrai, Nick Number, S77914767, Deipnosophista, Sean William, Dawnseeker2000, Natalie Erin, Escarbot, David D., KrakatoaKatie, Cyclonenim, Ialsoagree, Mattjblythe, AntiVandalBot, Mr Bungle, Ty146Ty146, Majorly, Luna Santin, Jeka911, Why My Fleece?, Opelio, Jayron32, 17Drew, Pro crast in a tor, TimVickers, Isilanes, Darklilac, Davegrupp, Storkk, Sjollema, Lklundin, Figma, Canadian-Bacon, JAnDbot, Deflective, Leuko, MER-C, Plantsurfer, Janejellyroll, DigitalGhost, Db099221, Alpinu, Andonic, Hut 8.5, Tstrobaugh, Clivedelmonte, Yahel Guhan, Efbfweborg, Sangak, Magioladitis, Pigmietheclub, WolfmanSF, Pedro, Bongwarrior, Hb2019, Carlwev, NighthawkJ, Bhar100101, Hullaballoo Wolfowitz, Jlh29, CattleGirl, Docjames, Scincesociety, GODhack~enwiki, Tobogganoggin, ThoHug, Avicennasis, WhatamIdoing, BatteryIncluded, Torchiest, David Eppstein, Emw, Gurko, Doctor Faust, Mr Meow Meow, Mika293, TheRanger, Stolsvik, Squidonius, Patstuart, E. Wayne, NatureA16, PsyMar, Yobol, MartinBot, NochnoiDozor, Sjjupadhyay~enwiki, ShaunL, Jonrunles, Pierceno, Dudewheresmywallet, Motley Crue Rocks, Rettetast, Fishingpal99, Zouavman Le Zouave, R'n'B, Test100000, CommonsDelinker, Nono64, Armored Ear, Impamiizgraa, Hairchrm, Fconaway, Jarhed, WelshMatt, Grandegrandegrande, 3dscience, Cinnamon colbert, DrKay, CFCF, Trusilver, Nate1028, Spathaky, UBeR, Nbauman, Boghog, Shmee47, Maurice Carbonaro, Vandelizer, WarthogDemon, Tdadamemd, Andrew wilson, BikA06, Ellis O'Neill, TheChrisD, Bmtbomb, DJRafe, Tonyrenploki, Nemo bis, Dr d12, MrErku, Lhenslee, Kemyou, Pyrospirit, (jarbarf), WHeimbigner, Aquaplus, NewEnglandYankee, Antony-22, Nwbeeson, SmilesALot, ArazZeynili, Rwessel, Master dingley, Touch Of Light, MKoltnow, Aatomic1, Potatoswatter, Scoterican, Cornacchia123, FOTEMEH, Aj123456, WJBscribe, YOUR DNA, Mleefs7, Dr.Kerr, Cainer91, Treisijs, Vaernnond, Etanol~enwiki, Idioma-bot, Plonk2, Hammersoft, VolkovBot, Preston47, DrMicro, Seldon1, Midoriko, Coolawesome~enwiki, Mstislavl, Daniel987600, Hersfold, Orthologist, Lia Todua, AlnoktaBOT, Fences and windows, The WikiWhippet, Loginbuddy, Philip Trueman, Director, TXiKiBoT, Oshwah, Muro de Aguas, Billmcgn189, CupOBeans, ElinorD, Qxz, Ocolon, Harianto, Sintaku, Michael H 34, Toninu, Chanora, Seb az86556, Ilia Kr., UnitedStatesian, Pristontale111, Chuck02, Wiae, Luuva, Johanvs, Hockey21dude, DJAX, Nighthawk380, Rajwiki123, Gilisa, Hannes Röst, Gwsrinme, Mentalmaniac07, MarvPaule, Flavaflav1005, Yomama9753, Madhero88, Amboo85, Ashnard, JamesMt1984, Usergreatpower, Synthebot, CephasE, Northfox, AlleborgoBot, Sly G, Timewatcher, Pediaknowledge, Isis07, EmxBot, Gustav von Humpelschmumpel, Kbrose, Safwan40, SieBot, Cubskrazy29, ShiftFn, Tiddly Tom, Graham Beards, BotMultichill, Sakkura, Gerakibot, Dawn Bard, Cwkmail, RJaguar3, Triwbe, Brockett, Execvator, Jerryobject, Keilana, Aaaxlp, Crowstar, Prestonmag, Hzh, Artoasis, Lightmouse, Nanettea, Escape Artist Swyer, Hairwheel, BenoniBot~enwiki, Cradlelover123, Diego Grez-Cañete, Vividonset, Juneythomas, Andrij Kursetsky, Taulant23, Bogwhistle, Randomblue, Paulinho28, Florentino floro, Lascorz, Forluvoft, SallyForth123, Stuart7m, Jimriz, ClueBot, GorillaWarfare, Artichoker, Wikievil666, Yikrazuul, Enthusiast01, TobyWilson1992, Niceguyedc, GoEThe, Llongland, Cross Reference, DragonBot, Excirial, Alexbot, Pumpkingrower05, Gulmammad, ThreeDaysGraceFan101, Estirabot, Coinmanj, NuclearWarfare, Jotterbot, Medos2, Highfly3442, Jackrm, OekelWm, Heyheyhack, Br shadow, Ardyn, Muro Bot, Gaara san, Aitias, Jonverve, Versus22, Josq, Johnuniq, Wnt, Wkboonec, Psymier, Rishi.bedi, DumZiBoT, InternetMeme, Jetsetpainter, Mandyj61596, Simultaneous, SilvonenBot, Priscilla 95925, Lemchesvej, Addbot, Giftiger wunsch, DOI bot, Medessec, TutterMouse, OBloodyHell, Low-frequency internal, Keepweek~enwiki, NjardarBot, Bernstein0275, DFS454, LinkFA-Bot, Kisbesbot, 84user, Numbo3-bot, Tide rolls, Luckas Blade, Zorrobot, WikiDreamer Bot, Jarble, Ettrig, Legobot, Drpickem, Luckas-bot, Yobot, Marcus.aerlous, Aquilla34, Nallimbot, KamikazeBot, LightFlare, AnomieBOT, Tryptofish, 1exec1, Jim1138, Galoubet, JWSurf, Mahmudmasri, Materialscientist, Citation bot, Bci2, ArthurBot, Quebec99, Xqbot, Sciencechick, Chaboura, Timir2, Kholdstare99, Khajidha, JWBE, P99am, Gap9551, GrouchoBot, Jhbdel, Ute in DC, Wizardist, RibotBOT, Wiki emma johnson, 7434be, Briland, GhalyBot, AustralianRupert, Metalindustrien, Hersfold tool account, Jack B108, Legobot III, FrescoBot, Paine Ellsworth, Dogposter, Rohitsuratekar, Jc3s5h, KerryO77, HJ Mitchell, Steve Quinn, Citation bot 1, Sebastian-Hawes, Scarce, Javert, Chenopodiaceous, Redrose64, DrilBot, Pinethicket, I dream of horses, Elockid, HRoestBot, Jonesey95, Tom.Reding, Supreme Deliciousness, RedBot, Wikiain, Curehd, Jujutacular, DixonDBot, Jann, Dinamik-bot, Clarkcj12, Sgt. R.K. Blue, HayleyJohnson21,

Earthandmoon, Jynto, Tbhotch, Jesse V., Minimac, Ugly Ketchup, DARTH SIDIOUS 2, Mean as custard, RjwilmsiBot, Cloakedyoshi, Salvio giuliano, Mandolinface, Toto Azéro, Mashin6, EmausBot, Orphan Wiki, Acather96, WikitanvirBot, Never give in, Jodon1971, Joeywallace9, GoingBatty, Black Yoshi, Winner 42, Dcirovic, K6ka, TeleComNasSprVen, Hhhippo, JSquish, Otterinfo, Empty Buffer, Dffgd, John Mackenzie Burke, SporkBot, AManWithNoPlan, Ocaasi, Thine Antique Pen, Hccc, JZuehlke, Brandmeister, Pkank, L Kensington, Perseus, Son of Zeus, SpikeballUnion, Rr2wiki, Donner60, Dnacond, Nanodance, Scientific29, Puffin, Venkatarun95, Nofatlandshark, LikeLakers2, Davidartois, Woodsrock, Mikhail Ryazanov, Teaktl17, Will Beback Auto, ClueBot NG, Jnorton7558, Ds2207, Gareth Griffith-Jones, MelbourneStar, Prathfig, Kaushlendratripathi, Zynwyx, O.Koslowski, Widr, Argionember, Username 772, Theopolisme, Vogel2014, Helpful Pixie Bot, Iyentra Rasonica, Aaronstonestrom, Calabe1992, Gob Lofa, Bibcode Bot, Lowercase sigmabot, BG19bot, BOK602, CatPath, Piguy101, Midnight Green, Astpurcell, JasonK33, TJRana, Jazzlw, Min.neel, Snow Blizzard, Uluru345, Achowat, Djihinne1, Comfr, AndroidOS, Biosthmors, TuringMachine17, Stigmatella aurantiaca, Jimw338, Cyberbot II, ChrisGualtieri, Saxophilist, Khazar2, IjonTichyIjonTichy, Dexbot, Webclient101, Mogism, Oliverrichardson, Laxative Brownies, THFC1996, Fox2k11, Zziccardi, Hopefuldonor, Mshamza112, Phil2793, Fjozk, RandomLittleHelper, Cor Ferrum, Joeinwiki, ProDawg5, 9FireStar, Diegomanzana, Keiyashi, Dbzhero5000, IncredibleWondersYes1, FamAD123, Youngdro2, AmericanLemming, Jodosma, ManofQueens, EvergreenFir, Evolution and evolvability, Bever, BruceBlaus, Anrnusna, Meteor sandwich yum, Andrewmhhs, Jlmalcos, Abitslow, Csutric, Chaya5260, Zettek95, Mahusha, Giancarlobasile, Monkbot, Virion123, Owais Khursheed, Acagastya, Mr. 1100100, Entitymasterblaster, HMSLavender, Cyntiamaspian, KH-1, Bhootrina, Soldier of the Empire, JordiYiman, DiscantX, MicroPaLeo, Jensberzelius, Forscienceonly, Azealia911, Craftwerker, 3 of Diamonds, Ira Leviton, COOL ANKUR CHOUDHURY, ProprioMe OW, Unmaterial scientist, MANEVIL 187, Rrwanga17, Maddog9962002, Christianmorasco, Underlyingboss3, CAPTAIN RAJU, S281305, Nn9888, Ryanross123, JoshMuirWikipedia, Lemondoge, Swissinator, MoonmanMCD, Majora, I like doors, Matiar Chris Brown, Mirenoula, Sweatypalmsggg, Joshhassweatylankles, Mikemorris123456789, Nafi172, TajBioinfo, MartinZ, Pro-Wiki21, Gantavya garg, PubSci, Lordmitchimus, CutiePie420, FormerBBC, Milkdudstruth, Alexander Rohner, Milkdudstruthback, Milkdudstruthagian, MariusOrion, I Study I Do, Harmonrasmu, Maria Cantana, InternetArchiveBot, Alexhaddad91, Holy Goo, Bender the Bot, Daniel.fedora, Shagil Kannur, Tornado0910, Heididoerr061, DennisPietras, L8 ManeValidus, Axiao12, Joyspark, Daisy.v.leon, UserDude, Magic links bot, BEades, Used name, KolbertBot and Anonymous: 1051

- **Palm print** *Source:* https://en.wikipedia.org/wiki/Palm_print?oldid=787471966 *Contributors:* Fishal, Reinyday, Mandarax, SmackBot, Sadads, Cydebot, Katharineamy, Mikael Häggström, Squids and Chips, EoGuy, Yobot, Paine Ellsworth, SIbuff, Lotje, SheepBot, Diannaa, SentinelAlpha, PrimeBOT and Anonymous: 1

- **Hand geometry** *Source:* https://en.wikipedia.org/wiki/Hand_geometry?oldid=736864860 *Contributors:* Conti, DocWatson42, Fdewaele, Shaddack, Phaedrus86, Boreas231, SmackBot, Tazmaniacs, Miguel.lima, Z22, DerHexer, STBot, SieBot, Viknesb, Jwilliams259, Addbot, AnomieBOT, Andriusval, Erik9bot, Trappist the monk, BG19bot, BattyBot, Dexbot and Anonymous: 7

- **Iris recognition** *Source:* https://en.wikipedia.org/wiki/Iris_recognition?oldid=800541536 *Contributors:* Kku, Tregoweth, Ahoerstemeier, Ronz, Kingturtle, Julesd, Ghewgill, Etaoin, Feedmecereal, Dysprosia, Zoicon5, Raul654, Irismeister, Qertis, Meelar, Seabhcan, HangingCurve, Markus Kuhn, Frencheigh, Chowbok, Freakofnurture, Bender235, Dewet, Art LaPella, Bontenbal, Interiot, Patrick Bernier, H2g2bob, Drbreznjev, JBellis, Tslocum, Btimlake, EbenVisher, Rjwilmsi, Alaney2k, AED, Old Moonraker, Bgwhite, UkPaolo, Wavelength, Fabricationary, Gaius Cornelius, Shaddack, Cpuwhiz11, Jpbowen, Furious Stormrage, Dbfirs, Graham Jones, Kkmurray, Phaedrus86, Pegship, Mendicott, Chrishmt0423, Shyam, Back ache, HereToHelp, SmackBot, Septegram, KYN, Gilliam, Chris the speller, RDBrown, Can't sleep, clown will eat me, Frap, Bc42, RoyalBlueStuey, Arif Zaman, Pen of bushido, Disavian, Roregan, Iridescent, Tawkerbot2, ShelfSkewed, Cydebot, Cahk, Theextract, Reywas92, Miguel.lima, Drogers.uk, Papajohnin, Thijs!bot, DwayneP, Nick Number, Escarbot, MikeLynch, MER-C, Z22, Magioladitis, Laburinthos, Adil zia, Roswell Crash Survivor, Jrmatey, CommonsDelinker, Nono64, Pekaje, EdBever, J.delanoy, Maurice Carbonaro, Jeepday, Joe 4641349, Dbjohnston, Jtyler111, Kiranwashindkar, Technopat, Király-Seth, Soumyadiprakshit, Turgan, Monty845, Adeelzafarbioid, Macchess, Egillsigurdur, Doc James, Michael Frind, Are2dee2, SieBot, 4wajzkd02, RJL Hartmans, Calatayudboy, Sfan00 IMG, Hurryswung, The Thing That Should Not Be, Lee1-LCRC, Mild Bill Hiccup, Iuhkjhk87y678, Designanddraft, Jcarroll26, Philip200291, BOTarate, Berean Hunter, Boitel, Michaeljthieme, Addbot, Escravoes, Willking1979, Download, AleistersCrow, Yobot, Kalirish, AnomieBOT, Jim1138, Soroosh129, Citation bot, Xqbot, Dithridge, Philipkoo2, Andriusval, Ashematian, A.amitkumar, JBBrask, Cannolis, Pinethicket, Canistabbats, Inamorata1, Serols, Trappist the monk, Onel5969, EmausBot, Timtempleton, Super48paul, GoingBatty, Playmobilonhishorse, Pyschobbens, Dcirovic, Smhossei, Shnako, Wanausha Khafaf, ClueBot NG, Jack Greenmaven, Rebwilli, Antiqueight, Helpful Pixie Bot, TechGeek70, BG19bot, JohnnyT22, Boodlemason, Sjeffcarter, Vanished user aij34tinwefj3jshylkqw1, BattyBot, Cyberbot II, Fly fly fly, FifthCrow, IridianTechnologies, FaradayLupin, Philipsaj, Abevac, Kind Tennis Fan, Whizz40, Fixuture, Rroizenblatt, Irisid3, Tony85poon, Monkbot, Lensiris, Nickytesla, Sona1994, KH-1, Elroyton, Jerodlycett, Moorestown2006, Chrisjohnmarshall, PhuongVuthuyanh, GreenC bot, Natureium, Bender the Bot and Anonymous: 156

- **Retinal scan** *Source:* https://en.wikipedia.org/wiki/Retinal_scan?oldid=800688395 *Contributors:* JohnOwens, Ciphergoth, Rl, Feedmecereal, Emperorbma, Giftlite, David Johnson, Discospinster, Rich Farmbrough, M3tainfo, Akuchling, Stefan Ivanovich, Tabletop, Spider1, AED, Bgwhite, UkPaolo, Phantomsteve, CambridgeBayWeather, Shaddack, Robchurch, Phaedrus86, Pegship, ASmartKid, Chrishmt0423, Matt Heard, SmackBot, Samij86, Imaginaryoctopus, The owner of all, L0rents, Coredesat, Jomsome, Tnorth, KimChee, Dgw, Phatom87, Miguel.lima, Red7jon, Lake54, Nicowalker, Escarbot, U-Mos, ClassicSC, ThomasO1989, Kacela, Nono64, Bilalis, Jeepday, Mikael Häggström, SieBot, Fullerenedream, Voldemore, Martarius, ClueBot, The Thing That Should Not Be, HiltonLange, Excirial, Gtstricky, Sophiecentaur, World, Rui Gabriel Correia, Berean Hunter, Addbot, Lightbot, Filip em, Yobot, AnomieBOT, Robertbuzzhill, Edrandall, Srr1053, Mweiss53, DrilBot, Spidey104, Serols, Tim1357, Ydontuleave, Mean as custard, BillyPreset, Cyber3d, Danmuz, Shnako, Shuffulo, ClueBot NG, Argha.hazra, This Is M4dn355 300, TheSnowLeppard, Cyberbot II, Grundy78, FoCuSandLeArN, BeautyOfChaos, Lakersfan0077, Tony85poon, BrayLockBoy, Master Tej, KH-1, Elroyton, Mr nerdy, GreenC bot, Bender the Bot, Camdennator11, Dorintosh, Wmdurand, Trinitysweeps and Anonymous: 86

- **Keystroke dynamics** *Source:* https://en.wikipedia.org/wiki/Keystroke_dynamics?oldid=798527071 *Contributors:* The Anome, Saqib (usurped)~enwiki, Chowbok, IlyaHaykinson, Bkdelong, BD2412, Rjwilmsi, Lockley, Bhadani, Gardar Rurak, Gadget850, Angoodkind, SmackBot, Hftf, FloNight, Gilliam, Deli nk, Toughpigs, WadeSchuette, JonHarder, Mion, Eldraco, Mr Stephen, Owen214, CmdrObot, Cydebot, Makwy2, Doom777, Equinexus, Outlook, Kiore, J.delanoy, Geek2006, Loungeapple, TopGun, Sterlingjones, Sintaku, Slysplace, SieBot, Jchecco, Tatterfly, Deanlaw, Arjayay, Kbacon101, Qwfp, Tprentice, MystBot, Addbot, Tide rolls, Allisonwarmstrong, Yobot, Fraggle81, AnomieBOT, Dithridge, GrouchoBot, Loyolaguy, Biometricguy, Milotoor, EmausBot, ZéroBot, Transformagic, Kmva, Zmax15, CocuBot,

26.3. TEXT AND IMAGE SOURCES, CONTRIBUTORS, AND LICENSES 205

Hermannh, Braincricket, Akilagreen, Cyberbot II, Theowenyin, Anonymous Random Person, AKS.9955, Probay wiki, AdrianPurley, Srednuas Lenoroc, InternetArchiveBot, Maitchison, GreenC bot, KolbertBot and Anonymous: 53

- **Gait analysis** *Source:* https://en.wikipedia.org/wiki/Gait_analysis?oldid=802983142 *Contributors:* Damian Yerrick, Fuzheado, Ed g2s, Donarreiskoffer, Altenmann, Doovinator, Jfdwolff, Chowbok, DragonflySixtyseven, User2004, HCA, Arcadian, Hooperbloob, Woohookitty, Hyung5kim, Graham87, Kbdank71, Koavf, Bgwhite, Wavelength, Shaddack, Skritek, Rbarreira, Sandstein, SmackBot, Varunbhalerao, Nihonjoe, Fngosa, Eurobikermcdog, JimmyWee, Powerload, Nicolharper, Pednm, Sphinxx, Cydebot, Dancter, BetacommandBot, John Barrett, Magioladitis, WhatamIdoing, CommonsDelinker, Bcooper1210, Motioncapture, Sunidesus, EMT1871, Jrcla2, Fathomharvill, CplDwayneHicks, SanitySolipsism, Dnjuls, AlleborgoBot, Glst2, Trackinfo, Jtlongwi, Callidior, ClueBot, Laudak, MasterXC, Undyne, Attaboy, Dthomsen8, Addbot, P0todd0p, Debresser, Lightbot, Will.M.Thompson, Luckas-bot, Yobot, Makesmanydays2, Rudolf.hellmuth, Drmaryannbell, J04n, Ruuddekeijzer, FrescoBot, Shadow scoob, Dger, Alpha plus (a+), Citation bot 1, HRoestBot, RedBot, Fama Clamosa, Conantwo, RjwilmsiBot, Semmendinger, John of Reading, WikitanvirBot, Wikipelli, Redmac54, RockMagnetist, Bmacwilliams, Jmvkrecords, Will Beback Auto, Amirtahataebi, Alba12 6, BarrelProof, WikiMSL, Helpful Pixie Bot, CoenLauwerijssen, BG19bot, Wifilocation, Sdueball, Biosensics, Meatsgains, Dustynrobots, ChrisGualtieri, TylerDurden8823, Jochen Burghardt, Buggiehuggie, BTSWIKI, Pttennessee, Crystallizedcarbon, Dbsdudahr, Yqm, Matthewcvbec, Bender the Bot, Magic links bot, Geomodelrailroader, KolbertBot and Anonymous: 67

- **Speaker recognition** *Source:* https://en.wikipedia.org/wiki/Speaker_recognition?oldid=800661666 *Contributors:* Edward, Mac, Nohat, Dbabbitt, BenFrantzDale, Finn-Zoltan, Mboverload, Quagmire, Beland, Jjk, Kjkolb, Brainy J, Grutness, Apoc2400, Stemonitis, Maartenvdbent, DeweyQ, Rjwilmsi, Alejo2083, Zotel, Nastajus, Simesa, Xaa, NawlinWiki, Dialectric, Yoninah, Tony1, DeadEyeArrow, Chriswaterguy, Allens, SmackBot, Willpower, Radagast83, Bejnar, C0pernicus, Shadowlynk, Beetstra, Bollinger, Dicklyon, Amitch, Iridescent, MightyWarrior, KyraVixen, Xous, Dancter, Alaibot, ChrisTomkinson, SusanLesch, Dawnseeker2000, Ekimd, AntiVandalBot, Mukake, Opelio, I'll bring the food, The Transhumanist, VoABot II, Bubba hotep, Galvanist, Gemuetlich, J.delanoy, Pgilmon, Elpasi, Martinevans123, SieBot, Oxymoron83, Evgeny Kapun, JuddFS, Jbening, Speechgrl, Three-quarter-ten, Hbeigi, Thingg, Versus22, Editorofthewiki, Rror, Addbot, AndrewHZ, Bultro, Josh1414, Jarble, Luckas-bot, Hippocrocopig, Yobot, SZEdit, Rjanag, Jim1138, Voiceverified, GrouchoBot, RibotBOT, Lolcatz90890809809809, FrescoBot, Gp2it, Skyerise, کاشف عقیل, Dewritech, Giuliopaci, Mbenayed, Tot12, Goulouc, Senator2029, ClueBot NG, Wrathkind, MerlIwBot, Helpful Pixie Bot, BG19bot, POOJAIN.888, Richterks, NotWith, Justincheng12345-bot, Cyberbot II, Dexbot, Me, Myself, and I are Here, Marco Ciaramella, Ed Oppty, Foisse, Monkbot, JonjoMaudsley, GeneralizationsAreBad, Blue rose red, Supereater14, GreenC bot, Bender the Bot, KolbertBot and Anonymous: 83

- **Electroencephalography** *Source:* https://en.wikipedia.org/wiki/Electroencephalography?oldid=802778966 *Contributors:* Fnielsen, Codeczero, Jtoomim, Heron, Rsabbatini, Llywrch, Bewildebeast, AFLastra~enwiki, DopefishJustin, Kku, TakuyaMurata, Karada, Ellywa, Ronz, JWSchmidt, BenKovitz, Hgamboa, Omegatron, Topbanana, Shantavira, Kizor, Rolando, Goethean, Pingveno, Ashdurbat, Clngre, Rhombus, Jondel, Geenah, Radomil, Wikibot, HaeB, Giftlite, DocWatson42, Philgp, Jfdwolff, Ding~enwiki, Eequor, Dfrankow, Andycjp, Sonjaaa, Beland, Glogger, Simoneau, Bk0, Sayeth, Geenah71, Indolering, Asbestos, Chris Howard, Pjacobi, Rama, Kndiaye, Gronky, Bender235, Jaberwocky6669, JustinWick, CanisRufus, RoyBoy, Muntfish, 2005, Army1987, Njyoder, Arcadian, Giraffedata, Forteanajones, Jumbuck, Jcsutton, Arthena, Sjschen, Riana, Hu, Caesura, Snowolf, Cburnett, Vuo, Ceyockey, Bobrayner, OwenX, Barrylb, Rvanschaik, Janbrogger, Robert K S, D.Right, MarcoTolo, Miroku Sanna, BD2412, Rjwilmsi, Nightscream, Heah, DoctorDog, Lyo, JuneD, EBlack, Brighterorange, FlaBot, Psydoc~enwiki, Intgr, Shooravi~enwiki, A.Warner, Chobot, Celebere, Bgwhite, Peterl, Vyroglyph, YurikBot, Wavelength, Sceptre, Chris Capoccia, Pi Delport, Monito, CanadianCaesar, Eleassar, Rsrikanth05, NawlinWiki, A314268, MrSandman, Deodar~enwiki, Banes, Daniel Mietchen, Balizarde, Supten, Wknight94, Rwxrwxrwx, D'Agosta, Esprit15d, Colin, Back ache, Dontaskme, AGToth, Axfangli, Eykanal, Macdorman, SmackBot, PSYBIRD1, Gilliam, Mikage31582, Sonicandfffan, Teemu08, Kxra, Tekhnofiend, Zsinj, Can't sleep, clown will eat me, Милан Јелисавчић, Lansey, Jumping cheese, Jdlambert, Hoof Hearted, Kieranfox, Xieliwei, Spiritia, SashatoBot, Zeraeph, General Ization, Simongraham, Gleng, JorisvS, AaronEJ, SandyGeorgia, PeterKonig, Th1alb, DabMachine, Chephyr, Iridescent, Michaelbusch, Hewn, AVJP619, Chirality, Dan1679, Wikifarzin, Mellery, XApple, CmdrObot, Ilphin, CBM, WeggeBot, Jefchip, Cydebot, Kanags, Anthonyhcole, Beefnut, Dancter, Deele, Kozuch, Thijs!bot, Epbr123, Headbomb, JustAGal, CharlotteWebb, Mmortal03, Escarbot, Ismailmohammed, KrakatoaKatie, Simbven, MER-C, SiobhanHansa, VoABot II, Arno Matthias, Tonyfaull, Fabrictramp, Thuglas, Alex Spade, LookingGlass, Middleman 77, Thibbs, Vssun, DerHexer, Hbent, Ashishbhatnagar72, Mmoneypenny, CliffC, BetBot~enwiki, Jim.henderson, Niclisp, CommonsDelinker, Gyro2222, PCock, Trusilver, Adavidb, Peter Chastain, Hans Dunkelberg, Tikiwont, Maurice Carbonaro, Mike.lifeguard, Extransit, Kpmiyapuram, Mrs.meganmmc, LordAnubisBOT, Longouyang, Notreallydavid, Mikael Häggström, Plasticup, LittleHow, Timokeefe, STBotD, DorganBot, Xetrov, Davecrosby uk, Poorman1, Mlewiss000, LLcopp, VolkovBot, Mbmaciver, Jones2, TXiKiBoT, Oshwah, Malinaccier, Kww, LabFox, Rei-bot, Ask123, Lradrama, Imasleepviking, Zeuszeus1122, KUutela, Mirasoledrecovery, Synthebot, Lova Falk, Temporaluser, Doc James, Simbamford, Jasontable, Nabinkm, SieBot, Fchapotot, Meightysix, Keilana, Berserkerus, Shwmtpf, Rena Silverman, OKBot, Correogsk, Savie Kumara, Michael Tangermann, Emptymountains, Tatterfly, ImageRemovalBot, Twinsday, Martarius, Elassint, ClueBot, Marleneklingeman, Fyyer, The Thing That Should Not Be, Journals88, Mild Bill Hiccup, Callumny, Goldkingtut5, Trivialist, Aorwing, Jumbolino, EeepEeep, Ryan.rakib, Keysanger, Lartoven, Sun Creator, M.O.X, Kaiba, ChrisHamburg, Roberrific, Sribulusu, Delldot on a public computer, Theo177, Elenaschifirnet, XLinkBot, Staticshakedown, EastTN, Rror, Legija, Facts707, Redhorseby, Addbot, Jackpickard1985, DOI bot, Zefryl, Jncraton, Fieldday-sunday, Ethanpet113, Leszek Jańczuk, Diptanshu Das, Dranorter, Looie496, MrOllie, Redheylin, Debresser, Quercus solaris, TangLab, Urness.sam, Numbo3-bot, Katharine908, Lightbot, Wojder, Filip em, Luckas-bot, Yobot, II MusLiM HyBRiD II, Sineenuchn, AnomieBOT, Tryptofish, Ciphers, Rjanag, Jim1138, Bluerasberry, Materialscientist, RobertEves92, Citation bot, Adnan niazi, Zad68, Capricorn42, Cambyzez nl~enwiki, Loveless, The Evil IP address, GrouchoBot, Taylorchas, RibotBOT, Stratocracy, Btait101, Uhhhhhno, Sirmikey, FrescoBot, Tobby72, Odissea, Dger, Steve Quinn, Endoran, Darrellx, Citation bot 1, Citation bot 4, Scidata, Dimo400, I dream of horses, Micromesistius, ImageTagBot, Veronica Roberts, BrandonSargent, Σ, TjeerdB, Corinne68, TobeBot, Trappist the monk, Datahaki, Namita123, Paiamshadi, Cronides2, Cp72, Solzhenitsyn1, RjwilmsiBot, Afowle, EmausBot, Kfederme, John of Reading, Yanglifu90, GoingBatty, Zagoury, Klbrain, TuHan-Bot, Dcirovic, Robertmabell, Scrane72, Ylwarrior, Schulze-bonhage, Midas02, Whet Under The Ears, Rcsprinter123, Δ, Srujan1001, Whelanrobwiki, Neuro11, Hazard-Bot, Alonker, GeoffreyMay, Mikhail Ryazanov, ClueBot NG, Horoporo, Colapeninsula, Beyondsquirrelly, Ariangiovanni, Jj1236, Cntras, Save me, Barry!, FiachraByrne, Alphalobe, Widr, Helpful Pixie Bot, XXLOLDAXx, Johnseiferth, Titodutta, Bibcode Bot, LizzardKitty, BG19bot, Virtualerian, MerryMilkMan, Bruce4949, Neøn, F Woodruff, Rstdenis, Exercisephys, KPFrerking, Glacialfox, Fernandopestana, BattyBot, Welpeo, Enirpmet, Darkcharmr, BrightStarSky, Dexbot, FoCuSandLeArN, Hmainsbot1, Mogism, Needcnest, CuriousMind01, TwoTwoHello, 93, Matt-in-a-hat-42, Mark viking, River2012wiki, Scarecrow, Etch6, Mzoltan24, Feydun, XABXCO, Evano1van, Dkz999, BallenaBlanca, Morozless, Dmscopio, Dalli 2013, SJ Defender, Fixuture, Omphalosskeptic, I3roly, Rhythm1140, Monkbot, BirthOfJesus, Lara comb, JaunJimenez, Clathrin, Globalglobes, Themulticaster, Minuit2400, Poiuytrewqvtaatv123321, Dr.Ashlesh.P, Zane Muir, BESA GmbH, Andre at besa, Samf4u, Rubbish computer, WhyPrivate?, Me.agmohit, Dfleur, Quinto Simmaco, KasparBot, VaibhavGandhi1, Ashfaquememon, Milly shepp, Wolk777, Haodong123, Blacknick, InternetArchiveBot, Muhammad Numan MN, RobbieIanMorrison, Yashrajyaguru999, Birjandtalab, Mungaisetta, PrimeBOT, Kleinhern, Damien Cameron, Mohkam78, Hameltion, TesLiszt, KolbertBot and Anonymous: 451

- **Electrocardiography** *Source:* https://en.wikipedia.org/wiki/Electrocardiography?oldid=801925848 *Contributors:* Kpjas, Bryan Derksen, The Anome, Jeronimo, Alex.tan, Andre Engels, Karen Johnson, Yaginuma, Kchishol1970, Ixfd64, Karada, Kosebamse, 168..., Haakon, Ronz, Jkanters, Statkit1, Julesd, Hashar, Lou Sander, Tpbradbury, Omegatron, Topbanana, Jerzy, Jeffq, Scalasaig, Carlossuarez46, Robbot, Gidonb, DHN, Kd4ttc, Wikibot, Seth Ilys, Diberri, GreatWhiteNortherner, Giftlite, Ksheka, Nunh-huh, Sampo, Ich, Michael Devore, Markus Kuhn, Bensaccount, Jfdwolff, ALargeElk, OldakQuill, Knutux, Rubik-wuerfel, Sonett72, Syvanen, Rich Farmbrough, Pjacobi, Roybb95~enwiki, Saintswithin, Bender235, Petersam, Glenlarson, Briséis~enwiki, Diamonddavej, Bobo192, Stesmo, Dean.jenkins, Arcadian, Scapermoya, CKlunck, Nsaa, HasharBot~enwiki, Poli, Espoo, Ranveig, Jumbuck, Zachlipton, Alansohn, Gary, BladeRunner99, Free Bear, Wouterstomp, Theodore Kloba, Wtmitchell, David Henderson, TaintedMustard, Iannigb, Twisp, Cburnett, Mauvila, JonSangster, ⁈, Robert K S, GregorB, Macaddct1984, 74s181, MarcoTolo, Graham87, Nirvelli, BD2412, Miss Pippa, Canderson7, Rjwilmsi, Koavf, Strait, MZMcBride, Brighterorange, Mortice, Alejo2083, WWC, Richdiesal, Kerowyn, Karelj, Stevenfruitsmaak, MithrandirMage, VolatileChemical, Bgwhite, YurikBot, Wavelength, Borgx, Sceptre, RussBot, Hede2000, Damato, Gaius Cornelius, CambridgeBayWeather, Dclapp, Wimt, Brian Crawford, Alison.philp, Ruhrfisch, SM, Htonl, Mysid, Karl Andrews, DRosenbach, Ozaru, CubicStar, Lt-wiki-bot, CWenger, Shawnc, Owain.davies, BarryH, Gwilz, Junglecat, Snaxe920, A bit iffy, SmackBot, Prodego, Unyoyega, Vanished user mdflkmweir234k56us3, Edgar181, Shai-kun, Gilliam, Skizzik, MPD01605, ERcheck, Prakashvankina, Armeria, Jmr30, Agateller, M0rt, Lennert B, J. Spencer, DHN-bot~enwiki, Hongooi, Oatmeal batman, Simpsons contributor, Can't sleep, clown will eat me, Chwats, Chlewbot, OrphanBot, J-Kama-Ka-C, OnixWP, Anazem, Dream out loud, Acdx, SanderB, Visium, Sbmehta, Hbachus, Vgy7ujm, Dddeoliveira, Mat8989, Mgiganteus1, IronGargoyle, Bilby, Rainwarrior, Beetstra, Vwozone, Kyoko, Johnvanzyl, MrDolomite, Hu12, Nehrams2020, Andthu, Madskile, Chirality, Billy Hathorn, Ghaly, Fvasconcellos, Kingishere, JForget, Robotsintrouble, 00110001, Mcstrother, 5-HT8, Cardsteam, Nathan Cole, Mpotse, John Yesberg, RelentlessRecusant, Gogo Dodo, Jkokavec, BlueAg09, Drur93, Lugnuts, Thom oost, Dancter, Ernstl, Vogey2002, FrancoGG, Thijs!bot, Yhevhe, Epbr123, Npatchett, Hazmat2, Who123, Headbomb, Marek69, Frank, Debbe, Uruiamme, ThomasPusch, Srvora, Stannered, Mentifisto, AntiVandalBot, Mr Bungle, Luna Santin, Nephlet, Grade4, Arx Fortis, Ironiridis, DrMacrophage, Leuko, MER-C, Patxi lurra, Robina Fox, Ph.eyes, Epinheiro, Tomskm, Z22, Burhan Ahmed, Lenny Kaufman, Thomas.Hedden, VoABot II, Appraiser, Lopkiol, Swpb, Peiter, SineWave, CharlieCLC, Jakeallenseo, Lošmi, WLU, Odje, Koska98, Scottalter, Yobol, Grandia01, CliffC, BetBot~enwiki, Dietzel65, Kitb, Mikr18, Fdixon, TheEgyptian, R'n'B, Lilac Soul, J.delanoy, CFCF, Nbauman, Cyanolinguophile, Boghog, Discott, Jerry, GoingThroughTheMotions, LordAnubisBOT, Avaron676, Mikael Häggström, Ephebi, DJ1AM, 97198, Danjeffers, Remember the dot, Wikquid, King Toadsworth, MoodyGroove, Halmstad, Danielhanlon, Idioma-bot, AndrewTJ31, Atom cz, Malik Shabazz, VolkovBot, Philip Trueman, TXiKiBoT, Oshwah, Tsuunen, Kychot, Vipinhari, Flashpoint145, Rei-bot, Sh111496, MuanN, Imasleepviking, Sirkad, LeaveSleaves, Tmarkopolo, KC Panchal, Brainiak4431, Lenborje, Madhero88, Tri400, Ashnard, Jon.j.henry, Jackryan, Cmcnicoll, Falcon8765, Doc James, AlleborgoBot, Symane, Guystout, Jpark4, SieBot, Doomgrr, Tresiden, Work permit, Intercontinental, Gerakibot, Tootenplop21, Yintan, Kenkku, LeadSongDog, Toddst1, Flyer22 Reborn, Oysterguitarist, Paolo.dL, Yerpo, Nk.sheridan, Lisatwo, Fleester, Mr Apple89, HendrixEesti, OKBot, Svick, Mike2vil, Silversin, Bogwhistle, Dabomb87, Guoshun2172, ImageRemovalBot, Mx. Granger, Celique, ClueBot, Sadm88, Mild Bill Hiccup, Shaun1045, Lbeben, Harland1, Trivialist, Youdiil, Walking Softly, Monobi, NuclearWarfare, Jotterbot, Razorflame, Den Hieperboree, Nawagaththegama, Kcallen78, Berean Hunter, A059970, XLinkBot, Staticshakedown, Facts707, WikHead, MystBot, Dnvrfantj, Toozdaygirl, Addbot, DOI bot, Fieldday-sunday, MagnusA.Bot, Diptanshu Das, MrOllie, Dapeda, Curap, LinkFA-Bot, Quercus solaris, Drjnk, Tahmmo, Tassedethe, Alphacolony, Medicellis, Cimiteducation, Tide rolls, Zorrobot, Ske, Legobot, Chaldor, Luckas-bot, Yobot, Mauler90, Donfbreed, Mmxx, Cardcop05, AnomieBOT, Piano non troppo, Bluerasberry, Materialscientist, Citation bot, Maxis ftw, Cjmike, Jmarchn, Obersachsebot, Zad68, JimVC3, Capricorn42, Biophysicsool, Hyjl, DSisyphBot, HNE3, EdithStarling, Abce2, RibotBOT, ViolaPlayer, Trafford09, GhalyBot, Amitrajpalanand, Captain-n00dle, FrescoBot, Vijaypinu9, Dger, Electrophys, Z0OMD, OgreBot, GunnarK, Pinethicket, Marinov84, RedBot, Serols, JamesGrimshaw, Jauhienij, Corinne68, عقیل کاشف, Arfgab, Ndkartik, Megumegun, Inferior Olive, PRINCE 1983, Reaper Eternal, Vanished user aoiowaiuyr894isdik43, Angelito7, Gene Omission, Solzhenitsyn1, Ripchip Bot, Krishnendu4776, Xdaedalus, Liquefier, EmausBot, WikitanvirBot, Reinhold Schäfer, GoingBatty, Dcirovic, Jasonanaggie, Lunagron, JP75, Ronk01, Jer5150, Bamyers99, H3llBot, Unreal7, Erianna, Card Zero, Jay-Sebastos, TyA, Rhynd, Brandmeister, MonoAV, Donner60, Noblesunny, DSITelemetry, Hazard-Bot, Williambq, Matthewrbowker, Dotmed, Liamc91, Mni9791, Pramicy, Jeffo1025, ClueBot NG, Gareth Griffith-Jones, Gosbear, Morgankevinj huggle, Stan3000, Yg12, Khalid Yousuf, ارائی1991, Saistmp, Widr, Abio87, Pluma, HS Offenburg MT, Helpful Pixie Bot, Asimlimbu, Overdamped, Curb Chain, Wbm1058, Tailor-tinker, BG19bot, Roberticus, JaimeTorchiana, Scorpian ad, Crocodilesareforwimps, Neøn, Raykwaku, NotWith, 220 of Borg, DanielChangMD, Scanbre, Rob Hurt, BattyBot, Yashovardhan Dhanania, Biosthmors, Millennium bug, Danluke007, Cyberbot II, Smagers, YFdyh-bot, Khazar2, Dexbot, Bop90, Kenneth.jh.han, Jakob-Steenberg, Frosty, Kkkchandu, Jamie bisson, Epicgenius, Scareccrow, Liuxv1986, BASA Спасимир, Muzzlenose, Davidirvan, Iztwoz, AmericanLemming, Tentinator, Peterjin3076, Ricktcb, HaveNoIdea, Sabrina Barton, Ashleyleia, Hamganu, Ugog Nizdast, Rbarraud, BruceBlaus, Schmidt23, Cbecc, Masterkielbaster, Anrnusna, Meteor sandwich yum, Robevans123, M dhanushkodi, CMcMillan13, Cooker2007, Monkbot, NewEnglandDr, The Last Arietta, AashishAashish, Shubhu dhanke, Poiuytrewqvtaatv123321, Goggledude, Dr.tanjil, Doctor cardiologist, Hayman30, Samf4u, Steffylorzy, Jerodlycett, KasparBot, Sutatstupni, Yahyansari, Gibbh027, Unenthusiastic, ECG niche, Pseudopig, CLCStudent, Composcompos12, Mar11, Wasiq 9320, InternetArchiveBot, Dramitvmmcsjh, Halynoor, GreenC bot, Mariodano, Andrea SI-91, PrimeBOT, Justeditingtoday, Zingvin, Chinchorro y atarraya, Stevenscholfield, Magic links bot, JCW-CleanerBot and Anonymous: 645

- **BioAPI** *Source:* https://en.wikipedia.org/wiki/BioAPI?oldid=791497906 *Contributors:* Edward, Pnm, Skysmith, Fugue~enwiki, BD2412, Rjwilmsi, NeilN, SmackBot, O keyes, Butko, Alaibot, Rei-bot, John Larmouth, Wetwarexpert, JL-Bot, Addbot, AnomieBOT, FrescoBot, Milostewart, Cyberbot II, Lugia2453, LauraALo, Shavaiz Shams, InternetArchiveBot, GreenC bot and Anonymous: 5

- **Biometric passport** *Source:* https://en.wikipedia.org/wiki/Biometric_passport?oldid=801222576 *Contributors:* Edward, Nealmcb, Pnm, Kku, Gabbe, Kingturtle, Darkwind, Pratyeka, Kaihsu, Silpol, LMB, Jusjih, Finlay McWalter, Robbot, Sander123, ChrisO~enwiki, Fredrik, Donreed, Pingveno, SchmuckyTheCat, Roscoe x, Xanzzibar, BenFrantzDale, WiseWoman, Everyking, Markus Kuhn, Joe Sewell, Elgaard, Steggall, Avala, Infinitysnake, Haggis, PFHLai, Klemen Kocjancic, Abdull, Edolen1, FT2, Roo72, Bender235, Swid, Zscout370, Walkiped, Benbread, Jeltz, Hipocrite, Garethhamilton, XLR8TION, Alinor, Danaman5, Max Naylor, Uceboyx, DrDaveHPP, Kelly Martin, UFu, Woohookitty, LoopZilla, Admrboltz, Dmol, Dzordzm, Maartenvdbent, Carlsmith, MD~enwiki, BD2412, Micga~enwiki, Josh Parris, Akashiii, Vegaswikian, JP Godfrey, Desmond Tsui, Caligvla, Wikiliki, Sky Harbor, McAusten, ViXx, IgorMagic, Gurch, Intgr, Idaltu, Bgwhite, Wavelength, Apancu, John Quincy Adding Machine, Gaius Cornelius, Shaddack, Rsrikanth05, Akhristov, Algaema, Nirvana2013, Martijno, Liastnir, Hanheng, Jpbowen, Laser2k, Clumsyone, Phaedrus86, SuperFlanker, Yeryry, Mendicott, Hanseichbaum, Excavator, Toddgee, Canley, GraemeL, Mike1024, NeilN, SmackBot, F, Reedy, McGeddon, Stifle, Apers0n, Xaosflux, Nixwrites, GoneAwayNowAndRetired, Chris the speller, Rakela, Nick4gwen, Lepetitvagabond, EquusAustralus, BBCWatcher, Vinay427, OrphanBot, Bc42, Hellimli, Muzi, Lra, Dantadd, Woodysee, Towsonu2003~enwiki, Ohconfucius, FunkyFly, Akendall, Ph89~enwiki, IronGargoyle, QDE-can, Trounce, Alex1111, Nabeth, Joseph Solis in Australia, Pimlottc, Goran.S2, FairuseBot, Pudeo, Little Grabbi, Makedonia, CmdrObot, Burlefot~enwiki, Jamesbrownontheroad, Francolee01, Silal, Cydebot, Cahk, Ntsimp, Rifleman 82, Travelbird, M.K, Nikopoley, Afinebalance, Emelpy, Edwardx, Dyslexik, Vetalius, Steve Appleton, BuffaloChip97, Zeuscho, Ithinkhelikesit, Harryzilber, CosineKitty, Hydrahead, Joshua, −1g, SiobhanHansa, VoABot II, RBBrittain, Cadsuane Melaidhrin, Felix Stember, Rafiqmahmood, Vaerner, Nyttend, Nineko, EagleFan, Tandras, Balazshubai, Khronos1, R.Schuster, 42croad, Marcusaffleck, Nunojpg, CommonsDelinker, Kaveish, Primetomas, J.delanoy, Mayatiita, Maurice Carbonaro, AgainErick, Encolpe, Repo stef, Adnanrao, Scforth, Jjhcap99, Ajfweb, RVJ, Gibmetal77, W. Frank, Funandtrvl, Moscatanix, VolkovBot, Praveentech, Linnartz, Rkt2312, Sweetness46, Alassiry, TXiKiBoT, Rizalninoynapoleon, Vipinhari, Dannyjnesh, Zaher1988, Abdullais4u, Seb az86556, Ilyushka88, Bonus bon, Majalinno, Billinghurst, Sherdim, Iapain wiki, Qlinz, Igor alexandrov, Noblesham, Rambo's Revenge, WereSpielChequers, Ori, Eagleal, Lucasbfrbot, ArsLongaVitaBrevisEst, GrooveDog, Infestor, DirectEdge, Xeltran, Sanya3, Mkt3000 dot com, Metsfan86, TaerkastUA, Dajes13, Lightpinkflower, HairyWombat, Robertscriva, ImageRemovalBot, EoGuy, Jagun, RashersTierney, Enthusiast01, Unbuttered Parsnip, Mild Bill Hiccup, Baksando, Niceguyedc, Auntof6, Du3357, Mfa fariz, Eu national, L.tak, M.O.X, Graavis, BalkanFever, DumZiBoT, OhioTrivium, Avrilko, Schengen sk, Normadic, Nordic Legend, AegeanGoddess, HolyCross4ever, Blackisblack, Fred the Oyster, Mitch Ames, Dindia, Passportistca, SilvonenBot, IngerAlHaosului, Mifter, Jk3000, Ds02006, MystBot, AndrewSamuel, Marc CAT, Haiderzainy, ThePutt, Addbot, X2X00, Auridia, Douglas the Comeback Kid, Svartkledd~enwiki, Tassedethe, Ivario, Lightbot, Captain, Shikuesi3, Luckas-bot, Yobot, Gabrielsouza15, Xeex, Aradic-es, Rubin16, Knownot, Bennyzhong, Kippi70, Xxxriainxxx, AnomieBOT, KDS4444, Hjlee84, 1exec1, Jim1138, Omarteacher, Bosonic dressing, Materialscientist, Clausekwis, Ozguroot, LilHelpa, Xqbot, Stoichkov8, Unique85, Anna Frodesiak, Keeper B, Guagd, Rownon, FaceOffic, Alblefter, Ping60637, Black Gold, Роман Валитов, TerraHikaru, Kubarp, Thayts, Luisitokj, SJJhang, Nakakapagpabagabag, Outback the koala, Sadfnjet, Asdfgaheashse, Overandaway, Ahmer Jamil Khan, PigFlu Oink, Ivangricenko, Iraqiedit, Nightsturm, SecurityBlackBelt, Blumonster, DrilBot, Desert pinguin, Ltmaneveld, Serongadamwiki, Philly boy92, Glany222, Full-date unlinking bot, Custen, Trappist the monk, TotoCZ, Lotje, Andre775, Tomi566, Adi4094, Faisal moon1, RjwilmsiBot, Neklat, Saftorangen, Kamran the Great, MarkHavel, Coolcole, EmausBot, John of Reading, Dolescum, MrFawwaz, Zollerriia, Dewritech, 3122WIKI, YassineWiki, Alexis85, Moyzes, Sudanese322, Ida Shaw, Slentee, Mattsacks, Tafounast, H3llBot, Foreone, Lexusuns, Ardagul87, Milostewart, Mahaztra, Dvacet, ClueBot NG, BarnabyJoe, Bogho, Awareinc, Stratforder, Jasonclarkmanager, Doma93, Prime409, Helpful Pixie Bot, Durrahbeach, MarkMysoe, BG19bot, Tiscando, Senaku, Albatalab, Mali81, Xtravel, TommyTu25, Alarbus, Asgew, Aisteco, Readybreakkkkkkk, BohemianRhapsody, BattyBot, 3Princip, Sayz, Cyberbot II, ChrisGualtieri, Arcandam, Dobie80, Adirbd, Jamshaidshaikh, Mogism, Cerabot~enwiki, Belowerapid Zhang, 2,4-dinitrophenylhydrazone, PassportCan, Aedagin, Дэвид, Tentinator, Xzesey, Deedat2009, Cooksey, Vincenty846, Twofortnights, Straback, CatcherStorm, Monkbot, Opencooper, Peruzulu, MaronitePride, Thefaiths, Developer15, Manuel.quijada.serrano, David8302, Wikola, DotStepOk, Khichar51, Ramosc55, RadhaadhaR, Zxs, CAPTAIN RAJU, Copperminekd, Porkyrocks, BushelCandle, InternetArchiveBot, JLaffy01, MelKrul, Janot, C-GAUN, GreenC bot, Ttansey, Bender the Bot, Junky Cat, Developer17, Larprice13, Jon Kolbert, KolbertBot and Anonymous: 548

- **Biometric voter registration** *Source:* https://en.wikipedia.org/wiki/Biometric_voter_registration?oldid=792852311 *Contributors:* Swister-Twister, Northamerica1000, Jupitus Smart, JJMC89 and E-DemSnoopy

- **Biometrics in schools** *Source:* https://en.wikipedia.org/wiki/Biometrics_in_schools?oldid=801750931 *Contributors:* Ronz, Topbanana, Bender235, Wikipedian231, Graham87, Rjwilmsi, GünniX, Wavelength, Grafen, SmackBot, OrientalHero, C.Fred, Ohnoitsjamie, Hmains, JonHarder, Tazmaniacs, Blue-Haired Lawyer, Mentifisto, TAnthony, Magioladitis, Bibliophylax, Steven J. Anderson, Wiae, Ltka, Lauracs, Torinfo, WaterBreak55, EoGuy, Niceguyedc, DumZiBoT, Mjsa, MrOllie, Teles, Yobot, RandyPng, AnomieBOT, Fmph, Surv1v4l1st, LittleWink, Trappist the monk, MrX, Connelly90, Diannaa, RjwilmsiBot, Strawfair, H3llBot, ClueBot NG, ColnelStewart, Cyberbot II, Sibby154, Sibby230, Whizz40, Ashley6891, Neilinely, Fingerman53, Ampletrails, Zombie45764, The Quixotic Potato, InternetArchiveBot, GreenC bot, Carl12345666789, Bender the Bot, B-identi, Jon Kolbert, Biometricsexpert, KolbertBot and Anonymous: 25

- **BioSlimDisk** *Source:* https://en.wikipedia.org/wiki/BioSlimDisk?oldid=791499319 *Contributors:* Stone, Orangemike, Guy M, Jeff3000, MadeYourReadThis, NeilN, SmackBot, Od Mishehu, Cydebot, Daniel J. Leivick, Doug Weller, Eastmain, ImageRemovalBot, Addbot, Ironholds, Crecy99, Vernonheng, GoingBatty, Mightymagicland, BattyBot, InternetArchiveBot and Anonymous: 2

- **Handwritten biometric recognition** *Source:* https://en.wikipedia.org/wiki/Handwritten_biometric_recognition?oldid=758312128 *Contributors:* Bearcat, Danhash, Rjwilmsi, De728631, Addbot, LittleWink, YFdyh-bot, MFZBCN, Fixuture and Monkbot

- **Private biometrics** *Source:* https://en.wikipedia.org/wiki/Private_biometrics?oldid=789011071 *Contributors:* Rjwilmsi, Malcolma, Dekushrub, Sadads, Robofish, Nick Number, Schmloof, Linnartz, Gunner11 05, MystBot, Addbot, Helpful Pixie Bot, BG19bot, PrimeBOT and Anonymous: 3

- **Signature recognition** *Source:* https://en.wikipedia.org/wiki/Signature_recognition?oldid=771608316 *Contributors:* Rjwilmsi, Gilliam, Dr Greg, Yobot, Ptbotgourou, Pink Bull, Trappist the monk, EmausBot, ClueBot NG, Satellizer, YFdyh-bot, Lone boatman, MFZBCN, Monkbot, Enrique argones and Anonymous: 5

- **Vein matching** *Source:* https://en.wikipedia.org/wiki/Vein_matching?oldid=801098391 *Contributors:* Gracefool, Geo Swan, Btimlake, Rjwilmsi, Pzerimars, RoyalBlueStuey, Ohconfucius, KimChee, Ixchup, PKT, Bplcse, Yobot, Materialscientist, Citation bot, Biometricse, Ed8r, Paine Ellsworth, I dream of horses, Schwede66, John of Reading, Puffin, BG19bot, Mogism, Rybec, Monkbot, Joerice50, Bender the Bot, Mariekevb, KolbertBot and Anonymous: 14

- **Voice analysis** *Source:* https://en.wikipedia.org/wiki/Voice_analysis?oldid=761563819 *Contributors:* The Anome, Tbackstr, SimonP, BigFatBuddha, Ganymead, Rich Farmbrough, Apoc2400, GregorB, RussBot, Shaddack, Leotohill, Joseph Solis in Australia, Magioladitis, VanishedUserABC, SieBot, WRK, Polishwonder74, Wnt, Addbot, Luckas Blade, AnomieBOT, FrescoBot, SporkBot, ClueBot NG, Frietjes, Wiki13, Casta947, AnaQy, Speech 33, Ystylianou and Anonymous: 10

- **Identity Cards Act 2006** *Source:* https://en.wikipedia.org/wiki/Identity_Cards_Act_2006?oldid=799022573 *Contributors:* The Anome, William Avery, Heron, Stevertigo, Michael Hardy, Gabbe, Philipdw, Arwel Parry, Kingturtle, Jll, Scott, Kaihsu, Greenrd, Morwen, Joy, Andrew

Yong, Trevor mendham, Chrism, Altenmann, Alan Liefting, Smjg, Seabhcan, Netoholic, Everyking, Djegan, Bobblewik, Deus Ex, ALargeElk, Avaragado, Haggis, Dvavasour, Antandrus, OwenBlacker, Bodnotbod, Necrothesp, Eddpayne, JulieADriver, Kingal86, Jim H, Bonalaw, JTN, Rich Farmbrough, Smyth, Martpol, Bender235, CanisRufus, Kiand, Acanon, BW, Cmdrjameson, Guidod, Honeycake, 119, Ben davison, Rd232, Andrewpmk, Andrew Gray, Hodg, Kel-nage, Benson85, Paul1337, LFaraone, H2g2bob, Nightstallion, Ceyockey, Pcpcpc, Megan1967, Alainna, Woohookitty, RHaworth, Xmp, Madchester, Pol098, BenSamples, Z303, Thruston, Tomhab, WikianJim, とある白い猫, Graham87, BD2412, Dearsina, CelticWonder, Rjwilmsi, George Burgess, LjL, SenorAnderson, Wikiliki, Ground Zero, Old Moonraker, JdforresterBot, Hiding, Fragglet, RobyWayne, Str1977, Benlisquare, Hairy Dude, Mark Ironie, Epolk, Gaius Cornelius, Kb1koi, AlbertR, Eurosong, Mholland, Sandstein, BMT, Zzuuzz, Silverhorse, Abune, Whobot, Mais oui!, NFH, Smurfy, Buybooks Marius, Ross UK, That Guy, From That Show!, AndrewWTaylor, SmackBot, Strangecolours, Jim62sch, Rrius, Kazuo62, Stifle, Flamarande, Mauls, Eiler7, Betacommand, HowFreeHmm, Chris the speller, Bluebot, Thumperward, Mdwh, Neo-Jay, ClarkF1, Brideshead, Duncancumming, Monsterkrumm, Grover cleveland, Valenciano, Wonkotsane, Rabidbob, Parrot of Doom, Bretonbanquet, Skinnyweed, ゼーロ, Ohconfucius, Globaltraveller, GiollaUidir, BrownHairedGirl, Guyjohnston, Stevenmc, Bob Whoops, Joseph 2166, Cnis, Jon186, Dean1970, Burto88, Iridescent, Joseph Solis in Australia, JoeBot, Lucy-marie, Courcelles, Mattbr, Ninetyone, Dracos, Icomb, Cydebot, Jpb1301, Mindjuicer, Pebb, David Moss, Joe 1987, Epbr123, Tartan Nutter, Hcobb, Smartse, Spartaz, Anne lenoir, Bzuk, Ecki~enwiki, Magioladitis, Strange but untrue, MattUK, Annasmee, Stephenchou0722, Pauly04, Timothy Titus, R'n'B, Francis Tyers, L337 kybldmstr, Cannonmc, PeterCrispin, M.allen.uk, Psidogretro, Ajfweb, Andy Marchbanks, Hugo999, TreasuryTag, Jhw38, Rwthplb, Willshaman, Christopher Connor, Bonus bon, Nedrutland, Larklight, Iceage77, Ltka, Kmfdmstud, Memo232, Lankyphil, Lmc169, Nootopian, Lightmouse, Millstream3, Dravecky, Jza84, ImageRemovalBot, Sfan00 IMG, EoGuy, RashersTierney, CaNNoNFoDDa, Mild Bill Hiccup, Niceguyedc, Auntof6, Copyeditor42, Excirial, Kelvin 101, L.tak, Promethean, Berean Hunter, DumZiBoT, Bud08, Jaymacdonald, Likelife, Qwerta369, Addbot, Willking1979, Mabdul, Gaafiw, Zen6172, Ondewelle, Lightbot, KevinBrydon, Yobot, Huylens, AnomieBOT, 1exec1, Jim1138, Materialscientist, James500, Date delinker, MJLRGS, Srich32977, J04n, Hneto, Bluemurder.rm, Kikodawgzzz, MerlLinkBot, Strangways, Mirrorme22, Beganlocal, André Devecserii, Cnwilliams, Tabletalker, User1497, Bahnfrend, RjwilmsiBot, John of Reading, Hasfg, H3llBot, Unreal7, SpikeballUnion, Vanished 1850, MarkWWW, TruckCard, EuropeanCowboy, ClueBot NG, Histree, MeanMotherJr, BattyBot, Pratyya Ghosh, Cyberbot II, Mogism, BurritoBazooka, Jodosma, Center life corska, Whizz40, Animal720, Zumoarirodoka, 93Dan, Tiptoethrutheminefield, Domdeparis, Interpuncts, Sizeofint, Neve-selbert, InternetArchiveBot, GreenC bot, Bender the Bot, Jon Kolbert, KolbertBot and Anonymous: 278

- **International Identity Federation** *Source:* https://en.wikipedia.org/wiki/International_Identity_Federation?oldid=717806582 *Contributors:* Edward, Cobaltbluetony, Beland, Neutrality, Wavelength, Mais oui!, SmackBot, Betacommand, Beetstra, Funandtrvl, Ianforrest, Ianmforrest, Addbot and BattyBot

26.3.2 Images

- **File:12leadECG.jpg** *Source:* https://upload.wikimedia.org/wikipedia/commons/b/bd/12leadECG.jpg *License:* Public domain *Contributors:* Original uploader was MoodyGroove at en.wikipedia *Original artist:* MoodyGroove

- **File:1st-eeg.png** *Source:* https://upload.wikimedia.org/wikipedia/commons/7/7e/1st-eeg.png *License:* Public domain *Contributors:* Berger H. Über das Elektrenkephalogramm des Menchen. *Archives für Psychiatrie.* 1929; 87:527-70. *Original artist:* Hans Berger

- **File:3DFingerprint.jpg** *Source:* https://upload.wikimedia.org/wikipedia/commons/b/be/3DFingerprint.jpg *License:* Public domain *Contributors:* Own work *Original artist:* Wamelculi

- **File:5-Methylcytosine.svg** *Source:* https://upload.wikimedia.org/wikipedia/commons/3/3a/5-Methylcytosine.svg *License:* Public domain *Contributors:* Own work *Original artist:* Yikrazuul (<a href='//commons.wikimedia.org/wiki/User_talk:Yikrazuul' title='User talk:Yikrazuul'>talk</a>)

- **File:A-DNA,_B-DNA_and_Z-DNA.png** *Source:* https://upload.wikimedia.org/wikipedia/commons/b/b1/A-DNA%2C_B-DNA_and_Z-DNA.png *License:* GFDL *Contributors:* Originally from en.wikipedia; description page is/was here. *Original artist:* Original uploader was Richard Wheeler (Zephyris) at en.wikipedia

- **File:ADN_animation.gif** *Source:* https://upload.wikimedia.org/wikipedia/commons/8/81/ADN_animation.gif *License:* Public domain *Contributors:* Own work *Original artist:* brian0918™

- **File:Access_control_door_wiring.png** *Source:* https://upload.wikimedia.org/wikipedia/commons/1/10/Access_control_door_wiring.png *License:* Public domain *Contributors:* Own work *Original artist:* Andriusval

- **File:Access_control_door_wiring_io_module.png** *Source:* https://upload.wikimedia.org/wikipedia/commons/f/f2/Access_control_door_wiring_io_module.png *License:* Public domain *Contributors:* Own work *Original artist:* Andriusval

- **File:Access_control_topologies_IP_controller.png** *Source:* https://upload.wikimedia.org/wikipedia/commons/5/5b/Access_control_topologies_IP_controller.png *License:* Public domain *Contributors:* Own work *Original artist:* Andriusval

- **File:Access_control_topologies_IP_master.png** *Source:* https://upload.wikimedia.org/wikipedia/commons/9/95/Access_control_topologies_IP_master.png *License:* Public domain *Contributors:* Own work *Original artist:* Andriusval

- **File:Access_control_topologies_IP_reader.png** *Source:* https://upload.wikimedia.org/wikipedia/commons/e/ef/Access_control_topologies_IP_reader.png *License:* Public domain *Contributors:* Midpoint Systems, UAB *Original artist:* Andriusval

- **File:Access_control_topologies_main_controller_a.png** *Source:* https://upload.wikimedia.org/wikipedia/commons/8/8f/Access_control_topologies_main_controller_a.png *License:* Public domain *Contributors:* Own work *Original artist:* Andriusval

- **File:Access_control_topologies_main_controller_b.png** *Source:* https://upload.wikimedia.org/wikipedia/commons/a/a5/Access_control_topologies_main_controller_b.png *License:* Public domain *Contributors:* Own work *Original artist:* Andriusval

- **File:Access_control_topologies_serial_controllers.png** *Source:* https://upload.wikimedia.org/wikipedia/commons/e/e0/Access_control_topologies_serial_controllers.png *License:* Public domain *Contributors:* Own work *Original artist:* Andriusval

26.3. TEXT AND IMAGE SOURCES, CONTRIBUTORS, AND LICENSES

- **File:Access_control_topologies_terminal_servers.png** *Source:* https://upload.wikimedia.org/wikipedia/commons/a/ac/Access_control_topologies_terminal_servers.png *License:* Public domain *Contributors:* Own work *Original artist:* Andriusval
- **File:Ambox_important.svg** *Source:* https://upload.wikimedia.org/wikipedia/commons/b/b4/Ambox_important.svg *License:* Public domain *Contributors:* Own work based on: Ambox scales.svg *Original artist:* Dsmurat, penubag
- **File:BASA-532K-1-2-15-Ran_Bosilek.jpg** *Source:* https://upload.wikimedia.org/wikipedia/commons/b/b7/BASA-532K-1-2-15-Ran_Bosilek.jpg *License:* Public domain *Contributors:*

Bulgarian Archives State Agency: <a href='http://www.archives.government.bg/' data-x-rel='nofollow'><img alt='Nuvola filesystems folder home.svg' src='https://upload.wikimedia.org/wikipedia/commons/thumb/8/81/Nuvola_filesystems_folder_home.svg/20px-Nuvola_filesystems_folder_home.svg.png' width='20' height='20' srcset='https://upload.wikimedia.org/wikipedia/commons/thumb/8/81/Nuvola_filesystems_folder_home.svg/30px-Nuvola_filesystems_folder_home.svg.png 1.5x, https://upload.wikimedia.org/wikipedia/commons/thumb/8/81/Nuvola_filesystems_folder_home.svg/40px-Nuvola_filesystems_folder_home.svg.png 2x' data-file-width='128' data-file-height='128' /></a> Home page

Original artist: ?

- **File:Base_pair_AT.svg** *Source:* https://upload.wikimedia.org/wikipedia/commons/d/db/Base_pair_AT.svg *License:* Public domain *Contributors:* Own work *Original artist:* Yikrazuul
- **File:Base_pair_GC.svg** *Source:* https://upload.wikimedia.org/wikipedia/commons/6/64/Base_pair_GC.svg *License:* Public domain *Contributors:* Own work *Original artist:* Yikrazuul
- **File:Benzopyrene_DNA_adduct_1JDG.png** *Source:* https://upload.wikimedia.org/wikipedia/commons/d/d8/Benzopyrene_DNA_adduct_1JDG.png *License:* CC-BY-SA-3.0 *Contributors:* ? *Original artist:* ?
- **File:Bifurcation.svg** *Source:* https://upload.wikimedia.org/wikipedia/commons/2/25/Bifurcation.svg *License:* Public domain *Contributors:* http://en.wikipedia.org/wiki/Image:Bifurcation.JPG *Original artist:* Nima Pirzadeh
- **File:Biometric_architecture.jpg** *Source:* https://upload.wikimedia.org/wikipedia/en/e/e5/Biometric_architecture.jpg *License:* PD *Contributors:*

Own work

Original artist:

John_Larmouth (talk) (Uploads)

- **File:Biometric_passports.png** *Source:* https://upload.wikimedia.org/wikipedia/commons/6/69/Biometric_passports.png *License:* CC BY-SA 3.0 *Contributors:* I (Alinor (talk)) created this work - based on the Ozguroot version and biometric passport article. *Original artist:* Alinor at English Wikipedia (Later version(s) were uploaded by Igor alexandrov, Albatalab, Mahaztra, Andyso, Nomi887 at en.wikipedia.)
- **File:Biometric_system_diagram.png** *Source:* https://upload.wikimedia.org/wikipedia/commons/3/3c/Biometric_system_diagram.png *License:* CC-BY-SA-3.0 *Contributors:* Own work *Original artist:* Alessio Damato
- **File:Biometrics.jpg** *Source:* https://upload.wikimedia.org/wikipedia/commons/b/b0/Biometrics.jpg *License:* CC-BY-SA-3.0 *Contributors:* ? *Original artist:* ?
- **File:Blood_vessels_in_the_human_arm.png** *Source:* https://upload.wikimedia.org/wikipedia/commons/2/28/Blood_vessels_in_the_human_arm.png *License:* Public domain *Contributors:* Scanned from Dagfinn Døhl Dybvig & Magne Dybvig (2003). *Det tenkende mennesket.* Oslo: Tapir akademisk forlag. ISBN 8251918642. Page 170. Which had it from Great Books of the Western World 26. Ancyclopedia Britannica 1990. *Original artist:* Not given.
- **File:Border-Control-Process.png** *Source:* https://upload.wikimedia.org/wikipedia/commons/8/8c/Border-Control-Process.png *License:* CC BY 3.0 *Contributors:* http://blog.protocolbench.org/2013/08/automatic-border-control-systems-egate/ *Original artist:* Holger Funke
- **File:Branch-DNA-multiple.svg** *Source:* https://upload.wikimedia.org/wikipedia/commons/0/08/Branch-DNA-multiple.svg *License:* CC BY-SA 3.0 *Contributors:* This vector image was created with Inkscape, and then manually replaced. *Original artist:* Otterinfo
- **File:Branch-dna-single.svg** *Source:* https://upload.wikimedia.org/wikipedia/commons/5/54/Branch-dna-single.svg *License:* CC BY-SA 3.0 *Contributors:* This vector image was created with Inkscape, and then manually replaced. *Original artist:* Otterinfo
- **File:BurglaryIsrael2.jpg** *Source:* https://upload.wikimedia.org/wikipedia/commons/b/b5/BurglaryIsrael2.jpg *License:* CC BY-SA 3.0 *Contributors:* Own work *Original artist:* Etan J. Tal
- **File:Chromosomal_Recombination.svg** *Source:* https://upload.wikimedia.org/wikipedia/commons/b/b2/Chromosomal_Recombination.svg *License:* CC BY 2.5 *Contributors:* Own work, Created using Inkscape, v0.44 *Original artist:* David Eccles (Gringer)
- **File:ColourIris.png** *Source:* https://upload.wikimedia.org/wikipedia/commons/d/d7/ColourIris.png *License:* CC BY 3.0 *Contributors:* Own work *Original artist:* Smhossei
- **File:Commons-logo.svg** *Source:* https://upload.wikimedia.org/wikipedia/en/4/4a/Commons-logo.svg *License:* PD *Contributors:* ? *Original artist:* ?
- **File:Contiguous_leads.svg** *Source:* https://upload.wikimedia.org/wikipedia/commons/3/33/Contiguous_leads.svg *License:* CC-BY-SA-3.0 *Contributors:* Own work in Inkscape *Original artist:* en:User:Cburnett
- **File:Cover_of_Mauritanian_Biometric_Passport.png** *Source:* https://upload.wikimedia.org/wikipedia/commons/f/f9/Cover_of_Mauritanian_Biometric_Passport.png *License:* CC BY-SA 3.0 *Contributors:* Own work *Original artist:* Noble
- **File:Cytosin.svg** *Source:* https://upload.wikimedia.org/wikipedia/commons/d/dd/Cytosin.svg *License:* Public domain *Contributors:* Own work *Original artist:* NEUROtiker
- **File:DNA-ligand-by-Abalone.png** *Source:* https://upload.wikimedia.org/wikipedia/commons/8/8b/DNA-ligand-by-Abalone.png *License:* CC BY-SA 3.0 *Contributors:* Own work *Original artist:* P99am

- **File:DNA_Structure+Key+Labelled.pn_NoBB.png** *Source:* https://upload.wikimedia.org/wikipedia/commons/4/4c/DNA_Structure%2BKey%2BLabelled.pn_NoBB.png *License:* CC BY-SA 3.0 *Contributors:* Own work *Original artist:* Zephyris

- **File:DNA_chemical_structure.svg** *Source:* https://upload.wikimedia.org/wikipedia/commons/e/e4/DNA_chemical_structure.svg *License:* CC-BY-SA-3.0 *Contributors:* <a href='//validator.w3.org/' data-x-rel='nofollow'><img alt='W3C' src='https://upload.wikimedia.org/wikipedia/commons/thumb/1/1a/Valid_SVG_1.1_%28green%29.svg/88px-Valid_SVG_1.1_%28green%29.svg.png' width='88' height='30' style='vertical-align: top' srcset='https://upload.wikimedia.org/wikipedia/commons/thumb/1/1a/Valid_SVG_1.1_%28green%29.svg/132px-Valid_SVG_1.1_%28green%29.svg.png 1.5x, https://upload.wikimedia.org/wikipedia/commons/thumb/1/1a/Valid_SVG_1.1_%28green%29.svg/176px-Valid_SVG_1.1_%28green%29.svg.png 2x' data-file-width='91' data-file-height='31' /></a>iThe source code of this SVG is <a data-x-rel='nofollow' class='external text' href='//validator.w3.org/check?uri=https%3A%2F%2Fcommons.wikimedia.org%2Fwiki%2FSpecial%3AFilepath%2FDNA_chemical_structure.svg,<span>,&,</span>,ss=1#source'>valid</a>.
  Original artist: Madprime (talk · contribs)

- **File:DNA_nanostructures.png** *Source:* https://upload.wikimedia.org/wikipedia/commons/5/55/DNA_nanostructures.png *License:* CC BY 2.5 *Contributors:* Strong M: *Protein Nanomachines.* PLoS Biol 2/3/2004: e73. http://dx.doi.org/10.1371/journal.pbio.0020073 *Original artist:* (Images were kindly provided by Thomas H. LaBean and Hao Yan.)

- **File:DNA_orbit_animated_static_thumb.png** *Source:* https://upload.wikimedia.org/wikipedia/commons/d/db/DNA_orbit_animated_static_thumb.png *License:* CC-BY-SA-3.0 *Contributors:* Derived from File:DNA orbit animated.gif originally from here. *Original artist:* 84user adapting file originally uploaded by Richard Wheeler (Zephyris) at en.wikipedia

- **File:DNA_replication_en.svg** *Source:* https://upload.wikimedia.org/wikipedia/commons/8/8f/DNA_replication_en.svg *License:* Public domain *Contributors:* Own work. Image renamed from File:DNA replication.svg *Original artist:* LadyofHats Mariana Ruiz

- **File:Dacty_poederen.JPG** *Source:* https://upload.wikimedia.org/wikipedia/commons/d/d3/Dacty_poederen.JPG *License:* CC-BY-SA-3.0 *Contributors:* Transferred from nl.wikipedia to Commons. *Original artist:* Arnij at Dutch Wikipedia

- **File:De-Modern_ecg_(CardioNetworks_ECGpedia).jpg** *Source:* https://upload.wikimedia.org/wikipedia/commons/2/23/De-Modern_ecg_%28CardioNetworks_ECGpedia%29.jpg *License:* CC BY-SA 3.0 *Contributors:* CardioNetworks: De-Modern_ecg.jpg *Original artist:* CardioNetworks: Googletrans

- **File:Dynamic_information_of_a_signature.jpg** *Source:* https://upload.wikimedia.org/wikipedia/commons/3/35/Dynamic_information_of_a_signature.jpg *License:* CC BY-SA 3.0 *Contributors:* MATLAB *Original artist:* MFZBCN

- **File:ECG_Paper_v2.svg** *Source:* https://upload.wikimedia.org/wikipedia/commons/9/96/ECG_Paper_v2.svg *License:* Public domain *Contributors:* This file was derived from: ECG Paper.jpg
  Original artist: User:Markus Kuhn modified trace by User:Stannered of original PowerPoint JPEG by User:MoodyGroove

- **File:ECG_Vector.svg** *Source:* https://upload.wikimedia.org/wikipedia/commons/9/94/ECG_Vector.svg *License:* CC-BY-SA-3.0 *Contributors:* Based on ECG Vector.jpg by MoodyGroove *Original artist:* Rick Manning

- **File:ECG_principle_slow.gif** *Source:* https://upload.wikimedia.org/wikipedia/commons/e/e5/ECG_principle_slow.gif *License:* CC-BY-SA-3.0 *Contributors:* selbst erstellt = Own work *Original artist:* Kalumet

- **File:EEG_cap.jpg** *Source:* https://upload.wikimedia.org/wikipedia/commons/b/bf/EEG_cap.jpg *License:* Public domain *Contributors:* Transferred from en.wikipedia to Commons by Sreejithk2000 using CommonsHelper. *Original artist:* Thuglas at English Wikipedia

- **File:EKG_Complex_en.svg** *Source:* https://upload.wikimedia.org/wikipedia/commons/3/34/EKG_Complex_en.svg *License:* CC BY-SA 3.0 *Contributors:* This file was derived from: EKG Komplex.svg
  Original artist:

- Derivative: Hazmat2

- **File:EKG_leads.png** *Source:* https://upload.wikimedia.org/wikipedia/commons/0/0e/EKG_leads.png *License:* CC BY-SA 4.0 *Contributors:* Own work *Original artist:* Npatchett

- **File:EPassport_logo.svg** *Source:* https://upload.wikimedia.org/wikipedia/commons/f/fb/EPassport_logo.svg *License:* Public domain *Contributors:* Uploaded to en-wp under the same name by en:User:Akhristov *Original artist:* en:User:Akhristov

- **File:EcoRV_1RVA.png** *Source:* https://upload.wikimedia.org/wikipedia/commons/d/dc/EcoRV_1RVA.png *License:* CC-BY-SA-3.0 *Contributors:* Transferred from en.wikipedia to Commons. *Original artist:* The original uploader was Zephyris at English Wikipedia

- **File:Edit-clear.svg** *Source:* https://upload.wikimedia.org/wikipedia/en/f/f2/Edit-clear.svg *License:* Public domain *Contributors:* The *Tango! Desktop Project*. *Original artist:*
  The people from the Tango! project. And according to the meta-data in the file, specifically: "Andreas Nilsson, and Jakub Steiner (although minimally)."

- **File:Eeg_SMR.svg** *Source:* https://upload.wikimedia.org/wikipedia/commons/b/be/Eeg_SMR.svg *License:* CC-BY-SA-3.0 *Contributors:* ? *Original artist:* ?

- **File:Eeg_alpha.svg** *Source:* https://upload.wikimedia.org/wikipedia/commons/e/ee/Eeg_alpha.svg *License:* CC-BY-SA-3.0 *Contributors:* Own work *Original artist:* Hugo Gamboa

- **File:Eeg_beta.svg** *Source:* https://upload.wikimedia.org/wikipedia/commons/2/28/Eeg_beta.svg *License:* CC-BY-SA-3.0 *Contributors:* Own work *Original artist:* Hugo Gamboa

- **File:Eeg_delta.svg** *Source:* https://upload.wikimedia.org/wikipedia/commons/5/54/Eeg_delta.svg *License:* CC-BY-SA-3.0 *Contributors:* Own work *Original artist:* Hugo Gamboa

- **File:Eeg_gamma.svg** *Source:* https://upload.wikimedia.org/wikipedia/commons/2/21/Eeg_gamma.svg *License:* CC-BY-SA-3.0 *Contributors:* Own work *Original artist:* Hugo Gambo

26.3. TEXT AND IMAGE SOURCES, CONTRIBUTORS, AND LICENSES

- **File:Eeg_raw.svg** *Source:* https://upload.wikimedia.org/wikipedia/commons/2/2a/Eeg_raw.svg *License:* CC-BY-SA-3.0 *Contributors:* ? *Original artist:* ?
- **File:Eeg_theta.svg** *Source:* https://upload.wikimedia.org/wikipedia/commons/3/33/Eeg_theta.svg *License:* CC-BY-SA-3.0 *Contributors:* This is raw eeg. The signal was acquired in the Oz position processed with scipy and saved with matplolib. *Original artist:* Hugo Gamboa
- **File:Electroencephalograph_Neurovisor-BMM_40_(close_view).jpg** *Source:* https://upload.wikimedia.org/wikipedia/commons/2/22/Electroencephalograph_Neurovisor-BMM_40_%28close_view%29.jpg *License:* CC BY-SA 3.0 *Contributors:* The photo was taken in educational-scientific center "Computer Diagnostic and Imaging" of Biomedical Systems Department of MIET *Original artist:* Юрий Петрович Маслобоев / Yury Petrovich Masloboev
- **File:Eukaryote_DNA-en.svg** *Source:* https://upload.wikimedia.org/wikipedia/commons/e/e2/Eukaryote_DNA-en.svg *License:* CC BY-SA 3.0 *Contributors:* This file was derived from Eukaryote DNA.svg: <a href='//commons.wikimedia.org/wiki/File:Eukaryote_DNA.svg' class='image'><img alt='Eukaryote DNA.svg' src='https://upload.wikimedia.org/wikipedia/commons/thumb/8/82/Eukaryote_DNA.svg/50px-Eukaryote_DNA.svg.png' width='50' height='31' srcset='https://upload.wikimedia.org/wikipedia/commons/thumb/8/82/Eukaryote_DNA.svg/75px-Eukaryote_DNA.svg.png 1.5x, https://upload.wikimedia.org/wikipedia/commons/thumb/8/82/Eukaryote_DNA.svg/100px-Eukaryote_DNA.svg.png 2x' data-file-width='1189' data-file-height='734' /></a>
Original artist: Eukaryote_DNA.svg: *Difference_DNA_RNA-EN.svg: *Difference_DNA_RNA-DE.svg: Sponk (talk)
- **File:Fingerprint_Arch.jpg** *Source:* https://upload.wikimedia.org/wikipedia/commons/c/c5/Fingerprint_Arch.jpg *License:* Public domain *Contributors:* ? *Original artist:* ?
- **File:Fingerprint_Identification.JPG** *Source:* https://upload.wikimedia.org/wikipedia/commons/f/fc/Fingerprint_Identification.JPG *License:* CC BY-SA 3.0 *Contributors:* Own work *Original artist:* aeroplanepics0112
- **File:Fingerprint_Identification_Room.JPG** *Source:* https://upload.wikimedia.org/wikipedia/commons/b/b4/Fingerprint_Identification_Room.JPG *License:* CC BY-SA 3.0 *Contributors:* Own work *Original artist:* aeroplanepics0112
- **File:Fingerprint_Loop.jpg** *Source:* https://upload.wikimedia.org/wikipedia/commons/0/06/Fingerprint_Loop.jpg *License:* Public domain *Contributors:* ? *Original artist:* ?
- **File:Fingerprint_Whorl.jpg** *Source:* https://upload.wikimedia.org/wikipedia/commons/4/49/Fingerprint_Whorl.jpg *License:* Public domain *Contributors:* http://web.archive.org/web/20050403155444/http://www.nist.gov/srd/fing_img.htm *Original artist:* NIST database
- **File:Fingerprint_cartridge.jpg** *Source:* https://upload.wikimedia.org/wikipedia/en/9/9b/Fingerprint_cartridge.jpg *License:* Cc-by-sa-3.0 *Contributors:* ? *Original artist:* ?
- **File:Fingerprint_detail_on_male_finger.jpg** *Source:* https://upload.wikimedia.org/wikipedia/commons/1/15/Fingerprint_detail_on_male_finger.jpg *License:* CC BY 3.0 *Contributors:* Own work *Original artist:* Frettie
- **File:Fingerprint_scanner_identification.jpg** *Source:* https://upload.wikimedia.org/wikipedia/commons/6/67/Fingerprint_scanner_identification.jpg *License:* CC BY-SA 3.0 *Contributors:* Own work *Original artist:* Rachmaninoff
- **File:Fingerprint_surfer.jpg** *Source:* https://upload.wikimedia.org/wikipedia/en/7/73/Fingerprint_surfer.jpg *License:* Cc-by-sa-3.0 *Contributors:* ? *Original artist:* ?
- **File:Fingerprintforcriminologystubs2.png** *Source:* https://upload.wikimedia.org/wikipedia/commons/2/2c/Fingerprintforcriminologystubs2.png *License:* CC-BY-SA-3.0 *Contributors:* Transferred from en.wikipedia to Commons. *Original artist:* Cyrillic at English Wikipedia
- **File:Fingerprinting_1928.jpg** *Source:* https://upload.wikimedia.org/wikipedia/commons/4/48/Fingerprinting_1928.jpg *License:* Public domain *Contributors:* Digital collections —UCLA Library *Original artist:* unknown Los Angeles Daily News staff photographer
- **File:Fingerprints_of_Anna_Timiriova_3.jpg** *Source:* https://upload.wikimedia.org/wikipedia/commons/3/39/Fingerprints_of_Anna_Timiriova_3.jpg *License:* Public domain *Contributors:* Последняя любовь Колчака снималась у Гайдая и Бондарчука [1] *Original artist:* Police Files
- **File:Fingerprints_taken_by_William_James_Herschel_1859-1860.jpg** *Source:* https://upload.wikimedia.org/wikipedia/commons/1/1b/Fingerprints_taken_by_William_James_Herschel_1859-1860.jpg *License:* Public domain *Contributors:* http://www.thamespilot.org.uk/asset_arena/image/2000/sl/sl/sl-sl-1526_herschelfprint-i-01-000.jpg *Original artist:* William James Herschel (1833-1917)
- **File:Flag_of_Albania.svg** *Source:* https://upload.wikimedia.org/wikipedia/commons/3/36/Flag_of_Albania.svg *License:* Public domain *Contributors:* ? *Original artist:* ?
- **File:Flag_of_Andorra.svg** *Source:* https://upload.wikimedia.org/wikipedia/commons/1/19/Flag_of_Andorra.svg *License:* Public domain *Contributors:* Llibre de normes gràfiques per a la reproducció i aplicació dels signes d'Estat per als quals el Govern és autoritat competent (Aprovat pel Govern en la sessió del dia 5 de maig de 1999) *Original artist:* HansenBCN
- **File:Flag_of_Bosnia_and_Herzegovina.svg** *Source:* https://upload.wikimedia.org/wikipedia/commons/b/bf/Flag_of_Bosnia_and_Herzegovina.svg *License:* Public domain *Contributors:* Own work *Original artist:* Kseferovic
- **File:Flag_of_Croatia.svg** *Source:* https://upload.wikimedia.org/wikipedia/commons/1/1b/Flag_of_Croatia.svg *License:* Public domain *Contributors:* http://www.sabor.hr/Default.aspx?sec=4317 *Original artist:* Nightstallion, Elephantus, Neoneo13, Denelson83, Rainman, R-41, Minestrone, Lupo, Zscout370,
<a href='//commons.wikimedia.org/wiki/User:MaGa' title='User:MaGa'>**Ma**</a><a href='//commons.wikimedia.org/wiki/File:Croatian_squares_Ljubicic.png' class='image'><img alt='Croatian squares Ljubicic.png' src='https://upload.wikimedia.org/wikipedia/commons/thumb/7/7f/Croatian_squares_Ljubicic.png/15px-Croatian_squares_Ljubicic.png' width='15' height='15' srcset='https://upload.wikimedia.org/wikipedia/commons/thumb/7/7f/Croatian_squares_Ljubicic.png/23px-Croatian_squares_Ljubicic.png 1.5x, https://upload.wikimedia.org/wikipedia/commons/thumb/7/7f/Croatian_squares_Ljubicic.png/30px-Croatian_squares_Ljubicic.png 2x' data-file-width='202' data-file-height='202' /></a><a href='//commons.wikimedia.org/wiki/User_talk:MaGa' title='User talk:MaGa'>**Ga**</a> (based on Decision of the Parliament)

- **File:Flag_of_Europe.svg** *Source:* https://upload.wikimedia.org/wikipedia/commons/b/b7/Flag_of_Europe.svg *License:* Public domain *Contributors:*
- File based on the specification given at [1]. *Original artist:* User:Verdy p, User:-xfi-, User:Paddu, User:Nightstallion, User:Funakoshi, User:Jeltz, User:Dbenbenn, User:Zscout370
- **File:Flag_of_Gibraltar.svg** *Source:* https://upload.wikimedia.org/wikipedia/commons/0/02/Flag_of_Gibraltar.svg *License:* Public domain *Contributors:* based on the original coat of arms of Gibraltar, granted in 1502 *Original artist:* Created on behalf of Isabella I of Castile in 1502; this version uploaded by Denelson83 (talk · contribs)
- **File:Flag_of_Guernsey.svg** *Source:* https://upload.wikimedia.org/wikipedia/commons/f/fa/Flag_of_Guernsey.svg *License:* CC0 *Contributors:* From the Open Clip Art website. *Original artist:* User:Nightstallion
- **File:Flag_of_Iceland.svg** *Source:* https://upload.wikimedia.org/wikipedia/commons/c/ce/Flag_of_Iceland.svg *License:* Public domain *Contributors:* **Proportions:** Forsetisráðuneyti Íslands **Colours:** Alþingi *Original artist:* Árni Dagur, and Magasjukur2
- **File:Flag_of_Jersey.svg** *Source:* https://upload.wikimedia.org/wikipedia/commons/1/1c/Flag_of_Jersey.svg *License:* Public domain *Contributors:* ? *Original artist:* ?
- **File:Flag_of_Liechtenstein.svg** *Source:* https://upload.wikimedia.org/wikipedia/commons/4/47/Flag_of_Liechtenstein.svg *License:* Public domain *Contributors:* ? *Original artist:* ?
- **File:Flag_of_Macedonia.svg** *Source:* https://upload.wikimedia.org/wikipedia/commons/f/f8/Flag_of_Macedonia.svg *License:* Public domain *Contributors:* Own work *Original artist:* User:SKopp, redrawn by User:Gabbe
- **File:Flag_of_Monaco.svg** *Source:* https://upload.wikimedia.org/wikipedia/commons/e/ea/Flag_of_Monaco.svg *License:* Public domain *Contributors:* ? *Original artist:* ?
- **File:Flag_of_Montenegro.svg** *Source:* https://upload.wikimedia.org/wikipedia/commons/6/64/Flag_of_Montenegro.svg *License:* Public domain *Contributors:* Own work *Original artist:* B1mbo, Froztbyte
- **File:Flag_of_Morocco.svg** *Source:* https://upload.wikimedia.org/wikipedia/commons/2/2c/Flag_of_Morocco.svg *License:* Public domain *Contributors:* Flag of the Kingdom of Morocco

<a data-x-rel='nofollow' class='external text' href='http://81.192.52.100/BO/AR/1915/BO_135_ar.PDF'>Moroccan royal decree (17 November 1915), BO-135-ar *page 6*</a>

Original artist: Denelson83, Zscout370

- **File:Flag_of_Norway.svg** *Source:* https://upload.wikimedia.org/wikipedia/commons/d/d9/Flag_of_Norway.svg *License:* Public domain *Contributors:* Own work *Original artist:* Dbenbenn
- **File:Flag_of_San_Marino.svg** *Source:* https://upload.wikimedia.org/wikipedia/commons/b/b1/Flag_of_San_Marino.svg *License:* Public domain *Contributors:* Own work: [/Users/bicio/Desktop/Cailungo logo 40°.jpg] *Original artist:* Zscout370
- **File:Flag_of_Scotland.svg** *Source:* https://upload.wikimedia.org/wikipedia/commons/1/10/Flag_of_Scotland.svg *License:* Public domain *Contributors:* http://kbolino.freeshell.org/svg/scotland.svg *Original artist:* none known
- **File:Flag_of_Serbia.svg** *Source:* https://upload.wikimedia.org/wikipedia/commons/f/ff/Flag_of_Serbia.svg *License:* Public domain *Contributors:* From http://www.parlament.gov.rs/content/cir/o_skupstini/simboli/simboli.asp. *Original artist:* sodipodi.com
- **File:Flag_of_Switzerland.svg** *Source:* https://upload.wikimedia.org/wikipedia/commons/f/f3/Flag_of_Switzerland.svg *License:* Public domain *Contributors:* PDF Colors Construction sheet *Original artist:* User:Marc Mongenet

Credits:

- **File:Flag_of_the_Faroe_Islands.svg** *Source:* https://upload.wikimedia.org/wikipedia/commons/3/3c/Flag_of_the_Faroe_Islands.svg *License:* Public domain *Contributors:* ? *Original artist:* ?
- **File:Flag_of_the_Isle_of_Man.svg** *Source:* https://upload.wikimedia.org/wikipedia/commons/b/bc/Flag_of_the_Isle_of_Man.svg *License:* CC0 *Contributors:* Sodipodi flag collection, OpenClipart *Original artist:* Edited by Reisio, Alkari, e.a.
- **File:Flag_of_the_Vatican_City.svg** *Source:* https://upload.wikimedia.org/wikipedia/commons/0/00/Flag_of_the_Vatican_City.svg *License:* CC0 *Contributors:* http://files.mojeeuro.meu.zoznam.sk/200000288-390ab3a04d/2_Commemorative_coin_Vatican_city_2010.jpg labelbasis *Original artist:* Unknown<a href='https://www.wikidata.org/wiki/Q4233718' title='wikidata:Q4233718'><img alt='wikidata:Q4233718' src='https://upload.wikimedia.org/wikipedia/commons/thumb/f/ff/Wikidata-logo.svg/20px-Wikidata-logo.svg.png' width='20' height='11' srcset='https://upload.wikimedia.org/wikipedia/commons/thumb/f/ff/Wikidata-logo.svg/30px-Wikidata-logo.svg.png 1.5x, https://upload.wikimedia.org/wikipedia/commons/thumb/f/ff/Wikidata-logo.svg/40px-Wikidata-logo.svg.png 2x' data-file-width='1050' data-file-height='590' /></a>
- **File:Fob-at-proximity-reader_532_130xauto.jpg** *Source:* https://upload.wikimedia.org/wikipedia/commons/c/cf/Fob-at-proximity-reader_532_130xauto.jpg *License:* CC BY-SA 3.0 *Contributors:* Own work Template:Self photograph Template:Http://www.accesscontrol.ie/products/category/pro/ *Original artist:* Mgavenda
- **File:Folder_Hexagonal_Icon.svg** *Source:* https://upload.wikimedia.org/wikipedia/en/4/48/Folder_Hexagonal_Icon.svg *License:* Cc-by-sa-3.0 *Contributors:* ? *Original artist:* ?
- **File:Front_cover_of_the_Panamanian_biometric_passport.jpg** *Source:* https://upload.wikimedia.org/wikipedia/commons/b/b6/Front_cover_of_the_Panamanian_biometric_passport.jpg *License:* CC BY-SA 4.0 *Contributors:* Own work *Original artist:* TommyTu25
- **File:Gait_laboratory.jpg** *Source:* https://upload.wikimedia.org/wikipedia/commons/2/22/Gait_laboratory.jpg *License:* CC BY-SA 3.0 *Contributors:* Own work *Original artist:* D. Gordon E. Robertson

26.3. TEXT AND IMAGE SOURCES, CONTRIBUTORS, AND LICENSES 213

- **File:Hand_Geometry_Reading_Device.jpg** *Source:* https://upload.wikimedia.org/wikipedia/commons/c/ce/Hand_Geometry_Reading_Device.jpg *License:* CC BY-SA 3.0 *Contributors:* Own work *Original artist:* Z22
- **File:Hand_Geometry_and_Measurements.jpg** *Source:* https://upload.wikimedia.org/wikipedia/commons/f/f3/Hand_Geometry_and_Measurements.jpg *License:* CC BY-SA 3.0 *Contributors:* Own work *Original artist:* Z22
- **File:HansBerger_Univ_Jena.jpeg** *Source:* https://upload.wikimedia.org/wikipedia/commons/6/69/HansBerger_Univ_Jena.jpeg *License:* Public domain *Contributors:* http://www.psychiatrie.uk-j.de/Geschichte.html *Original artist:* Unknown<a href='https://www.wikidata.org/wiki/Q4233718' title='wikidata:Q4233718'><img alt='wikidata:Q4233718' src='https://upload.wikimedia.org/wikipedia/commons/thumb/f/ff/Wikidata-logo.svg/20px-Wikidata-logo.svg.png' width='20' height='11' srcset='https://upload.wikimedia.org/wikipedia/commons/thumb/f/ff/Wikidata-logo.svg/30px-Wikidata-logo.svg.png 1.5x, https://upload.wikimedia.org/wikipedia/commons/thumb/f/ff/Wikidata-logo.svg/40px-Wikidata-logo.svg.png 2x' data-file-width='1050' data-file-height='590' /></a>
- **File:Holliday_Junction.svg** *Source:* https://upload.wikimedia.org/wikipedia/commons/8/83/Holliday_Junction.svg *License:* Public domain *Contributors:*
- Holliday Junction.png *Original artist:*
- derivative work: Mouagip (talk)
- **File:Holliday_junction_coloured.png** *Source:* https://upload.wikimedia.org/wikipedia/commons/9/92/Holliday_junction_coloured.png *License:* CC-BY-SA-3.0 *Contributors:* Transferred from en.wikipedia to Commons. *Original artist:* Zephyris at English Wikipedia
- **File:Human_EEG_artefacts.png** *Source:* https://upload.wikimedia.org/wikipedia/commons/b/b1/Human_EEG_artefacts.png *License:* CC BY-SA 4.0 *Contributors:* Own work *Original artist:* Andrii Cherninskyi
- **File:Human_EEG_with_prominent_alpha-rhythm.png** *Source:* https://upload.wikimedia.org/wikipedia/commons/8/87/Human_EEG_with_prominent_alpha-rhythm.png *License:* CC BY-SA 4.0 *Contributors:* Own work *Original artist:* Andrii Cherninskyi
- **File:Human_EEG_without_alpha-rhythm.png** *Source:* https://upload.wikimedia.org/wikipedia/commons/1/1c/Human_EEG_without_alpha-rhythm.png *License:* CC BY-SA 4.0 *Contributors:* Own work *Original artist:* Andrii Cherninskyi
- **File:IRiris.jpg** *Source:* https://upload.wikimedia.org/wikipedia/commons/f/f4/IRiris.jpg *License:* CC BY-SA 3.0 *Contributors:* Own work *Original artist:* Boodlemason
- **File:Id_card.jpg** *Source:* https://upload.wikimedia.org/wikipedia/commons/e/ea/Id_card.jpg *License:* Public domain *Contributors:* Walter Jeffs (1937-2010) ID card *Original artist:* UK Government
- **File:India_1952_2r_stamped_paper.jpg** *Source:* https://upload.wikimedia.org/wikipedia/commons/7/74/India_1952_2r_stamped_paper.jpg *License:* Public domain *Contributors:* Scan of original *Original artist:* British Government
- **File:Info_numeros.jpg** *Source:* https://upload.wikimedia.org/wikipedia/commons/4/4c/Info_numeros.jpg *License:* CC BY-SA 3.0 *Contributors:* Own work *Original artist:* MFZBCN
- **File:Intelligent_access_control_door_wiring.PNG** *Source:* https://upload.wikimedia.org/wikipedia/commons/5/53/Intelligent_access_control_door_wiring.PNG *License:* Public domain *Contributors:* Own work *Original artist:* Andriusval
- **File:IriScan_model_2100_iris_scanner_1.jpg** *Source:* https://upload.wikimedia.org/wikipedia/commons/1/1b/IriScan_model_2100_iris_scanner_1.jpg *License:* CC BY-SA 2.5 *Contributors:* Own work *Original artist:* Mark Pellegrini
- **File:IrisGuard-UAE.JPG** *Source:* https://upload.wikimedia.org/wikipedia/en/d/d4/IrisGuard-UAE.JPG *License:* PD *Contributors:* ? *Original artist:* ?
- **File:Iris_Recognition_Enabled_ATM.jpg** *Source:* https://upload.wikimedia.org/wikipedia/commons/f/f2/Iris_Recognition_Enabled_ATM.jpg *License:* CC BY-SA 4.0 *Contributors:* Own work *Original artist:* Jcarroll26
- **File:Lambda_repressor_1LMB.png** *Source:* https://upload.wikimedia.org/wikipedia/commons/8/8f/Lambda_repressor_1LMB.png *License:* CC-BY-SA-3.0 *Contributors:* Transferred from en.wikipedia to Commons. *Original artist:* Zephyris at English Wikipedia
- **File:Limb_leads.svg** *Source:* https://upload.wikimedia.org/wikipedia/commons/c/c9/Limb_leads.svg *License:* Public domain *Contributors:*
- Limb Leads.jpg *Original artist:* Twisp
- **File:Limb_leads_of_EKG.png** *Source:* https://upload.wikimedia.org/wikipedia/commons/1/19/Limb_leads_of_EKG.png *License:* CC BY-SA 4.0 *Contributors:* Own work *Original artist:* Npatchett
- **File:Lock-green.svg** *Source:* https://upload.wikimedia.org/wikipedia/commons/6/65/Lock-green.svg *License:* CC0 *Contributors:* en:File:Free-to-read_lock_75.svg *Original artist:* User:Trappist the monk
- **File:Locksmiths-11211.jpg** *Source:* https://upload.wikimedia.org/wikipedia/commons/7/7a/Locksmiths-11211.jpg *License:* CC0 *Contributors:* http://nyclocksmithbrooklyn.com/security-systems/fast-locksmith-anytime-locks/ *Original artist:* Locksmithwilli
- **File:Loudspeaker.svg** *Source:* https://upload.wikimedia.org/wikipedia/commons/8/8a/Loudspeaker.svg *License:* Public domain *Contributors:* New version of Image:Loudspeaker.png, by AzaToth and compressed by Hautala *Original artist:* Nethac DIU, waves corrected by Zoid
- **File:Maclyn_McCarty_with_Francis_Crick_and_James_D_Watson_-_10.1371_journal.pbio.0030341.g001-O.jpg** *Source:* https://upload.wikimedia.org/wikipedia/commons/e/ed/Maclyn_McCarty_with_Francis_Crick_and_James_D_Watson_-_10.1371_journal.pbio.0030341.g001-O.jpg *License:* CC BY 3.0 *Contributors:* Lederberg J, Gotschlich EC (2005) A Path to Discovery: The Career of Maclyn McCarty. *PLoS Biol* 3(10): e341 doi:10.1371/journal.pbio.0030341 *Original artist:* Marjorie McCarty
- **File:NIRIris.png** *Source:* https://upload.wikimedia.org/wikipedia/commons/a/a6/NIRIris.png *License:* CC BY 3.0 *Contributors:* Own work *Original artist:* Smhossei
- **File:NewYorkCitySubwayEntranceInterior.jpg** *Source:* https://upload.wikimedia.org/wikipedia/commons/5/55/NewYorkCitySubwayEntranceInterior.jpg *License:* CC BY 3.0 *Contributors:* Own work *Original artist:* Canadaolympic989

- **File:Nucleosome1.png** *Source:* https://upload.wikimedia.org/wikipedia/commons/b/be/Nucleosome1.png *License:* CC BY-SA 3.0 *Contributors:* Own work *Original artist:* Thomas Splettstoesser
- **File:Numeros_dinamicos.jpg** *Source:* https://upload.wikimedia.org/wikipedia/commons/a/a8/Numeros_dinamicos.jpg *License:* CC BY-SA 3.0 *Contributors:* Own work *Original artist:* MFZBCN
- **File:Online_signture.jpg** *Source:* https://upload.wikimedia.org/wikipedia/commons/4/44/Online_signture.jpg *License:* CC BY-SA 3.0 *Contributors:* Own work *Original artist:* MFZBCN
- **File:Parallel_telomere_quadruple.png** *Source:* https://upload.wikimedia.org/wikipedia/commons/9/9e/Parallel_telomere_quadruple.png *License:* CC-BY-SA-3.0 *Contributors:* self-made using the open source vizualization program PyMol *Original artist:* Thomas Splettstoesser (www.scistyle.com)
- **File:Pencil_sketch_of_the_DNA_double_helix_by_Francis_Crick_Wellcome_L0051225.jpg** *Source:* https://upload.wikimedia.org/wikipedia/commons/a/ab/Pencil_sketch_of_the_DNA_double_helix_by_Francis_Crick_Wellcome_L0051225.jpg *License:* CC BY 4.0 *Contributors:* http://wellcomeimages.org/indexplus/obf_images/a2/0c/23cf65b022b84aa38d0dad401daf.jpg *Original artist:* ?
- **File:Physical_security_access_control_with_a_fingerprint_scanner.jpg** *Source:* https://upload.wikimedia.org/wikipedia/commons/8/8b/Physical_security_access_control_with_a_fingerprint_scanner.jpg *License:* CC BY 3.0 *Contributors:* Own work *Original artist:* Lgate74
- **File:Police_CCTV_Van,_Comer_Crescent_-_geograph.org.uk_-_1263396.jpg** *Source:* https://upload.wikimedia.org/wikipedia/commons/9/92/Police_CCTV_Van%2C_Comer_Crescent_-_geograph.org.uk_-_1263396.jpg *License:* CC BY-SA 2.0 *Contributors:* From geograph.org.uk *Original artist:* J Taylor
- **File:Precordial_leads_in_ECG.png** *Source:* https://upload.wikimedia.org/wikipedia/commons/4/41/Precordial_leads_in_ECG.png *License:* CC0 *Contributors:* Own work *Original artist:* Mikael Häggström
- **File:Proce.jpg** *Source:* https://upload.wikimedia.org/wikipedia/commons/2/23/Proce.jpg *License:* CC BY-SA 3.0 *Contributors:* Own work *Original artist:* Alba12 6
- **File:Q-Lane_Turnstiles.jpg** *Source:* https://upload.wikimedia.org/wikipedia/commons/8/83/Q-Lane_Turnstiles.jpg *License:* CC BY-SA 4.0 *Contributors:* Own work *Original artist:* Fabtron
- **File:Question_book-new.svg** *Source:* https://upload.wikimedia.org/wikipedia/en/9/99/Question_book-new.svg *License:* Cc-by-sa-3.0 *Contributors:*
Created from scratch in Adobe Illustrator. Based on Image:Question book.png created by User:Equazcion *Original artist:* Tkgd2007
- **File:Real_fingerprints_on_fake_crime_scene.JPG** *Source:* https://upload.wikimedia.org/wikipedia/commons/a/aa/Real_fingerprints_on_fake_crime_scene.JPG *License:* CC BY-SA 3.0 *Contributors:* Own work *Original artist:* Alchemica
- **File:Retina_camera_controls.jpg** *Source:* https://upload.wikimedia.org/wikipedia/commons/5/5c/Retina_camera_controls.jpg *License:* CC BY-SA 3.0 *Contributors:* Own work *Original artist:* User:Jason7825
- **File:Ridge_ending.svg** *Source:* https://upload.wikimedia.org/wikipedia/commons/f/f1/Ridge_ending.svg *License:* CC-BY-SA-3.0 *Contributors:* http://en.wikipedia.org/wiki/Image:Ridge_ending.JPG *Original artist:* Nima Pirzadeh
- **File:RitechBSDLogo.jpg** *Source:* https://upload.wikimedia.org/wikipedia/en/d/db/RitechBSDLogo.jpg *License:* Fair use *Contributors:*
The logo is from the https://www.bioslimdisk.com/ website. *Original artist:* ?
- **File:Royal_Coat_of_Arms_of_the_United_Kingdom_(HM_Government).svg** *Source:* https://upload.wikimedia.org/wikipedia/commons/1/1a/Royal_Coat_of_Arms_of_the_United_Kingdom_%28HM_Government%29.svg *License:* CC BY-SA 3.0 *Contributors:* Own work *Original artist:* Sodacan
- **File:Short_ridge.svg** *Source:* https://upload.wikimedia.org/wikipedia/commons/7/78/Short_ridge.svg *License:* Public domain *Contributors:* http://en.wikipedia.org/wiki/Image:Short_ridge.JPG *Original artist:* Nima Pirzadeh
- **File:Sound-icon.svg** *Source:* https://upload.wikimedia.org/wikipedia/commons/4/47/Sound-icon.svg *License:* LGPL *Contributors:* Derivative work from Silsor's versio *Original artist:* Crystal SVG icon set
- **File:Surveillance_equipment_5411.jpg** *Source:* https://upload.wikimedia.org/wikipedia/commons/8/85/Surveillance_equipment_5411.jpg *License:* Public domain *Contributors:* Transferred from en.wikipedia *Original artist:* Original uploader was Maraparacc at en.wikipedia
- **File:Surveillance_equipment_5413.jpg** *Source:* https://upload.wikimedia.org/wikipedia/commons/c/ca/Surveillance_equipment_5413.jpg *License:* Public domain *Contributors:* Transferred from en.wikipedia to Commons. *Original artist:* Maraparacc at English Wikipedia
- **File:Symbol_book_class2.svg** *Source:* https://upload.wikimedia.org/wikipedia/commons/8/89/Symbol_book_class2.svg *License:* CC BY-SA 2.5 *Contributors:* Mad by Lokal_Profil by combining: *Original artist:* Lokal_Profil
- **File:Symbol_list_class.svg** *Source:* https://upload.wikimedia.org/wikipedia/en/d/db/Symbol_list_class.svg *License:* Public domain *Contributors:* ? *Original artist:* ?
- **File:Symbol_question.svg** *Source:* https://upload.wikimedia.org/wikipedia/en/e/e0/Symbol_question.svg *License:* Public domain *Contributors:* ? *Original artist:* ?
- **File:T7_RNA_polymerase.jpg** *Source:* https://upload.wikimedia.org/wikipedia/commons/e/e3/T7_RNA_polymerase.jpg *License:* CC BY-SA 3.0 *Contributors:* Own work *Original artist:* Thomas Splettstoesser
- **File:TPI1_structure.png** *Source:* https://upload.wikimedia.org/wikipedia/commons/1/1c/TPI1_structure.png *License:* Public domain *Contributors:* based on 1wyi (http://www.pdb.org/pdb/explore/explore.do?structureId=1WYI), made in pymol *Original artist:*
→<a href='//commons.wikimedia.org/wiki/User:AzaToth' title='User:AzaToth'>Aza</a><a href='//commons.wikimedia.org/wiki/User_talk:AzaToth' title='User talk:AzaToth'>Toth</a>

26.3. TEXT AND IMAGE SOURCES, CONTRIBUTORS, AND LICENSES

- **File:Tented_arch.jpg** *Source:* https://upload.wikimedia.org/wikipedia/commons/5/5b/Tented_arch.jpg *License:* Public domain *Contributors:* NIST.
  Original artist: NIST
- **File:Text_document_with_red_question_mark.svg** *Source:* https://upload.wikimedia.org/wikipedia/commons/a/a4/Text_document_with_red_question_mark.svg *License:* Public domain *Contributors:* Created by bdesham with Inkscape; based upon Text-x-generic.svg from the Tango project. *Original artist:* Benjamin D. Esham (bdesham)
- **File:TheEaglePub-Cambridge-BluePlaque.jpg** *Source:* https://upload.wikimedia.org/wikipedia/commons/f/f1/TheEaglePub-Cambridge-BluePlaque.jpg *License:* Public domain *Contributors:* ? *Original artist:* ?
- **File:Thymin.svg** *Source:* https://upload.wikimedia.org/wikipedia/commons/1/15/Thymin.svg *License:* Public domain *Contributors:* Own work *Original artist:* NEUROtiker
- **File:UK_National_Identity_Card.png** *Source:* https://upload.wikimedia.org/wikipedia/en/3/3e/UK_National_Identity_Card.png *License:* Fair use *Contributors:*
  http://www.mirror.co.uk/news/uk-news/calais-migrant-crisis-means-id-6243331 *Original artist:* ?
- **File:USMC_Sergeant_identifies_Baghdaddi_city_council_member_with_iris_scanner.jpg** *Source:* https://upload.wikimedia.org/wikipedia/commons/f/f6/USMC_Sergeant_identifies_Baghdaddi_city_council_member_with_iris_scanner.jpg *License:* Public domain *Contributors:* defenselink.mil *Original artist:* Gunnery Sergeant Michael Q. Retana, U.S. Marine Corps
- **File:US_Navy_050308-N-2385R-029_Master-at-Arms_Seaman_Carly_Farmer_checks_an_identification_card_(ID)_before_allowing_a_driver_to_enter_the_gate_at_U.S._Fleet_Activities_Sasebo,_Japan.jpg** *Source:* https://upload.wikimedia.org/wikipedia/commons/4/42/US_Navy_050308-N-2385R-029_Master-at-Arms_Seaman_Carly_Farmer_checks_an_identification_card_%28ID%29_before_allowing_a_driver_to_enter_the_gate_at_U.S._Fleet_Activities_Sasebo%2C_Japan.jpg *License:* Public domain *Contributors:*
  This Image was released by the United States Navy with the ID 050308-N-2385R-029 <a class='external text' href='//commons.wikimedia.org/w/index.php?title=Category:Files_created_by_the_United_States_Navy_with_known_IDs,<span>,&,</span>,filefrom=050308-N-2385R-029#mw-category-media'>(next)</a>.
  This tag does not indicate the copyright status of the attached work. A normal copyright tag is still required. See Commons:Licensing for more information.
  Original artist: U.S. Navy photo by Photographer's Mate 3rd Class Yesenia Rosas
- **File:Unbalanced_scales.svg** *Source:* https://upload.wikimedia.org/wikipedia/commons/f/fe/Unbalanced_scales.svg *License:* Public domain *Contributors:* ? *Original artist:* ?
- **File:Wiki_letter_w_cropped.svg** *Source:* https://upload.wikimedia.org/wikipedia/commons/1/1c/Wiki_letter_w_cropped.svg *License:* CC-BY-SA-3.0 *Contributors:* This file was derived from Wiki letter w.svg: <a href='//commons.wikimedia.org/wiki/File:Wiki_letter_w.svg' class='image'><img alt='Wiki letter w.svg' src='https://upload.wikimedia.org/wikipedia/commons/thumb/6/6c/Wiki_letter_w.svg/50px-Wiki_letter_w.svg.png' width='50' height='50' srcset='https://upload.wikimedia.org/wikipedia/commons/thumb/6/6c/Wiki_letter_w.svg/75px-Wiki_letter_w.svg.png 1.5x, https://upload.wikimedia.org/wikipedia/commons/thumb/6/6c/Wiki_letter_w.svg/100px-Wiki_letter_w.svg.png 2x' data-file-width='44' data-file-height='44' /></a>
  Original artist: Derivative work by Thumperward
- **File:Wikiquote-logo.svg** *Source:* https://upload.wikimedia.org/wikipedia/commons/f/fa/Wikiquote-logo.svg *License:* Public domain *Contributors:* Own work *Original artist:* Rei-artur
- **File:Wiktionary-logo-en-v2.svg** *Source:* https://upload.wikimedia.org/wikipedia/commons/9/99/Wiktionary-logo-en-v2.svg *License:* CC-BY-SA-3.0 *Contributors:* ? *Original artist:* ?
- **File:Willem_Einthoven_ECG.jpg** *Source:* https://upload.wikimedia.org/wikipedia/commons/1/1c/Willem_Einthoven_ECG.jpg *License:* Public domain *Contributors:* http://en.wikipedia.org/wiki/Image:Willem_Einthoven_ECG.jpg *Original artist:* ?

26.3.3 Content license

- Creative Commons Attribution-Share Alike 3.0

www.ingramcontent.com/pod-product-compliance
Lightning Source LLC
Chambersburg PA
CBHW082325220526
45470CB00008B/2403